Zoo Animal Learning and Training

Zoo Animal Learning and Training

Edited by

Vicky A. Melfi
Hartpury University,
Hartpury,
Gloucestershire,
UK

Nicole R. Dorey
Department of Psychology,
University of Florida,
Gainesville, FL,
USA

Samantha J. Ward
School of Animal, Rural and Environmental Sciences,
Nottingham Trent University,
Brakenhurst Campus,
Southwell,
UK

WILEY Blackwell

Registered Office(s)
John Wiley & Sons, Inc., 111 River Street, Hoboken, NJ 07030, USA
John Wiley & Sons Ltd, The Atrium, Southern Gate, Chichester, West Sussex, PO19 8SQ, UK

Editorial Office
The Atrium, Southern Gate, Chichester, West Sussex, PO19 8SQ, UK

For details of our global editorial offices, customer services, and more information about Wiley products visit us at www.wiley.com.

Wiley also publishes its books in a variety of electronic formats and by print-on-demand. Some content that appears in standard print versions of this book may not be available in other formats.

Library of Congress Cataloging-in-Publication Data

Names: Melfi, Vicky A., editor | Dorey, Nicole R., editor | Ward, Samantha J., editor.
Title: Zoo animal learning and training / edited by Vicky A. Melfi, Nicole R. Dorey, Samantha J. Ward.
Description: First edition. | Hoboken, NJ : Wiley-Blackwell, 2020. |
    Includes bibliographical references and index.
Identifiers: LCCN 2019024551 (print) | LCCN 2019024552 (ebook) | ISBN
    9781118968536 (cloth) | ISBN 9781118968567 (adobe pdf) | ISBN
    9781118968550 (epub)
Subjects: LCSH: Zoo animals–Training.
Classification: LCC SF408 .M45 2020 (print) | LCC SF408 (ebook) | DDC
    636.088/9–dc23
LC record available at https://lccn.loc.gov/2019024551
LC ebook record available at https://lccn.loc.gov/2019024552

Cover Design: Wiley
Cover Image: © Jeroen Stevens

Set in 10/12pt Warnock by SPi Global, Pondicherry, India
Printed and bound by CPI Group (UK) Ltd, Croydon, CR0 4YY

C9781118968536_081223

This book is dedicated to bridging the gap between those working within the zoo profession and academia; all of whom share a common goal to improve the welfare of animals.

We have been lucky to study animal learning principles and work to practically apply this information in practice within animal training programmes in zoos. With this perspective and respect for the importance of both the art and science of animal training, we hope this book will facilitate greater cooperation and support evidence-based practice.

Contents

Modality Boxes
Consideration of what modalities animals use to communicate with one another; as training programmes are based on good communication between the zoo professional and animal they are working with.

Bigger Training Consideration Boxes
For some of the topics we felt it might be helpful to provide a box outside to the general text to consider some bigger concepts in animal training. These include 'training multiple animals' by Kirstin Anderson-Hansen and a concluding positive note on including training within zoo animal management programmes by Gary Priest.

Notes on Contributors

Kirstin Anderson-Hansen is currently a postdoc at the University of Southern Denmark and in cooperation with the University of Veterinary Medicine Hannover, where she trains grey seals and aquatic birds, currently cormorants and common murres to investigate the effects of underwater noise on marine life. She started her career at the University of California in Santa Cruz over 25 years ago, working with cetaceans as a research assistant and trainer. Later, she followed some of the dolphins to the Shedd Aquarium in Chicago, where she worked and trained Pacific white-sided dolphins, beluga whales, harbour seals, sea otters, and penguins. In 1998, she was offered a trainer position at the Fjord and Baelt in Denmark, training harbour porpoises and harbour seals for research and public demonstrations. From 2003 to 2013, she was the training coordinator and zoological curator at Odense Zoo, where she had the opportunity to expand her training experience to all types of animals, including lions, tigers, giraffes, tapirs, birds, and manatees. Kirstin is the co-chair for the Training Committee at the Danish Association for Zoos and Aquariums (DAZA), as well as an expert advisor for the Animal Training Working Group at the European Association of Zoos and Aquariums (EAZA), where she is also an instructor for the animal training and management courses for both DAZA and EAZA.

Cristiano Schetini de Azevedo is a Brazilian biologist interested in zoo animals, animal behaviour, and animal conservation. This author has been studying the effects of environmental enrichment on the behaviour and welfare of zoo animals, especially birds. In addition, antipredator training techniques for captive-born animals are being tested with the aim of increasing reintroduction success. Finally, aspects of animal personalities are being investigated to increase the efficiency of the environmental enrichment and for conservation purposes.

Kathy Baker works for the Whitley Wildlife Conservation Trust and her main role is to manage the zoo-based research and higher education delivery for Newquay Zoo. She coordinates and supervises student projects from FdSc to MSc level. Her research focus covers a wide range of behaviour and welfare related topics. In particular, cross species comparisons of animal personality, the evaluation of personality as a management tool for captive animals, and multi-institutional research to inform management practices for captive animals. Kathy's also a committee member of SHAPE – UK-Ireland, the regional division of The Shape of Enrichment, Inc. that is dedicated to furthering enrichment efforts in the UK and Ireland, and she is also a member of the Primate Society of Great Britain.

Gordon B. Bauer is professor emeritus of psychology at New College of Florida where he held the Peg Scripps Buzzelli chair in psychology until his retirement. He received an MS from Bucknell University and a PhD from the University of Hawaii, where he studied under Louis M. Herman at the Kewalo Basin Marine Mammal Laboratory. He is a fellow of the American Psychological Association (Division 6) and Association for Psychological Science. He has studied animal senses, cognition, and behaviour in a variety of species including manatees, bottlenose dolphins, humpback whales, sea turtles, honeybees, and humans. In recent years he has focused on a broad survey of manatee senses, including vision, hearing, and touch. He has also investigated magnetoreception by cetaceans and sea turtles, hearing by dolphins and sea turtles, imitation and synchronous behaviour of dolphins, humpback whale behaviour, and memory in honeybees.

Sabrina Brando is director of animal welfare consulting company AnimalConcepts and 247 Animal Welfare. Sabrina is trained as a human psychologist, has a MSc in animal studies, and is currently completing a PhD in human and non-human animal welfare. Sabrina's research interests are welfare, behaviour, effecting change, advocacy, and storytelling. She has presented extensively as an invited and keynote speaker at animal welfare and advocacy conferences globally and is a reviewer for various animal behaviour and welfare journals. Sabrina is passionate about animals and the natural world, and focuses on promoting positive animal welfare and good human–animal interactions and relationships, to facilitate excellent animal care and protection and with the aims to effect behaviour change and challenge the status quo. Sabrina uses stories and storytelling as a means to connect, share, encourage, and empower.

Culum Brown is an associate professor at Macquarie University and has made a significant contribution to the study of behavioural ecology of fishes over his research career. His research niche lies in the study of fish behaviour with his most significant contribution being enhancing our understanding of fish cognition and behaviour. Culum is a well-known champion of fish intelligence and welfare.

Gordon M. Burghardt is alumni distinguished service professor in the Departments of Psychology, and Ecology and Evolutionary Biology at the University of Tennessee. He received his PhD in biopsychology from the University of Chicago and his research focus has been on comparative studies of behavioural development in species as diverse as turtles, bears, lizards, stingrays, spiders, crocodilians, and, especially, snakes. He has worked on many topics involving snakes including sensory perception, foraging, and prey capture, antipredator behaviour, sociality, multiple paternity, sexual dimorphism, colour and pattern variation, environmental enrichment, learning, genetics, conservation, ethical treatment, and mating systems. He has served or is serving as editor or editorial board member of numerous journals including, *Ethology, Herpetologica, Herpetological Monographs, Journal of Comparative Psychology, Animal Learning and Behaviour, Zoo Biology, Society and Animals, Journal of Applied Animal Welfare Science,* and *Evolutionary Psychology.* He is a past president of the Animal Behaviour Society and Division 6 (Neuroscience and Comparative Psychology) of the American Psychological Association. He has edited or co-edited 7 books, including *The Cognitive Animal: Empirical and Theoretical Perspectives on Animal Cognition* (MIT Press, 2002) and the *APA Handbook of Comparative Psychology* (APA, 2017) and authored *The Genesis of Animal Play: Testing the Limits* (MIT Press, 2005). Besides his reptile research, his current research involves play in animals and responses of primates and other animals to snakes.

Jackie Chappell's research interests focus on the ways in which the environment shapes intelligence through evolution, the ways in which animals (including humans) understand their physical environments, and how this changes during development. For example, how do animals integrate information about their physical environments and properties of objects discovered during exploration with their pre-existing knowledge? Her current research as head of the Cognitive Adaptations Research Group primarily focuses on great apes (in collaboration with Dr Susannah Thorpe), but she is also interested in avian and human cognition, and the design of behaviourally flexible, interactive robots, able to explore and learn about their environment.

Fay Clark is an animal welfare scientist based at Bristol Zoo Gardens, who specialises in the assessment and enhancement of captive animal welfare in traditional zoos, safari parks, sanctuaries, and aquariums. Fay received her PhD from the Royal Veterinary College and Institute of Zoology (Zoological Society of London/University of Cambridge) after an MPhil at the University of Cambridge. Prior to joining Bristol Zoological Society in 2013, Fay worked for and studied at the Zoological Society of London for six years, examining gorilla welfare and how the welfare of bottlenose dolphins and chimpanzees could be enhanced by providing them with cognitively challenging activities. Fay has a special interest in how technology and advanced statistical techniques can be used to improve the validity of zoo-based research. Fay project leads internal (zoo-based) health and welfare research at Bristol Zoo Gardens, including two longitudinal studies on primate cognition and enrichment, and animal welfare assessments.

Cynthia Fernandes Cipreste is a biologist in the Animal Welfare Sector of the Belo Horizonte Zoo in Brazil and is responsible for the environmental enrichment and animal conditioning (training) activities. She has a keen interest in animal behaviour and animal learning as well as in laws and ethics applied to animal welfare.

Nicole R. Dorey is a Senior Lecturer at the University of Florida, where she teaches courses in psychology and animal behaviour. In addition to her teaching, Nicole is founder of the undergraduate Animalia research

laboratory, which has published many peer reviewed papers on a variety of species and has earned her a faculty/mentoring award. Dr. Dorey is a Certified Applied Animal Behaviourist (CAAB), has served as a board member for professional organisations, consulted on animal research and training at a variety of zoos, and has been an invited speaker to a number of national and international conferences and workshops. Nicole holds a BS degree in both zoology and psychology from the University of Florida, an MS degree in behaviour analysis with a minor in biology from the University of North Texas and a PhD in animal behaviour from the University of Exeter (UK).

Richard Gibson has vast herpetology experience from a variety of zoos across the world. He began in Jersey Zoo, UK where he was responsible for several world-first breeding successes, and for the planning and delivery of numerous field research and conservation initiatives. He then moved to Mauritius where he took up the role of fauna manager for the Mauritian Wildlife Foundation. In 2003, he became curator of herpetology for the Zoological Society of London. He implemented significant upgrades to the herpetological facilities and husbandry practices and over-saw the first UK breeding of Komodo dragons and the subsequent discovery and publication of parthenogenesis in this species. Between 2005 and 2011, Richard devoted much of his time to helping establish and develop the Amphibian Ark (AARK). When he relocated to Chester Zoo to take up the position of curator of lower vertebrates and Invertebrates. Currently Richard is the curator of ectotherms and birds at Auckland Zoo where he is responsible for a department of 18 full-time staff dedicated to the care of the native species-focused ex situ collection and to a growing portfolio of in situ field conservation programmes.

Heidi Hellmuth has been involved in the animal care and training field for over 30 years, since graduating from the University of Cincinnati in 1987. She started out as a marine animal trainer, working for facilities including Mystic Aquarium, Sea World of Florida, and the Brookfield Zoo. She then spent several years as a zoo keeper with carnivores and large hoofstock at Zoo Atlanta. Heidi managed education animal collections and a wildlife education programme, as well as gaining experience as a supervisor and zoo curator. She spent over six years as the curator of enrichment and training for Smithsonian's National Zoo and is currently the curator of primates at the Saint Louis Zoo. During her career Heidi has worked with a wide array of species and diverse taxa. She has been actively involved in the fields of animal training, enrichment, and animal welfare and is a founding director and past president of the Animal Behaviour Management Alliance. Heidi has authored numerous articles and presented at a wide variety of conferences, including instructing at several animal training, enrichment, and welfare-related workshops, and being an invited keynote speaker at the 1st International Animal Training Conference hosted by Twycross Zoo.

Betsy Herrelko is the assistant curator of animal welfare and research at the Smithsonian's National Zoological Park (NZP). Within the 'WelfareLAB' (Welfare Laboratory of Animal Behaviour), she focuses on research, practice, outreach, and compliance. As a behavioural scientist, Betsy's interests focus on the pursuit of advancing animal welfare science with an emphasis on animal management and how animals think. She started her tenure at NZP as the David Bohnett cognitive research fellow studying primate cognition (cognitive bias, a measure of emotional affect) in zoo-housed apes and husbandry and welfare topics with various species around the zoo.

Geoff Hosey was principal lecturer in biology at the University of Bolton until his retirement in 2005, and is now honorary professor there. His experience of undertaking research and supervising students has mostly been in

behavioural biology, animal welfare, and primatology, and he is still involved in research on zoo animal welfare, particularly about human-animal relationships in the zoo. He is a member of the BIAZA Research Committee and is one of the authors of the textbook *Zoo Animals: Behaviour, Management and Welfare* (Oxford UP, 2nd edition 2013).

Sarah L. Jacobson is a PhD student in cognitive and comparative psychology at the Graduate Center of the City University of New York. She received her BA in neuroscience from Colorado College in 2013. She is interested in the behaviour and cognition of social species including elephants, and the application of that knowledge to conservation and wildlife management.

Neil Jordan is a lecturer in the Centre for Ecosystem Science, University of New South Wales (Sydney) and conservation biologist at Taronga Conservation Society Australia. His current research focus is in applying behavioural ecology to conservation management problems, particularly in using animal signals to resolve human–wildlife conflicts involving large carnivores in Botswana and Australia.

Jim Mackie was appointed The Zoological Society of London's (ZSL's) first animal training and behaviour officer in 2012 having previously demonstrated the value of trained behaviours to improve husbandry and welfare in the zoo's living collections. Jim's interest in animal behaviour began when he trained his own raptors for educational demonstrations 25 years ago. This led to an opportunity to join ZSL's animal display department where he worked for 10 years developing the zoo's visitor education programme. Jim's passion for sharing information in the field of operant learning and behavioural enrichment led to the formation of ZSL's Behaviour Management Committees, at both London and Whipsnade Zoo and later the BIAZA British and Irish Association of Zoos and Aquaria (BIAZA) Animal Behaviour and Training Working Group which he chairs.

Khyne U. Mar is a veterinarian and conservation biologist, with more than 30 years of experience at academic, research, and administrative positions. Recipient of various research grants, fellowships, and scholarships. Specially trained for elephant management, elephant breeding, conservation medicine, and biology. Extensive work experience in South-east Asian countries as a consultant veterinarian.

Steve Martin is president of Natural Encounters, Inc. a company of over 50 professional animal trainers who teach animal training strategies and produce educational animal programmes at zoological facilities around the world. He spends over 200 days each year on the road serving as an animal behaviour and visitor experience consultant at zoological facilities worldwide. Steve teaches several training workshops each year and is an instructor at the AZA Animal Training School, an instructor at the Recon – Elephant Training Workshop, a trustee with the World Parrot Trust, and a member of the AZA Animal Welfare Committee. He is also president of Natural Encounters Conservation Fund, Inc. a non-profit company that has raised and donated over $1.3 million to in situ conservation programmes. As a core team member of the California Condor Recovery Team, Steve helped guide the release of the captive-bred condors back into the wild.

Lindsay R. Mehrkam is an applied animal behaviourist, animal welfare scientist, and doctoral-level board certified behaviour analyst (certificate number: 1-15-17919, certified February 2015). Her primary research interests focus on the benefits of human–animal interaction, enrichment, and training for improving the welfare of both animals and people in society. As the director of the Human & Animal Welfare Collaboratory (HAWC) and the faculty representative for the Six Flags Field Experience Programme, her teaching and research programmes in applied animal behaviour, learning, and well-being have led to grants, publications, workshops, internships, and service learning opportunities in animal shelters, zoos, aquariums, and animal sanctuaries, amongst others. Dr Mehrkam's research has been published widely in peer-reviewed scientific journals and has been presented at national and international conferences. She has been recognised through popular media outlets, grants, and scholarly and industry awards, including the Association for Professional Dog Trainers, Maddie's Fund, and the Animal Behaviour Management Alliance. In addition to teaching and research, Dr Mehrkam serves as the vice president for the Applied Animal Behaviour Special Interest Group for the Association for Behaviour Analysis International (ABAI) and as an advisory board member for Big Oak Wolf Sanctuary.

Vicky A. Melfi is a Professor at Hartpury University, Gloucestershire, UK. She has almost 30 years' experience working within the zoo profession, focusing on animal welfare and conservation, in appointments within the UK, Ireland, and Australia. She has also held various academic appointments, notably at the Universities of Exeter, Plymouth, and Sydney. Vicky is a passionate advocate of professional–academic collaborations, which serve to gather data which can underpin evidence based practice, and improve our understanding of animal and human behaviour and their interactions, both of which can lead to great animal welfare and conservation outcomes.

Erik Miller-Klein is a founding partner of A3 Acoustics, a licenced acoustical engineer, and is board certified by the Institute of Noise Control Engineering in the United States. His background in science and music gives him a finely tuned ear to noise issues and a general curiosity to understand the fundamental impact of noise and vibration. His work focuses on designing environments that acknowledge the complex nature of acoustics, whilst integrating in solutions that address the role sound plays in the way we communicate, relax, and interpret our world.

Joshua M. Plotnik is a comparative psychologist and conservation behaviour researcher who has studied elephant cognition since 2005. Recently, Josh has been working in Thailand to understand how research on animal thinking can be applied directly to the mitigation of human/wildlife conflict. He is a faculty member in the Department of Psychology and the animal behaviour and conservation programme at Hunter College, and in the cognitive and comparative psychology graduate programme at the Graduate Center of the City University of New York. He is also the founder and executive director of Think Elephants International, a US non-profit charity working to bridge the gaps between research, education, and conservation by using elephants as a conduit. Dr Plotnik was previously a Newton international fellow at the University of Cambridge, and has earned degrees from both Emory University (MA and PhD) and Cornell University (BS).

Gary Priest 'Growing up on a ranch in Southern California, I have been around animals all my life. For as long as I can remember, I have been fascinated by the ways both animals and people learn.' For over four decades, Gary Priest has enjoyed a career that has taken him all over the world to work with all types of animals and train all sorts of people to understand and use operant conditioning to improve animal care and welfare. A senior manager with San Diego Zoo Global for 35 years, Priest is curator of animal care training with San Diego Zoo Global Academy www.sdzglobalacademy.org. Gary received his BA (natural science) from Western Illinois University and his MA (management) from National University in San Diego. Gary's patient wife has managed to expertly shape his behaviour from their first encounter at 14.

Ken Ramirez is the EVP and chief training officer for Karen Pryor Clicker Training where he helps to oversee the vision, development and implementation of training education programmes. Previously, Ken served as EVP of animal care and training at Chicago's Shedd Aquarium. A 40+ year veteran of animal care and training, Ramirez is a biologist and behaviourist who has worked with many zoological organisations and dog programmes throughout the world. He hosted two successful seasons of the TV series *Talk to the Animals*. Ramirez authored the book *Animal Training: Successful Animal Management through Positive Reinforcement* in 1999 and *Better Together: The Collected Wisdom of Modern Dog Trainers* in 2017. He taught a graduate course on animal training at Western Illinois University for 20 years. He currently teaches at ClickerExpo every year, offers hands on courses and seminars at the Karen Pryor National Training Center (the Ranch), and teaches online courses through Karen Pryor Academy.

Marty Sevenich-MacPhee is an animal behaviour and husbandry professional who has devoted her career to animal training, enrichment, and welfare. Since earning her bachelor's degree from the University of Illinois in 1984, Marty has held leadership positions responsible for programming in behavioural husbandry at Chicago's Brookfield Zoo and Walt Disney World. In addition to her contributions to the betterment of animal husbandry, Marty also specialises in the education of animal trainers as an active consultant and is the author of publications on animal husbandry. Marty has also served as a board member and/or on committees for IMATA, IAATE, and AZA. She has been an instructor for various AZA courses on training and enrichment of animals that include managing animal enrichment and training programmes, crocodilian biology and management, and animal training applications in zoos and aquariums.

Andrew Smith's main areas of expertise are behavioural ecology and primate colour vision. His work looks at how animals, from aardvarks to goldfish, interact with each other and their environment. Much of his work involves primates and the ecological influences of perceptual capabilities and resource partitioning. Using a mixed approach of captive and field studies he investigated the evolution and advantages of trichromacy.

Sarah Spooner has been working in zoos and zoo education for the past decade. She is currently education and research manager at Flamingo Land, UK, where she is working on updating and implementing the zoo's conservation education strategy. Her current research investigates the impact of animal encounters and handling animals on visitor attitudes and knowledge. Her doctoral research at the University of York, 'Evaluating the effectiveness of education in zoos', examined the roles of live animal shows, zoo signage, and educational theatre as a means of educating zoo visitors about animal facts and conservation messages. She holds post graduate degrees from the University of York and the University of Cambridge and an undergraduate degree from the University of Cambridge. She has worked in formal education as a primary school teacher and as a lecturer in evolutionary biology, ecology, and statistics. She has taught in informal settings including museums and zoos to a wide range of ages and abilities. Additionally, she has experience as a zoo keeper and as an animal trainer, predominantly working with parrots and birds of prey.

Tim Sullivan has been employed by the Chicago Zoological Society at Brookfield Zoo for the last 38 years. He spent 16 years as a keeper in the marine mammal department training and caring for the zoo's dolphins, walrus, sea lions, and seals. In 1997, Tim was asked to implement an elephant protected-contact behaviour management programme in the Pachyderm department. In 1998, he was offered his present position as the zoo's curator of behavioural husbandry. Tim's primary responsibilities are to manage the zoo's animal training and environmental enrichment programmes. Tim oversees the behaviour of the zoo's large and diverse animal collection – from aardvarks to zebras – and is responsible for developing the skills of over 100 animal keepers who care for them. Tim consults on animal training and environmental enrichment at other zoological institutions and conducts international training workshops. Tim is currently on the Instructor team of the Association of Zoo and Aquariums (AZA) annual animal training applications course. He is active in international training organisations and has been an officer on the board of directors of both the International Marine Animal Trainers Association and the Animal Behaviour Management Alliance; an organisation he cofounded.

Greg A. Vicino is the curator of applied animal welfare, San Diego Zoo Global. Mr Vicino studied biological anthropology at UC Davis where he focused on non-human primate husbandry, behaviour, welfare, and socialisation. Mr Vicino focuses on integrated management strategies, in which all animals receive the benefit of every specialty at each facility. By emphasising the frequency and diversity of behaviour, he and his team have worked on developing integrated management strategies that exploit the adaptive relevance of behaviour and making behaviour meaningful for managed populations. This strategy is designed to be applicable to all species both captive and wild and he has extensive experience in the Middle East and East Africa applying these concepts to in situ conservation programmes and rehab/re-release sites. Greg has continued to work towards his institute's mission of ending extinction, and has staunchly stood by the idea that all animals should be given an opportunity to thrive.

Samantha J. Ward is a Senior Lecturer in Animal Science at Nottingham Trent University. Previously Sam worked as a zoo animal keeper of various hoofstock, primate, and macropod species and was a zoo conservation and research manager with animal record (ZIMS), animal transportation, and studbook responsibilities. Sam completed an MSc in Animal Behaviour and a PhD in Animal Behaviour and Welfare. Sam's research focuses on zoo animal welfare and the impacts that human–animal interactions, human–animal relationships and zoo animal husbandry and management techniques impact and improve captive animal welfare. Sam is a member of the BIAZA Research Committee and also sits as the Welfare Expert on Defra's Zoo Executive Committee.

Gerard (Gerry) Whitehouse-Tedd has over 35 years' international experience working in zoos, private and government animal collections, and theme parks. He is a professional animal manager, trainer, and zoo supervisor, with specialisation in the husbandry, training, and management of exotic and domestic species. This includes management of animals for captive breeding programmes, medical husbandry, rehabilitation, public display, and presentations, as well as visitor experience, encounter, and outreach programmes with zoo animals. Gerry has held senior zoo management roles for the past 20 years and is currently the operations manager for the Kalba Bird of Prey Centre in the United Arab Emirates, a position he has held since 2012.

Katherine Whitehouse-Tedd is a senior lecturer at Nottingham Trent University, specialising in human–wildlife conflict, zoo conservation education, and carnivore husbandry. Kat teaches on undergraduate and postgraduate degrees in zoo biology and conservation, and supervises a range of student research projects both locally in the UK and abroad. Kat's research interests are currently concentrated on the use of livestock guarding dogs as mitigating tools for promoting human–carnivore coexistence, as well as vulture conservation, and the use of ambassador animals in zoos for promoting conservation behaviours in visitors. Kat holds a PhD in nutritional science (investigating bioactive dietary compounds in large

felid zoo diets) from Massey University in New Zealand, and has previously worked as a conservation manager, research scientist, and education programme coordinator for zoos in the UK, South Africa, and the United Arab Emirates.

Jonathan Webb is an associate professor in conservation biology with a background in applied research focusing on actions for recovering threatened fauna and mitigating the impacts of invasive species. His recent research has centred on using behavioural techniques to reduce the impacts of cane toads on northern quolls. Jonathan has published more than 120 scientific papers and he currently teaches wildlife ecology at the University of Technology Sydney. He believes that scientists should communicate their research to the wider public, and his research on northern quolls was featured extensively in the media, including the BBC documentary, *Attenborough's Ark*. Jonathan is a keen wildlife photographer and he lives with his wife and two children on a wildlife friendly acreage on the NSW central coast.

Heather Williams began SCUBA diving at 14 and first helped in an aquarium at 16 with work experience at Scarborough Sealife Centre. She has volunteered with pilot whale photo-ID work (Projecto Ambiental, Tenerife), and cetacean distribution research (Organisation Cetacea). She studied ocean science with marine biology at the University of Plymouth, during which time she volunteered at the National Marine Aquarium (NMA), completing her honours project there focusing on seahorses (*Hippocampus kuda* specifically). After university, she was offered a job within the animal husbandry department at the NMA. Following this, she moved to Grand Cayman to work in a relatively new tourist attraction – Boatswain's Beach. This is where she first became interested in training the animals in her care. A move to The National Sealife Centre in Bray, Ireland for a couple of years meant she could implement a simple training programme there for some of the animals. she was lucky enough to be able to spend a month volunteering at the Monterey Bay Aquarium where she got to see how important training was in encouraging feeding in an ocean sunfish (*Mola mola*) with only one eye in a large exhibit. Moving back to the NMA in 2010 meant she had the experience and knowledge to push forwards with a training programme and now the aquarium has a large number of animals trained for various purposes. She was also lucky enough to be a founding committee member of the BIAZA (British and Irish Association of Zoos and Aquariums) Animal Behaviour and Training Working Group as an aquarium liaison officer.

Robert John Young is a British professor of wildlife conservation who divides his research time between working with animals in captivity (principally the welfare of zoo animals) and conservation fieldwork in countries such as Brazil. He is the author of the book, *Environmental enrichment for captive animals*.

Foreword

I was delighted to be asked to write the forward for this important book and, as I read through the chapters, that delight turned to sincere appreciation for the wealth of knowledge presented by all the authors and editors. Time and time again I would write in my notes 'clearly explained', 'really great applied examples', and 'good use of relatable real-life situations'. What makes this even more impressive is that the experiences and examples are drawn from across the world. This not only demonstrates the widespread utilisation of training but, also where there are commonalities and differences in approaches. I cannot recommend this book highly enough and hope the readers will allow me a few paragraphs to expand on why ....

Many years ago, at the start of my career I was employed to be a course manager for a new BSc (Hons) in Applied Animal Behaviour and Training. This degree was one of the first of its kind and reflected the increasing interest in not only understanding why animals behave in the ways that they do, but also how we can use that knowledge to shape and train different behaviours. Although this course focused mainly on training of companion animals it was also forward thinking enough to include modules on wild animals as well; and oh, what myself and my fellow course tutors would have given for this book to be available at that time!

Managing that course not only reinforced my deep fascination with animal behaviour (and my enjoyment at having opportunities to be able to pass that fascination onto others) but also opened up the brand new world of animal training to me. Through my interactions with other tutors and our students it quickly became apparent to me that the theoretical knowledge of animal learning and how it could best be practically applied to training was still at an early stage. I was also made painfully aware of my own lack of training ability; put bluntly, although I had a good grasp of the theoretical knowledge I struggled with the practical skills to quickly see nuances in behaviour and then adjust my timing to respond effectively and provide a good learning environment for the animal. Having now read this book, I have been given new hope and inspiration that I would be able to do better if I tried again.

The authors provide an excellent range of applied examples across varied species and situations. They are able to take complex topics and write about them in an easily accessible way. We are quite rightly reminded that as the field of animal training expanded so did the range of terms and techniques being used. This has the potential to be at the least confusing, and at the worst conflicting. In turn this will result in poor technique and negative welfare, and that is before you even get into the complexities of translation into different languages. This provides me with yet another reason to recommend this book; in addition to consistency of terms used and clear explanation of them (the differences between positive and negative reinforcement and punishment being just one example), there is also an excellent glossary that I am sure will be utilised again and again by readers.

There is a clear value in reading this book chapter by chapter as each builds on the other and opens up a new and relevant aspect of animal learning and training. However, I can also see that each chapter can be taken on its own merit, either to introduce the reader to a new topic, act as a refresher for a topic already known, or as a handy reference guide when you want to check back on x or y or search for solutions. There is such a lot to know about learning and training that it is difficult to keep it all in ones' head. Having it here in handy reference format allows the trainer to get on with the important task of noticing small changes in the behaviour of their animals and adapting aspects of training to this.

I will leave the reader to discover each chapter for themselves however, I couldn't resist the opportunity to share a few aspects that particularly caught my attention. Having used Pavlov as an example in various 'introduction to animal behaviour' modules for more years than I can count, it was lovely to be reminded of Twitmyer's related work classically conditioning students to exhibit a knee jerk reaction at the sound of a bell. Chapters 2 and 3 reminded me of all the fascinating and complex ways animals learn and gain information over and above classical and operant conditioning. They also served to provide a clear reminder of the benefits that can be gained by studying animals in zoos and aquariums compared to the wild ones, whilst also giving reasons why we shouldn't always directly compare one to the other. At a more individual level, there was also some thought-provoking research shared on how early environments and individual experiences shape adult behaviours. Chapter 4 is excellent at making the reader think about ethical considerations related to training and just because we can train something doesn't necessarily mean we should. Chapter 5 has some fascinating examples about all the different learning opportunities that are available to animals in human care from the embryonic stage through all subsequent phases and life experiences.

As my career developed I found my way to the European Association of Zoos and Aquaria (EAZA) and was employed in the newly developed position of EAZA Academy Training Officer; sadly, my role did not involve training wild animals but the much more challenging species of 'zoo and aquarium staff'! All jokes aside, what became very apparent was the strong desire from the community for the development of a course on animal training. To this day, this course is consistently one of our most popular and has expanded to additional courses looking at different applied aspects. Humans are an essential ingredient in the training of the animals in our care and so I was especially interested in Chapters 7, 8, 9, and 13 which all touch on different aspect of the human(s) involved in learning and training interactions. Whether this be the influence, on both sides, of motivation, trust, and control (Chapter 7); how to engage staff in training programmes so that they see it as an opportunity rather than a burden (Chapter 8); human animal interactions (Chapter 9); or ways to make sure staff and animals stay safe (Chapter 13). Chapter 10 on the different ways animals are used in 'shows' left me with a clear understanding of the varied situations and nuances that are often grouped together under this one title. It also left me with a strong desire to go back and cross check our EAZA Guidelines for the use of animals in public demonstrations with the information given here.

The authors of Chapter 11 couldn't have said it better when they wrote 'considering the impact of training on animal welfare is an epic challenge' however, they prove they are more than up to this challenge. Their chapter provides some excellent queries about the terms we use, as well as serving as an important reminder that we should always come back to considering welfare from the individual animals' perspective and not how we might perceive the situation. Challenges between welfare and training are further considered in Chapter 12. We are asked to consider the ethics of situations where we might use our knowledge of learning and training to actively compromise short-term welfare for a long-term welfare gain when returning animals to the wild. As the threats to wildlife continue to increase we will need to be more and more sure of our ability to understand and balance ethical and welfare decisions relating to aiding increased survival post-release.

We have come an awfully long way since my early days managing that course on behaviour and training. This book provides a great overview of how far we have come, yet there is still much to learn, especially when it comes to some taxa and situations. Consequently, I wholeheartedly share the desires expressed by the authors and editors throughout this book that it will act to encourage more evidenced based research and publications to help our expanding knowledge. I encourage anyone interested in the field of animal learning, training, and welfare, from student, to zoo professional, to academic, to read this book. There is so much to learn from, and be inspired by, here.

Myfanwy Griffith, Executive Director,
European Association of Zoos and Aquaria
(EAZA)
March 2019

Preface

Zoo Animal Learning and Training (ZALAT) has been inspired to fill a need: a need to bridge the gap between those studying animal learning theory and those using these principles to implement training programmes in zoos. In creating ZALAT we hoped, and believe we have succeeded, in bringing together expert academics and zoo professionals, so that you the reader can benefit from their shared wisdom. This unified approach brings together the art and science of animal training. Consequently, we're proud to present a book written by experts in their fields, with academic and professional knowledge and experience operating worldwide.

ZALAT provides a clear, easy to read and concise introduction to basic animal learning theory, alongside tangible application of this theory when training animals in zoos and aquaria. We hoped to clarify the jargon, the different terms used between academic disciplines, and different professional settings to bridge the barriers to implementing and understanding the consequences of training in zoos.

We invited academic and professional experts working in zoo animal learning and training to provide content, from the principles of animal learning theory (Chapters 1–4), to the application of this theory (Chapters 5–9), all the way through to practical considerations of implementing zoo animal training programmes (Chapters 10–13). Given the magnitude of the subject and implications for translating theory into practice, we soon realised that it would also be great to include text boxes. The first set of text boxes consider the modalities animals employ to communicate, as communication is key to successful training. Thereafter follow a series of boxes which focus on taxa specific information, providing information about what is known about a taxa's cognitive abilities and what considerations might need to be thought about when training those animals in zoos; thus we organised the boxes into two groups, those including an academic and professional perspective. For elephants there is an exception to this pattern, as we have three boxes relating to them, as with the other taxa we detail the current knowledge of their cognitive abilities, but we have included two boxes related to their captive management, one relates to zoos and one relates to their training and management in elephant camps. We then have the final set of boxes, which relate to 'other' aspects of training, which were not explicit within the chapters. In this section, we consider training more than one animal at a time; often a necessary practical consequence of socially housed animals and/or few staff allotted to training. Finally a box is included which provides a general view of zoo animal training, where the art started, and what it has been able to achieve; a bite size introduction, overview, and view to the future.

We have referred to zoos throughout, even in the title, not to exclude aquaria but as shorthand for a profession which operates to care for and conserve species in captivity whether that be a zoo, aquaria, or sanctuary. To improve accessibility of the theoretical background to animal learning theory, a rouge guide for translation has been included in the form of a handy glossary.

This indispensable collation of terms recognised for their scientific foundation, along with definitions, will hopefully provide clarity in an area which can sometimes be mindboggling; even to the seasoned trainer. Our no-nonsense approach based both in science and practical experience, will hopefully demystify the complexities of training zoo animals and enable all readers to clearly understand how to effectively train animals and understand the consequences implementing these training actions might have on the lives of the animals in your care.

Vicky A. Melfi, Nicole R. Dorey, and
Samantha J. Ward

Acknowledgements

As editors we are extremely grateful to the contributors of this book who have shared and translated their knowledge to further the understanding and improve implementation of zoo animal training. We'd also like to thank those anonymous reviewers of the publication proposal, as well as, Geoff Hosey and Lucy Bearman-Brown for reviewing some materials for this book, we're grateful for their time and expertise.

The book has been brought to life not only by the talent and enthusiasm of our authors, but also by the photographs provided by Kirstin Anderson-Hansen, Gordon M. Burghardt, Sabrina Brando, Elisabeth S. Herrelko, Katherina Herrmann, Jim Mackie, Steve Martin, Joshua Plotnik, Jereon Steven, Heather Williams, and Ray Wiltshire.

We're grateful to Myfanwy Griffith (CEO EAZA) for taking the time to support this publication by writing our foreword; as someone who has a personal connection and passion for animal learning theory and training, Myfanwy has supported its scientific study and professional implementation.

We'd also like to acknowledge the support of the staff at Wiley, especially David McDade who supported the publication proposal and Andrew Harrison who was part of the publication team in its final year, during which time Athira Menon facilitated the editorial needs and the publication requirements. We are also grateful to the production team for making this project into a book.

From a personal perspective Vicky Melfi would like to thank Edward Pickersgill and Isabel MP for their love support. Nicole Dorey would like to thank Michael for supporting her "garage band" as well as Xander and Aria for all the hugs, kisses, and cuddles as well as only hiding under her desk a few times. Samantha (Sam) Ward wishes to thank Jonathan Hatton and Connie for always being there and supporting her when taking on "these exciting projects".

Part A

Demystifying Zoo Animal Training

In this opening section, we hope to provide an overview of the principles needed to approach and understand animal learning theory. Knowledge of zoo animal learning principles is essential for those interested in setting-up, managing, and implementing zoo animal training programmes. We have worked to provide overviews of the scientific understanding of this field, in a clear and interesting text. Too often animal learning theory is portrayed as a complex and difficult subject. Though there are intricacies and debates about the learning theory, there are some rules and principles, which can be easily understood, which also make it clear what can and cannot be achieved by different training programmes. How learning might benefit the animals taking part is considered. And finally, the basic training methods commonly used and their scientific basis is explored.

1

Learning Theory

Nicole R. Dorey

1.1 Introduction

Before really diving into learning theory, we must first define learning. Scientists define *learning* as a change in an organism's behaviour or thought resulting from experience. This can be the standard definition found in any general psychology textbook. As a student, I heard this definition a million times, and I'm sure you have too. Sometimes when you hear something over and over again you tend to not pay attention to it anymore. But I think we should. Not only should we pay attention to the definition we should also start to break it apart and not just take the definition at face value. Questions like, why is it *change* in behaviour and not *acquisition* of *knowledge*? What does 'due to experience' mean? This section will guide you through the answers to those important questions.

Let's look at the first of these questions. Why do we use the phrase 'change in behaviour'? One might instead like to think of learning as an *acquisition* of knowledge rather than a *change* in behaviour. Chance (1988) explains that the word *change* is preferred over *acquisition* because 'learning does not always appear to involve acquiring something, but does always involve some sort of change' (p. 24). Furthermore, we use the word *behaviour* instead of *knowledge* because behaviour can be observed. We can't see what someone knows and psychologist know that knowledge isn't enough to change

behaviour. For example, when a person stops performing a problem behaviour, like biting their nails, a behaviour is not acquired but a change occured. In addition they might know that these behaviours are bad for them, but that knowledge doesn't change the observable behaviour.

Learning also must be 'due to experience'. Behaviour change is sometimes, but not always, a result of experience. For example, consider zebra A and zebra B who are both being trained to target a frisbee on command. After zebra A acquires the targeting behaviour, s/he will touch its nose to the frisbee when you the trainer say 'target'. Zebra B who also acquired the behaviour is now sick and refuses to touch the target with its nose. Although zebra B changed their behaviour when cued for 'target' their behaviour change is not a result of learning. So zebra A and B both learned the targeting behaviour from experience. However, zebra B changed their behaviour due to an illness.

What is experience? 'Experience' refers to events that occur in the animal's environment or surroundings (Chance 1988). Thus, changes in the animal's environment can cause learning. For example, consider an elephant in a zoo. In the evening, the elephant might start walking towards the night enclosure (where it is fed at night) when it hears keys jingle. In the past, keys jingling have signalled mealtime and thus the animal has learned from experience that the jingle

Zoo Animal Learning and Training, First Edition. Edited by Vicky A. Melfi, Nicole R. Dorey, and Samantha J. Ward.
© 2020 John Wiley & Sons Ltd. Published 2020 by John Wiley & Sons Ltd.

of keys at that time means food is being provided.

Now that we have a better sense of the term learning, let's move on to an overview of animal learning theory. In the rest of the chapter you should expect the discussion of the history and expansion on the types of learning that will be important as we continue further into the book.

1.2 Individual Learning

A great place to start to explain individual learning is with the work of Edward Thorndike. Although there were scientists before him who referred to animal behaviour (such as Darwin, Romanes, and Morgan), Thorndike was the first to study animal behaviour empirically. Before Thorndike, scientists observed animal behaviour in naturalistic settings. They collected anecdotal accounts from a variety of people (pet owners, amateur nature enthusiasts, military officers, etc.) and compiled them as evidence for a hypothesis or theory.

This anecdotal approach was not just anthropocentric but anthropomorphic. *Anthropocentrism* is when a person regards human beings as the central or most significant species on the planet. An anthropocentric person may believe that humankind has unique abilities superior to those of other animals. For example, some scientists, like Romanes, used animal models to prove animals could think and solve problems the way people do. Instead of studying animal behaviour in its own right, he tried to model their behaviour to meet that of human standards. For example, he had a few anecdotal 'experiments' where he trapped ants and watched to see if their mound-mates helped free them. He wrote 'I next covered one up with a piece of clay, leaving only the ends of its antennae projecting. It was soon discovered by its fellows, which set to work immediately, and by biting off pieces of the clay soon liberated it' (Romanes 1888, p. 48).

Anthropomorphism refers to the attribution of human traits, such as language or emotion, to other animals. It is often used to maintain anthropocentric biases. For example, Dr Seuss's *Cat in the Hat* features a cat that walks upright, speaks perfect English and wears human cloths in a similar way to a human. Anthropomorphism also characterises many Disney movies, for example *Finding Nemo, The Incredible Mr. Fox, Bambi*. All these movies give human characteristics to animal species. Animal professionals know that although an animal's behaviour mimics a behaviour engendered by a human in a similar setting, it may be explained in a different way. One reason for this is parsimony. Scientists look for the most parsimonious, or simple, explanations of observed behaviour. Between two equally good explanations, choose the simpler. For example, you may come home to the living room covered in what used to be the stuffing of very expensive pillows. Your sweet playful dog got bored and found a way to entertain himself. You look at your dog only to see those big droopy eyes, tucked tail, dropped ears, and him backing up very slowly. You think to yourself 'he looks guilty and knows that what he has done is wrong'. However, the most parsimonious explanation is that he is actually reacting to your body language (tone in your voice, staring in your dog's eyes, moving erratically, not greeting your dog as you normally would, etc.). Indeed, experiments have shown this body-language explanation to be the better explanation (Horowitz 2003).

Thorndike was the first to introduce experimental methods with animals in 1898. What were these methods ... to put a cat in a box of course!

Thorndike was a very unconventional scientist for his time. He didn't have a laboratory, but kept his subjects (chickens, cats, dogs, fish, and monkeys) in his own home 'until the landlady's protests were imperative' (Thorndike 1936, p. 264). He also felt that previous scientists paid too much attention to animal intelligence and although he took on a conventional title for his thesis (Animal intelligence), his goal was to investigate animal stupidity (Walker 1983). The apparatus that Thorndike used to study animal stupidity was called a puzzle box. To picture a puzzle box, think of a small wooden crate with wooden slats, in the box was some kind of device (a sequence of strings that when pulled in the correct order would release a pin and open the box). The experiment started with Thorndike placing a subject in the puzzle box. Thorndike put a hungry cat into the box, with a piece of food visible outside the box. He found that the cats did not understand the sequence of strings to open the door. However, if that was the end of the story, this chapter could end here. In fact, the cats did get out and could get out again and again. He credited the cats' escape to trial-and-error learning, pointing out that the animals weren't using insight, inference, or any other signs of 'intelligence' (Walker 1983). Out of these experiments came Thorndike's *law of effect*. The law of effect states that if a stimulus is followed by a behaviour that results in a reward, the stimulus is then more likely to give rise to the behaviour in the future. According to proponents of the law of effect, most of the behaviour we emit is due to the association between the behaviour and its consequence. We will look at this in more detail in the operant conditioning section when we talk about the psychologist B.F. Skinner.

1.3 Classical Conditioning

Thorndike had a strong methodological and theoretical impact on animal behaviour research influencing key concepts in learning theory that are still relevant today. However, in the early 1900s, scientists had a new leader in the area of animal learning theory. His name was Ivan Pavlov.

Pavlov's experiments furthered the doubt, raised by Thorndike, that animals had much, if any, cognitive ability. This uncertainty was due to how the responses were made by the animal. Pavlov designed experiments involving unintentional behaviours. For example, in one classic experiment Pavlov elicited salivation in a dog by presenting the sound of a metronome. The dog didn't need to think about his salivation; it was an automatic response to the sound. Few people know that Pavlov was an established scientist who studied processes of digestion. He made major advances in the study of digestion by developing surgical techniques which in 1904, won him a Nobel Prize. Fewer people know that although he is credited for discovering classical conditioning, this title should perhaps go to a man named Edwin Twitmyer.

Twitmyer's PhD dissertation tested the knee-jerk reflexes of college students by sounding a bell half a second before hitting the patellar tendon. After repeatedly doing this, he found that the sound of the bell alone caused the knee-jerk reflex. In 1904, Twitmyer presented his findings at the American Psychological Association meeting, where it drew no interest. From here, history gets a little muddled. Some historians state that Pavlov saw this talk, and noticed when he got back that his dogs salivated in the presence of lab coats before the food was present, and it was this that caused him to change the focus of his lab. Others state that Twitmyer and Pavlov were working on this subject independently of each other. No matter what the real story was, we all know that Pavlov is the one that is credited for classical conditioning.

As we know, Pavlov's process of association is called *classical conditioning* (or *Pavlovian conditioning*). It should also be noted that during his research, Pavlov coined the term *reinforcement* to describe the strengthening of the association between an unconditioned and conditioned stimulus. In classical conditioning, the animal learns to respond to a previously neutral stimulus that had been paired with another stimulus that elicits an automatic response. Let's look at a real world example that I mentioned earlier and one you may have witnessed yourself. Keepers at most zoos and aquariums wear keys on their belt loops. When they walk, the keys make a distinct sound. An animal that is naive to the zoo (just weaned from its mother and now eating with the adults for example) might not notice the sound of keys jingling. Thus, the sound of keys is a *neutral stimulus*: one that elicits no response and thus no meaning. Food is the *unconditioned stimulus*. It requires no conditioning to elicit a response (in this case, salivation). The animal's salivation is the *unconditioned response*, because if you present the animal with food, it will salivate automatically (without training). After multiple pairings of the jingling of keys with feeding, the once-neutral stimulus (the sound of keys) is now the conditioned stimulus and causes a conditioned response (salivating at the sound of the keys).

Let's look at a second example. This time try to label the neutral stimulus, unconditioned stimulus, unconditioned response, conditioned response, and the conditioned stimulus.

Monkeys, in general, are great animals to watch because they are so active. They are always involved in a fight of some sort, sometimes one that includes weapons. Once, I was commissioned to watch a group of capuchins for a behaviour project not related to conditioning. I noticed that two individuals were fighting; George picked up a tree branch and used it to hit Declan on the head. At first, Declan did not react to the tree branch coming towards him. However, after George hit him a few times with it, Declan started to flinch in pain and even would flinch in pain when the wind would move the branches on trees.

Do you think you labelled them correctly? Here is the answer:

The tree branch started out as a *neutral stimulus* because it had no meaning to Declan prior to conditioning. Tree branches are normally benign items that dangle from trunks. The *unconditioned stimulus* is getting hit. The *unconditioned response* is the pain the individual felt because there is no conditioning required to make an animal react to pain. After pairing the tree branch with the pain

(conditioning), the once-neutral stimulus becomes the conditioned stimulus and causes a conditioned response (just seeing a tree branch can cause the capuchin to feel pain).

This process also occurs in humans. Let's look at a classic study.

In 1913, a farm boy from South Carolina who had a passion for animals, outlined his thoughts after about a decade of research in the area of psychology. His conclusion was that psychology should be the study of behaviour and not the study of consciousness or mental processes. The farm boy was John Watson and the paper that outlined his belief and coined the term 'behaviourism' was 'Psychology as the Behaviourist Views it'. Although Watson spent over a decade doing research at John's Hopkins University, a single study received the most attention: the study with Little Albert. In this study, Watson and his assistant, Rosalie Rayner, showed a healthy nine-month-old infant, 'Little Albert', a rat, a rabbit, a dog, and a monkey. Because he didn't have exposure to these animals, Albert didn't cry. To show that one can condition fear, Watson paired these animals with a loud noise (a hammer that he would strike against a steel bar). For example, when Albert touched the rat, they made the clanging sound. After multiple pairings, the infant cried and avoided the rat even when there wasn't a loud noise.

After the experiments, Watson wrote, 'Give me a dozen healthy infants, well formed, and my own specified world to bring them up in and I'll guarantee to take any one at random and train him to become any type of specialist I might select – doctor, lawyer, artist, merchant-chief, and, yes even beggar-man and thief, regardless of his talents, penchants, tendencies, abilities, vocations, and race of his ancestors' (Watson 1924, p. 104). In the next sentence, he writes, 'I am going beyond my facts and I admit it, but so have my advocates of the contrary and they have been doing it for many thousands of years' (Watson 1924, p. 104). Watson believed that introspective psychologists, like Freud, made untenable assumptions about behaviour. Watson and Rayner (1920) even poked fun at psychoanalysis, saying that, 'When they [Freudians] come to analyse [Little] Albert's fear of a [white] seal skin coat, assuming that he [Little Albert] comes to analysis at that age, [Freudians] will probably tease him on the recital of a dream which upon their analysis will show that Albert at three years of age attempted to play with the pubic hair of the mother and was scolded violently for it' (p. 14). Clearly, Watson believed behaviour could be better explained by ones environment and the consequences encountered during his lifetime and not on unconscious thoughts.

Almost every semester, my students would ask me where Little Albert is today, and whether he's still afraid of white rats. Finally, in 2010, I was able to answer. Beck et al. (2009) searched medical records and historical documents and worked with facial recognition experts for seven years, finally finding a baby and mother that matched Little Albert. The puzzle pieces lead them to a boy named Douglas (not Albert) Merritte. Sadly, we will never know if Little Douglas is still scared of rats because he died at the age of six from hydrocephalus.

Nowadays classical conditioning is regularly discussed when training zoo-housed animals, often when a clicker is involved. If you are an animal trainer, then you may wonder why I don't discuss clickers in this section. For those non-animal trainers, *Clickers* are hand-held devices that, when pressed, make a clicking sound. Clickers and similar devices (like whistles) are discussed by animal trainers as conditioned reinforces, because they are paired with food. However, this has recently been questioned (see Dorey & Cox 2018 for a discussion on the topic) and more research needs to be conducted to make this claim.

1.4 Operant Conditioning

Watson, Pavlov and Thorndike set the tone for future research in animal behaviour. They were major influences on our next scientist, B.F. Skinner.

As a graduate student at Harvard University, Skinner worked in both the psychology and the physiology departments. Each department assumed the other was supervising Skinner, which wasn't the case. Without supervision, Skinner had the freedom to do whatever he wanted! One of the first things he did was build new equipment. Skinner found Thorndike's experimental setup to be lacking, because the researcher had to place the cat in the puzzle box after every trial. To fix this flaw, Skinner invented the *Skinner box*, or *operant chamber*. The operant chamber contained a bar or key that the animal pressed to obtain food. Food was delivered by a dispenser called a *hopper* and a light signalled the start of an experiment. Data from in the operant chamber were collected electronically on a *cumulative record*. This recorded every response the animal made as an upward movement on a horizontally moving line. This record allowed Skinner to collect data on the effects of contingencies and on the subject's rate of responding. With this new equipment, Skinner showed that the rate of responding depended on what *followed* the bar press and not what *preceded* it. This differed from the original work of Watson and Pavlov.

To further distinguish his findings from Pavlov's classical conditioning, Skinner coined the term *operant conditioning* (also known as instrumental conditioning). Skinner used 'operant conditioning' to refer to how an organism operates on the environment. In other words, according to Skinner, an animal's own behaviour can cause events in the environment. For example, if the key is turned then the car starts; if the tail is pulled then the dog bites; if the target is touched then food is delivered; if a leash is pulled then the dog is choked; if the electric fence is touched then the animal is shocked. Unlike classical conditioning, operant conditioning involves the consequence occurring only if the animal engages in a particular behaviour, making the behaviour more likely to occur in the future.

1.4.1 Reinforcement and Punishment

Although Pavlov originally coined the term *reinforcement* to describe the strengthening of the association between an unconditioned and a conditioned stimulus, it was Skinner in the 1930s who found that if a reward is given for performing a target behaviour, the rate of that target behaviour increased the next time it is performed. This was a very important discovery and although it might seem simple, it is not. For example, by giving your dog a treat, you are not using the concept of reinforcement. For you to know if you are using reinforcement, you must know what is happening to a behaviour after giving the reward. If you don't know what is happening then you don't know if the reward is reinforcing the target behaviour or not. So, reinforcement is defined by increasing behaviour. For the sake of my favourite undergraduate professor, Dr Hank Pennypacker, I should also point out that when we speak about reinforcement, we should state that we are reinforcing the target *behaviour* and that we are not reinforcing the *subject*. Dr Pennypacker would always say, 'Reinforcing a subject would entail a metal rod and concrete'. You never forget that image.

On the opposite side of the coin is punishment. In an earlier version of his law of effect, Thorndike mentioned that behaviours which are punished were eliminated. In experiments on human verbal learning, he found that if someone said 'right', the subject had an increase in rate of responding. However, if the experimenter said 'wrong', the effects were similar to if the experimenter said nothing. Thorndike took these results as evidence against punishment being an effective process to stamp out behaviour (Catania 1998). Thus, he replicated the second half of the law of effect, keeping the reinforcement

part but discarding the punishment part (Catania 1998).

Today we know that punishment decreases the rate of behaviour. Thus, like reinforcement, you need to know/look at the behaviour afterwards to define punishment. For example, if I slap someone for grabbing me, am I punishing them? It's possible, but it's also possible that I could be reinforcing it, just watch the movie 50 Shades of Grey! Only if that slap decreases the likelihood of grabbing can one call it punishment. In summary, reinforcement increases the likelihood of a behaviour occurring in the future and punishment decreases the likelihood of a behaviour occurring in the future.

Skinner (1953) identified four basic procedures of operant conditioning. Table 1.1 is helpful for understanding the types of operant conditioning. You will notice the words positive and negative. These words do not mean 'good' and 'bad'. Instead, you should think of these words as mathematical terms; 'positive' means *adding* a stimulus to the situation, and 'negative' means taking *away* a stimulus.

In *positive reinforcement*, a stimulus following a behaviour is added which increases the likelihood of that behaviour. For example, it is positive reinforcement when someone gives an elephant a piece of watermelon for touching her/his foot to a target (operant behaviour), and then that elephant touches her/his foot when the target is presented more often (increases in response strength).

Negative reinforcement is when a response results in the removal of an event, and the response rate increases. This is ordinarily

Table 1.1 Overview of the four basic procedures of operant conditioning.

	Effect on behaviour	
Consequence	Increases the likelihood of future behaviour	Decreases the likelihood of future behaviour
Stimulus is added (+)	Positive reinforcement	Positive punishment
Stimulus is removed (−)	Negative reinforcement	Negative punishment

something the organism tries to avoid or escape, such as electric shock from an electric fence. For example, when you want a horse to turn left, pressure is applied to the left rein. The horse, to avoid the pressure of the bit, will turn its head left and the behaviour is likely to increase. A horse that is used to being ridden will take the slightest cue of the reins to avoid any pressure. Negative reinforcement confuses a lot of people, so let's look at a second example. A person with a headache may take an aspirin. If the headache goes away, the person is likely to take an aspirin the next time a headache occurs. The response (taking aspirin) results in the removal of an event (headache) and the response rate increases (taking aspirin increases).

In *positive punishment,* a stimulus is added, but the rate of the behaviour decreases over time. For example, when someone puts a shock collar on a dog to stop the dog from barking, they are adding a stimulus (shock) in hopes of decreasing a behaviour (barking). If the dog does decrease its rate of barking in the future, then positive punishment was used.

The last basic procedure is *negative punishment.* In this procedure, the removal of a stimulus decreases the target behaviour. For example, you might remove your attention by walking away or turning your back to an animal during a training session if they did something incorrect (i.e. a timeout). If the rate of that incorrect behaviour decreases in the future, then negative punishment was used.

To assess which of these four basic procedures is being used, you must know *what behaviour* you are analysing. For example, a parent and a child are in a grocery store. As they pass the candy aisle, the child grabs a bag of candy. The parent, noticing this behavior, asks the child to put the candy back on the shelf. Immediately, the child has a tantrum in the store yelling that they want the candy. The parent, embarrassed by the child's behavior, gives the child the candy back. If the rate of tantrums increase in the future when the child wants something, which procedure was used? If you said positive reinforcement, you are correct. But what if the behaviour under analysis is the *parent's?* What procedure is being used if their behaviour of giving the child candy for decreasing the tantrum? This one isn't easy, but you already know everything you need to answer this question correctly. The parent did not like the tantrum and the embarrassing situation it was creating. This aversive situation was presented before the parent allowed the child to have the candy. After the parent gave the child the candy, the child stopped tantruming and the aversive situation was gone. Therefore, the parent's behaviour was negatively reinforced.

1.4.2 Schedules of Reinforcement

Inarguably, Skinner's most influential contribution to the science of behaviour is his 1957 publication on schedules of reinforcement with his colleague Charles Ferster. However, this idea wasn't created by formal empirical methods or theoretical questioning. The idea was serendipitous, caused by necessity to have some time off and lack of resources. In explaining the origins of the schedules of reinforcement theory, Skinner wrote:

> Eight rats eating a hundred pellets each day could easily keep up with production. One pleasant Saturday afternoon I surveyed my supply of dry pellets, and, appealing to certain elemental theorems in arithmetic, deduced that unless I spent the rest of that afternoon and evening at the pill machine, the supply would be exhausted by ten-thirty Monday morning …. It led me to apply our second principle of unformalized scientific method and to ask myself why every press of the lever had to be reinforced. (Skinner 1956, p. 226)

Skinner's data showed interesting patterns of behaviour when the pellets were provided at different intervals and decided that 'when you run into something interesting, drop everything else and study it' (Skinner 1956, p. 363). Schedules of reinforcement were born that day.

But what are schedules of reinforcement?

Generally, schedules of reinforcement can be continuous or intermittent. On a *continuous* reinforcement schedule, every emitted target behaviour is followed by a reinforcer. An example of this in the natural environment would be a baby learning to drink from their mother's teat. Every time they suck on the teat they will get milk, which increases their chances of sucking on the teat in the future.

However, schedules of natural contingencies are often *intermittent*: the behaviour isn't reinforced every time it occurs. A classic example of an intermittent schedule in nature is bees' foraging behaviour. A single bee will visit several different flowers to find nectar. However, if they visit the same flowers, they may not find nectar every time because the flower needs time to fill. We also see this in zoos/aquariums if you vary whether food is put in environmental enrichment items. The animal might check the environmental enrichment item every time it is placed into the enclosure, but will only manipulate the item if it is filled with food.

Continuous and intermittent reinforcement can be broken down into four schedules or reinforcement: fixed ratio, variable ratio, fixed interval, and variable interval (Table 1.2).

Schedules of reinforcement can differ in two ways. Firstly, they can differ based on whether they come from the number of responses or the amount of time passed. In *ratio* schedules, reinforcement depends on the number of responses made. Ratio schedules are set to deliver reinforcement following a particular number of responses. *Interval* schedules are set to deliver reinforcement when one response is made after some amount of time has passed. In *fixed* schedules, the number of responses needed to obtain reinforcement is the same every time. The number of responses can be 1 or 1000, but that number is fixed. In *variable* schedules, the number of responses required for reinforcement varies around some average.

Let's go over some examples. A keeper training an elephant to touch a target delivers food on a variable ratio 5 (written as VR 5). This means that on average, every fifth response, the elephant will receive food. So the elephant might receive a piece of sweet potato on the first response, sixth response, second response, eighth response, fifth response, and the eighth response and so on. If we were to train that same elephant

Table 1.2 Reinforcement schedules.

Reinforcement schedule	Definition	Example
Fixed interval	Reinforcement is delivered at a predictable time interval.	Turning out the animals to the yard: every morning at 10 am the keeper opens the night enclosure door, but the animal's behaviour of checking the door to go outside isn't reinforced until they check the door after 10 am.
Variable interval	Response is reinforced after an interval of time which varies but centres around some average amount of time.	Animal feedings: the time of feeding an animal may vary from day to day, but on average a keeper gives food every 4 hours. Therefore, the animal's response of checking the bowl will not be reinforced until an average of 4 hours has passed.
Fixed ratio	Response is reinforced only after a specified number of responses.	Multiple repetitions: you want the animal you are training to do multiple repetitions of the same behaviour. Therefore, you deliver reinforcement after every 2 correct responses.
Variable ratio	Response is reinforced after an average number of responses.	Laboratory study: the lever in a Skinner box gives a pellet on average after 20 pulls. Thus the rat might receive a pellet after 2 pulls or after 15 pulls, but on average it is 20 pulls of the lever to receive a pellet.

and behaviour with a fixed ratio 5 (written as FR 5), then we would give him a piece of sweet potato after every fifth response: the number of required responses is *fixed* at 5.

1.4.3 Extinction

Extinction was discovered through another serendipitous event. During one of Skinner's experiments on satiation, a rat was pressing the lever when the pellet dispenser jammed. Skinner was not in his lab at the time and thus couldn't simply fill the dispenser. However, when he returned he found an interesting curve. Skinner states in his book, 'The change was more orderly than the extinction of a salivary reflex in Pavlov's setting and I was terribly excited. It was a Friday afternoon and there was no one in the laboratory whom I could tell. All that weekend I crossed streets with particular care and avoided all unnecessary risks to protect my discovery from loss through my death' (Skinner 1979, p. 95).

Extinction is defined as withholding reinforcement from a previously reinforced response. As a training process, during extinction, there is a zero probability of reinforcement for the response you were previously reinforcing. As a behavioural process, extinction is a decline in the rate of responding caused by withdrawal of reinforcement.

The interesting curve that Skinner found was probably an *extinction burst*. When the extinction procedure has started, a sharp increase in responding occurs, followed by a slow decline in responding as the trials continue. An example that you have probably witnessed is when you wait for an elevator. If you press the button and the doors don't open, you press it again. You are then frustrated so you press it 10 times in a row (extinction burst). If the doors still don't open, you might press it one or two times more before taking the stairs.

1.4.4 Shaping

The story of Skinner's discovery of shaping is quite interesting so I thought I would include it in this chapter. Skinner was commissioned

by the military in World War II to train pigeons to guide missiles (Skinner 1960). During this time, Skinner took a leave of absence from the University of Minnesota and moved his lab into a flour mill provided by General Mills. Taking breaks from training the soon to be missile guiding pigeons, Skinner and his students noticed a wild flock of pigeons hanging around the mill. One day in 1943, Skinner and his students decided to teach a pigeon to bowl (Skinner 1958). To achieve this behaviour the pigeon needed to first swipe a ball with its beck. The group got prepared to reinforce the first swipe, but nothing happened. In the hopes of speeding up the process they decided to reinforce simply looking at the ball, then a behaviour 'which more closely approximated the final form' (Skinner 1958, p. 94). Although Skinner had used shaping before, this was the first time he had hand-shaped an animal to perform a behaviour (Peterson 2004). It wouldn't be until 1951 when an article by *LOOK* magazine was published that showed Skinner training a Dalmatian to jump, that the procedure of shaping was disseminated to the general public.

Today the term shaping is widely known to anyone that has tried to train their dog or has taken an introductory psychology course. Shaping is training closer and closer approximations to the target behaviour. For example, a rough shaping plan for getting a dog into a crate on its own using shaping may look like this:

Start with the dog 2 ft (60 cm) from the crate. Decrease the distance between the dog and the crate. When the dog is close enough, have the dog touch the crate with its paw. After the dog is touching the crate you might ask the dog for one paw in the crate and then increase the distance into the crate until the entire dog is in the crate. Finally you would work on closing the crate door whilst the dog is inside.

Shaping allows animal trainers to set out a plan that will bring the target behaviour to fruition. However, more importantly, shaping is an essential process in teaching because a behaviour cannot be rewarded and thus increased unless it first occurs. Although not all behaviour can be trained by using shaping (e.g. a sit behaviour), it still plays an important role in animal training. Shaping provides animal trainers with guidance and direction in training new behaviours.

1.5 Conclusion

In this chapter we discussed learning theory which is at the heart of animal training. To be a great trainer you will need to understand the basics so that you can apply these basics outside of a laboratory or operant conditioning chamber. Learning theories emphasise the role of external events in changing observable behaviour, so you too should focus on the antecedents and the consequences of the behaviour you observe. In addition to correctly labelling the antecedents and consequences related to the behaviour you observe, I hope, after you read this chapter, that you will also have a better idea on the differences between operant and classical conditioning; reinforcement and punishment; extinction and shaping, as well as the schedules of reinforcement. All of these concepts will be important as you move forward through this book.

Another thing you might have noticed in this chapter is that there was a focus on individual learning. This is because historically investigators have focused on the individual and this tradition carries into the current research in the area of behavioural science. This too is an important concept for those training animals. Professionals know that one size does not fit all and that training programmes need to be individualised even when working with groups of animals. I hope that you keep this in mind as you continue to read the rest of this book.

The remainder of the book will not focus on definitions, but instead will focus on the applications of these theories in both captivity and in nature giving you tons of examples to help you learn exactly how these learning theories are relevant to those working in zoos/aquariums. We will discuss more applied aspects of these terms in Chapter 4.

References

Beck, H.P., Levinson, S., and Irons, G. (2009). Finding little Albert: a journey to John B. Watson's infant laboratory. *American Psychologist* 64 (7): 605–614.

Catania, A.C. (1998). *Learning*, 4e. Upper Saddle River, NJ: Simon & Schuster.

Chance, P. (1988). *Learning and Behavior*, 2e. Belmont, CA: Wadworth Inc.

Dorey, N.R. and Cox, D.J. (2018). Function matters: a review of terminological differences in applied and basic clicker training research. *Peer J* 6: e5621. https://doi.org/10.7717/peerj.5621.

Horowitz, A.C. (2003). Do humans ape? Or do apes human? Imitation and intention in humans and other animals. *Journal of Comparative Psychology* 117: 325–336.

Peterson, G.B. (2004). A day of great illumination: B. F. Skinner's discovery of shaping. *Journal of the Experimental Analysis of Behavior* 82 (3): 317–328.

Romanes, G. (1888). *Animal Intelligence*. New York: N.Y. Appleton.

Skinner, B.F. (1953). *Science and Human Behavior*. New York: Macmillan.

Skinner, B.F. (1956). A case history in scientific method. *American Psychologist* 11 (5): 221.

Skinner, B.F. (1958). Reinforcement today. *American Psychologist* 13 (3): 84–99.

Skinner, B.F. (1960). Pigeons in a pelican. *American Psychologist* 15: 28–37.

Skinner, B.F. (1979). *The Shaping of a Behaviorist*. New York: Knopf.

Thorndike, E.L. (1936). Autobiography. In: *A History of Psychology in Autobiography*, vol. 3 (ed. C. Murchison), 263–270. Worcester, MA: Clark University Press.

Walker, S. (1983). *Animal Thoughts*, International Library of Psychology Series, 437. London: Routledge & Kegan Paul.

Watson, J.B. (1924). *Behaviorism*. New York: W. W. Norton.

Watson, J.B. and Rayner, R. (1920). Conditioned emotional reactions. *Journal of Experimental Psychology* 3: 1–14.

2

The Cognitive Abilities of Wild Animals

Lindsay R. Mehrkam

An overview of the cognitive abilities of wild animals: what can they learn in the wild and what have we been able to demonstrate experimentally?

The importance of learning for animals living in the wild should not surprise us; the physical and social environments that free-roaming animals live in are very changeable, so we would expect behaviour to be flexible, and differences in learning abilities to have adaptive significance and thus have evolutionary consequences. In this chapter, we will review the diversity of cognitive abilities among a range of animal species that have been observed in their natural habitat.

2.1   Classical Conditioning in the Wild

Just as behavioural ecologists have moved in recent years towards including learning as a function of behaviour, so too have researchers in animal learning been more willing to consider the learning abilities of animals within an evolutionary perspective. This has been particularly prominent in the field of classical conditioning. You should remember reading about the very 'classic' classical conditioning laboratory experiments of scientists such as Pavlov (see Chapter 1). While the importance of these controlled laboratory experiments cannot be understated, it is also important to recognise that conditioning is not only a pro-cess that occurs in the lab; quite the contrary, there are many examples of classical conditioning that occur in the wild as well.

In the wild, classical conditioning is widely observed in predator recognition and avoidance. Many animal species produce alarm calls when they detect a predator, and in some species, different alarm calls may be given to denote different categories of predator. For example, vervet monkeys (*Chlorocebus aethiops*) emit specific alarm calls for different categories of predators (e.g. eagles flying overhead, terrestrial snakes) that in turn elicit different responses among conspecifics in the group (Seyfarth and Cheney 1986). Alarm calls have also been studied in other species such as ground squirrels (*Spermophilus beecheyi*), which have also been shown to have specific alarm calls for snakes but not for other predators (Owings and Leger 1980). The 'meaning' of these calls (i.e. which category of predator each type of call relates to), and the appropriate responses to those calls, are often learned or acquired through social processes (see section 'social cognition' below). Stryjek et al. (2018) found that free-living Norway rats do not avoid predator odours or display other fear-related behaviour, such as freezing or increased grooming in the presence of predator odours when foraging in a well-known territory and in relative proximity to burrows and other shelters, suggesting that, although an association exists, this also depends on the context of

Zoo Animal Learning and Training, First Edition. Edited by Vicky A. Melfi, Nicole R. Dorey, and Samantha J. Ward.
© 2020 John Wiley & Sons Ltd. Published 2020 by John Wiley & Sons Ltd.

other unconditional stimulus or conditioned stimuli. Predator-avoidance acquisition appears to be a case of classical conditioning, where the alarm behaviours of a demonstrator are the unconditional stimulus, and the conditioned stimulus are cues about the predator and environment, to which the animal acquires avoidance responses (Griffin 2004). This is somewhat different from other examples of classical conditioning in that, if the conditioned stimulus predicts a biologically important event (the unconditional stimulus), then it should work best if it precedes that event ('forward conditioning', as described in Chapter 1), whereas in alarm call learning the conditioned stimulus follows the unconditional stimulus ('backward conditioning' as described in Chapter 1; Griffin and Galef 2005). The possibility that socially acquired predator avoidance is less sensitive to forward conditioning than other cases of classical conditioning may reflect an example of learning processes being shaped by the unique demands of a species' environment (Griffin 2008). Thus we can see how the ability to form conditioned stimulus–unconditional stimulus associations in the wild can ultimately have adaptive significance to the animals, in this case by promoting success in behaviours that lead to a fitness benefit for the animal or species.

One of the greatest threats facing wildlife today is conflict with humans. For many species, behaviours that cause conflict (such as overgrazing, crop destruction, or predation on livestock) are often learned behaviours (Much et al. 2018). Furthermore, these behaviours become more likely as animals become habituated to people and to stimuli paired with people. Through classical conditioning processes, we know that non-lethal aversive stimuli can be at least temporarily effective in reducing learning potential. Most recently, Found et al. (2018) discovered that the frequency of aversive conditioning (specifically in the form of subjecting marked individuals to predator-resembling chases by people over a period of three months) had an effect on the wariness of elk (*Cervus canadensis*). During this conditioning period, overall wariness in elk increased significantly for elk in both high and low frequency groups. However, some animals habituated to the stimuli depending on their behavioural flexibility and so it was concluded that this method might be further increased with proactive assessment of the elk's individual personality.

The ability to learn is not just possible for the larger, mammalian species. Previously, it was thought that smaller, 'simpler' animals (notably invertebrates) were guided predominantly by genetically-mediated behaviours, since it was believed that their central nervous systems were just too small and simple, and their life histories too short, for learning to have much significance (Tierney 1986). This view is wrong. The honeybee (*Apis mellifera*), for example, learns the characteristics and whereabouts of different seasonal flowers, navigation to and from flower patches, some aspects of the 'waggle dance', characteristics of hivemates, and a lot of other things about its social and physical environment (Menzel and Müller 1996). It does all of this with a brain with a volume of about $1\,\text{mm}^3$ and containing just 960 000 neurons (Menzel and Giurfa 2001) as compared with 100 billion in the human brain. As an example, it is possible to train bees through classical conditioning to extend their proboscis in response to different solutions or odours, the normal unconditioned stimulus being sucrose solution, the conditioned stimulus being whatever solution is paired with sucrose. As well as permitting investigation of the bees' abilities to discriminate different tastes and odours, this experimental paradigm has also been used to investigate the neural substrates of learning (Menzel and Giurfa 2001).

Until recently, investigations of classical conditioning took place in laboratory settings, and even now there are few demonstrations of this form of learning in free-living animals. Nevertheless this is a powerful way by which animals learn about the relationship between different events in their environment.

2.2 Operant Conditioning in the Wild

As you learned in Chapter 1, animals also learn about relationships in their environment based on the events that precede (i.e. antecedents) and follow a particular behaviour (i.e. its consequences). When this type of learning occurs, is it termed operant conditioning. As discussed in Chapter 4, zoos are undoubtedly very well versed in operant conditioning procedures in the training of their collection animals, but how is operant conditioning useful to an animal that lives in the wild?

2.2.1 Reinforcement

Reinforcement is a process by which a stimulus change increases the future likelihood of a behaviour. Animals are reinforced in many situations in their natural environment and reinforcers can take many forms; For example, the provision of food following a successful hunt, obtaining fresh water after a long trek to a communal watering hole, a positive interaction with a conspecific, a safe place to rest or hide from predators, or an opportunity to mate with a conspecific. Many of these examples are primary reinforcers that are biologically relevant to the organism, such as food, water, shelter, and sex. As these stimuli are biologically relevant, they do not require any conditioning or learning to become a reinforcer. In contrast, secondary reinforcers are stimuli that *do* require conditioning or pairing with a primary reinforcer to become rewarding to an organism. Some examples of secondary reinforcers in the wild could be something such as finding mating signals or characteristics of quality habitat.

Reinforcement can also occur as a continuum, for example, positive reinforcement occurs when a behaviour is followed by the presentation of an (usually) appetitive stimulus and results in an increased likelihood of the behaviour being performed in the future. For example, if stalking silently in underbrush allows a lion to successfully pounce on a gazelle, the likelihood of the lion stalking during a hunt will increase in the future. If a chimpanzee is able to more easily extract ants from a log using a long stick, (as we will see later in this chapter), the chimpanzee may be more likely to manipulate sticks in future foraging tasks. If gathering moss for a nest is more likely to result in a warmer nest for a bald eagle, then the eagle may be more likely to gather moss during nest construction in the future. In contrast, negative reinforcement occurs when a behaviour results in the removal of a (usually aversive) stimulus, and thus, increases the future likelihood of that behaviour. Let's imagine we are observing a pack of wolves feeding on a carcass, where the dominant male of the pack growls at a subordinate pack member that is also attempting to feed from the carcass. If the subordinate wolf ceases to approach the carcass or retreats, we would expect the dominant male to stop growling as well; thus, for the subordinate wolf, retreating is negatively reinforced because the behaviour results in the removal of the aversive stimulus (i.e. growling from the dominant male). Seen from the perspective of the dominant wolf, a case could also be made for growling to be negatively reinforced, because growling also presumably results in the removal of the presence of the subordinate wolf. Furthermore, individuals can learn throughout their lifetime that, as pack hierarchies shift, these contingencies will apply to different individuals and in different situations. Maternal behaviours provide many excellent examples of negative reinforcement, as providing maternal care to offspring in many species may result in the removal of species-typical distress and contact vocalisations emitted by offspring. These examples help to demonstrate that it is important to remember that negative reinforcement, though it may not be preferred in training contexts, is a natural learning process that allows animals to adapt to their natural environments under certain circumstances.

Although we have just provided some very clear examples of positive and negative reinforcement that would be likely to occur in free-roaming animals, it can sometimes be difficult to determine whether a behaviour is increasing due to positive reinforcement or negative reinforcement. If we observe a grizzly bear swimming in a river on a hot summer day, one might wonder whether the bear is swimming because the behaviour (i.e. swimming) is providing coolness or opportunities for foraging (and is thus positively reinforcing) or because the behaviour is resulting in the removal of discomfort from overheating (and thus negatively reinforcing)? In a mobbing situation, do European bee-eaters exhibit aggression toward snakes because mobbing results in the removal of an aversive predator near their eggs (i.e. negatively reinforcing) or because mobbing provides increased opportunities for social cohesion among bee-eaters or a higher number of intact eggs (i.e. positively reinforcing). Although these are empirical questions that require further scientific study to tease apart, it is nonetheless interesting to consider exactly what processes are contributing to the learning history that modifies an individual's behaviour over its lifetime.

In addition to different types of reinforcement, the schedule at which an animal may obtain reinforcers in the wild varies tremendously. Broadly speaking, animals may obtain reinforcers *continuously* or *intermittently*. An animal that is receiving continuous reinforcement for a behaviour is obtaining reinforcement every time it exhibits that behaviour, whereas behaviours that are intermittently reinforced are not reinforced every time. Interestingly, behaviours that are *not* reinforced every time are actually more likely to persist longer in the absence of reinforcement than behaviours that are reinforced every time. As you might suspect, most behaviours that an individual animal exhibits in the wild are intermittently reinforced. For example, stalking or hunting by a male jaguar does not always result in successfully taking down prey, male peacocks that engage in sexual displays in soliciting for a potential mate do not always achieve copulation, and intrasexual competitions by bighorn sheep do not always result in a won contest for that individual animal.

2.2.2 Punishment

Just like both positive and negative reinforcement, animals experience punishment in many situations in their natural environment. Within social groups of primates, for example, subordinate animals who attempt to access preferred resources (e.g. food, reproduction with potential mates) may often be punished by dominant members of the group. Another example can be thought of in terms of mobbing – an antipredator behaviour – in birds toward other avian predators. The function of mobbing behaviour is described as a means to bring together conspecifics – usually within a social group – to remove intruders (Caro 2005). A hawk or eagle that is mobbed by a flock of crows might be less likely to visit that spot because it resulted in the aversive consequence of being attacked by multiple blackbirds simultaneously (i.e. positive punishment) or losing a preferred food item in the process (i.e. negative punishment). Individuals who leave a food source unguarded without caching or storing their food, may find that as a consequence, other individuals – either from within or outside a social unit – can run off with these resources. This would be a form of negative punishment, in which the removal of appetitive stimuli (i.e. resources) can follow from overly aggressive or passive behaviour.

Scientific evidence of negative punishment is much less prevalent than positive punishment in wild animals. However, there are some extreme examples that can be seen in cases of infant mortality and disappearances in male takeovers in free-ranging howling monkeys (*Alouatta palliata*) in Costa Rica (Clarke 1983), a species in which dominant males have exclusive access to mate with receptive and high-estrous females. Thus, punishment contingencies, although not strictly programmed as in a captive setting, are observed frequently in competitive and aggressive interactions in many species. Interestingly, in contrast to reinforcement processes, the efficacy of an animal to stop engaging in a behaviour is often higher when the behaviour is punished continuously (i.e. every time) rather than intermittently.

2.2.3 Stimulus Control

The behaviour of animals clearly changes in ways that are adaptive in their natural environments. Stimulus control occurs when a stimulus exerts discriminative control over an organism's behaviour. Many aspects or features of an animal's environment may come under stimulus control, and as such, there are infinite examples of stimulus control in the wild. For example, African hoofstock species tend to avoid areas in their natural habitat that contain stimuli that has been associated with predators (Griffin 2004; see Figure 2.1). Many ungulate species avoid grazing in habitat patches that, although filled with plentiful and high levels of dense vegetation, can conceal predators quite well.

Thus, engaging in grazing in these areas may lead to an aversive consequence of seeing a lion or being chased, injured, or killed by a predator. For example, the richness of a flower's hue may signal for the availability of nectar for foraging hummingbirds, thus serving as a discriminative stimulus.

2.3 Cognitive Abilities

Cognitive abilities in animals are often not simply conditioning, but the result of higher-level or complex forms of learning, which may nonetheless involve similar processes. However, it is important to remember that demonstrations of higher cognitive abilities in animals also often interact with classical and operant conditioning processes. In this section of the chapter, I will focus on several well-known higher cognitive abilities including tool use, spatial learning, discrimination, social learning, and cultural transmission; and examples of how these abilities have been demonstrated observationally and experimentally in wild animals.

Figure 2.1 Wild African herbivores, like zebra, have been seen to actively avoid grazing in areas associated with a predator. *Source:* Vicky Melfi.

2.3.1 Tool Use

Tool use may be defined as behaviours performed by altering a target object by mechanical means and behaviours that mediate information between the tool user and the surrounding environment (St Amant and Horton 2008). Like many cognitive skills, tool use was once considered a strictly human ability; however, it is now acknowledged that tool use occurs in primates, some other mammalian taxa, some avian species, and even reptilians. Tools can be used to achieve strictly physical tasks or can be used to achieve social goals as well. Tool use can also be transmitted socially through communities in the wild. Furthermore, specific features of an animal's environment (e.g. barriers, natural disturbance) can facilitate the use of inanimate objects as tools.

In wild animals, the creation and use of tools in a variety of adaptive ways has been observed in a wide range of species. Perhaps the most well-cited and obvious use of tools is to obtain food that does not appear directly accessible. Apes, and particularly chimpanzees (*Pan troglodytes*), are arguably the most skilful and flexible users of tools in the wild (Biro et al. 2003). Chimpanzees use sticks and poles to extract insects (e.g. ants, termites) from wooden logs and stumps (Suzuki et al. 1995). Wild orangutans (*Pongo pygmaeus*) exhibit flexible tool use in their native habitat (van Schaik et al. 1996). Western lowland gorillas (*Gorilla gorilla*) use shrub stumps as both bridges and as stabilisers during food processing (Breuer et al. 2005). Chimpanzees and bonobos use a wide variety of tool types, including twigs (see Figure 2.2), grasses, and stones, for different functions (Boesch and Boesch 1990; Inoue-Nakamura and Matsuzawa 1997), but specifically use large, flat 'anvil' stones and smaller 'hammer' stones to crack oil palm nuts (Inoue-Nakamura and Matsuzawa 1997). Similarly, tufted capuchins (*Cebus apella*) use stones to crack open Syragus

Figure 2.2 Bonobos are trying to access food in a closed basket by using a twig to 'fish' for food: bonobos have been observed to use a wide variety of tool types. *Source:* Jeroen Stevens.

nuts (Ottoni and Mannu 2001) and black-handed spider monkeys also engage in tool use with detached sticks in self-directed behaviours (Lindshield and Rodrigues 2009).

There are several examples of tool use in free-ranging non-primate terrestrial species that have been documented as well. A classic example is the use of stones by Egyptian vultures (*Neophron percnopterus*) to break open the shells of ostrich eggs (van Lawick-Goodall and van Lawick-Goodall 1966). California sea otters (*Enhydra lutris*) position stones against another stone or on their chest while supine and pound mussels repeatedly against them in order to open the mussels (Hall and Schaller 1964). Tool use has also been demonstrated experimentally in many corvids, including New Caledonian crows and ravens, which bend hooks from leaf stems to obtain larvae from logs (Bluff et al. 2010).

Although the social behaviour of aquatic species can be particularly difficult to study in the wild, there are several reports of tool use in marine mammals and fishes as well. Free-ranging bottlenose dolphins (*Tursiops* spp.) are particularly well-represented in observations of tool use, with sponging being the primary tool (Krützen et al. 2005), where a dolphin breaks a marine sponge off the seafloor and wears it over its closed rostrum to apparently probe into the substrate for fish. South American fresh water stingrays use water to extract food from an experimental test apparatus (Kuba et al. 2010). It is quite clear that tool use can be highly adaptive for foraging purposes in many species of distinct taxonomic groups.

In addition to different species' biology and characteristics, certain features of an animal's habitat can facilitate tool use as well, such as the availability of potential tools or the presence of barriers to a site with additional resources. Western gorillas have been observed to use branches as walking sticks to test water deepness in elephant pools. African elephants (*Loxodonta africana*) have also been observed to use a variety of tools in the wild; elephants will swat and scratch their bodies with vegetation held in their trunks, possibly to remove insects or alleviate itching from insect bites; toss sand or dirt in threats or play; and siphon mud through their trunk onto their bodies for thermoregulation and cooling (Chevalier-Skolnikoff and Liska 1993). Chimpanzees used leaves to aid in collecting and drinking water (Sousa et al. 2009). Boinski (1988) reported observations of a white-faced capuchin using a large tree branch as a club to attack a venomous snake. Many more examples of a variety of tool use in wild animals exist both in the scientific literature as well in anecdotal reports.

2.4 Spatial Learning, Navigation, and Migration

Another way animals learn from their environment is how to move through it. To date, most experimental demonstrations of how animals use maps and landmarks have been conducted in laboratory settings with rats, meadow voles (Gaulin and Fitzgerald 1989), and insects, and even in species with very large home ranges. Interestingly, male meadow voles (which have home ranges up to ten times larger than females) were superior to female voles on spatial learning tasks.

2.4.1 Optimal Foraging

Wild animals of all species need to be able to successfully navigate adaptively through their environment for a variety of reasons, but perhaps the most essential reason is to locate food. But it is not enough to know simply the location of food; it is also vital to be able to obtain food efficiently in the wild. One of the best-known concepts in animal behaviour and learning is optimal foraging theory (OFT), which comes from influences of ethology, ecology, evolution, psychology, and economics (Dugatkin 2013). OFT is a mathematical theory to predict various aspects of animal foraging behaviour under certain conditions (Sih and Christensen 2001). Because animals have to make a number of decisions about foraging under a given set of circumstances, those that have studied OFT might ask what food items are available and valuable and can help predict how long an animal may remain in a specific food patch.

OFT has been demonstrated with numerous species in the wild. Given what we have discussed earlier about the unpredictability of obtaining food during foraging, animals must decide what type of food to eat, based on not only size and caloric intake of the food item, but also the energy cost of obtaining that food item; that is, animals need to decide about the *profitability* of a foraging choice. This is especially true for predators that must engage considerable effort to locate and successfully take down a prey animal. For example, the depth and length of blue whales' (*Balaenoptera musculus*) foraging dives compensate for longer transit times between food patches and optimise resource acquisition; short shallow dives yielded the highest feeding rates (Doniol-Valcroze et al. 2011). Bottlenose dolphins use various types of synchronous behaviours in the wild in order to catch prey; in Shark Bay, Australia, for example, dolphins perform 'kerplunking' by making a loud splash with their tails, as a foraging strategy that may stun prey (Connor et al. 2000). Cheetahs (*Acinonyx jubatus*) prefer to kill prey species that are most abundant, medium-sized, and can be consumed before kleptoparasites arrive; furthermore, the morphological adaptations of the cheetah appear to have evolved to specifically hunt species like Thompson's and Grant's gazelles and impala with minimal risk of injury (Hayward et al. 2006). For prey animals such as ungulates, optimal foraging must also involve minimising predation risk (Kie 1999). Plains bison (*Bos bison bison*) search for food and move between suitable food patches in ways that are highly influenced by environmental conditions such as snowfall (Fortin 2003). Mountain gorillas (*Gorilla gorilla beringei*) have also been shown to alter their ranging patterns based on both the distribution and abundance of food, and these patterns change at different times of the year that influence the quality of the food in their habitats

(Vedder 1984); specifically, gorillas revisited regions more often that were associated with a greater renewal rate of foods. Other species have included great tits (Cowie 1977), white-throated sparrows (Schneider 1984), little penguins (Ropert-Coudert et al. 2006), and bluegill sunfish (Werner and Hall 1974).

2.5 Learning from Others: Social Cognition and Learning

Animals learn a great deal of valuable information from conspecifics in the wild. Imagine if one had to learn everything by trial and error or through direct manipulation of the environment; in such cases, animals could only learn how to hunt, forage, construct habitats, communicate and navigate through their surroundings, all things required for survival, by successfully engaging in that task themselves. While some individuals may successfully learn these behaviours through trial and error, we would suspect that many individuals do not succeed in performing behaviours as well or as quickly as we might expect (if at all). Learning information or behaviours by observing others is therefore adaptive for many species.

It is important to keep in mind, that although there are many ways in which an animal can use information from other individuals (e.g. social facilitation, stimulus and local enhancement), it is important to be able to distinguish these cases from formal types of social learning (e.g. observational learning, imitation, emulation, and cultural transmission) (Galef 2012).

2.5.1 Social Facilitation

Social facilitation is considered to be an involuntary or automatic process in which an increase in responding occurs due to simply being in close proximity to other individuals (usually, but not always, a conspecific) (Zentall and Galef 2013). When a herd of wildebeest collectively flee in response to an approaching lion, birds feeding in flocks, or a school of fish swim collectively together through coral reefs, you are likely observing social facilitation.

Often highly adaptive, social facilitation is not considered a type of social learning. First, the animal need not learn something new in social facilitation. When all members of a wildebeest herd flee from a predator on sight, those individuals are not learning a new behaviour nor are they likely to be learning any new information about the predator; wildebeest, like many ungulates, are predisposed to recognise that predators signal the threat of an attack, injury, or death. Although individuals in groups are less vigilant, grouping makes it less likely that any individual will be preyed upon (this is termed the *dilution effect*). Field studies are rich with examples of social facilitation exhibited by animals in the wild. For example, eggs may hatch simultaneously from the same clutch (Vince 1964), sea turtle hatchlings follow one another in their initial migration to the sea (Carr and Hirth 1961), and male tropical frogs' mating choruses are strongly facilitated in leks (Brooke et al. 2000).

2.5.2 Stimulus and Local Enhancement

Social enhancement is an increase in the tendency to interact with an object (stimulus enhancement) or approach a location (local enhancement) because of the presence and actions of another individual (Zentall and Galef 2013; see Figure 2.3). A very well-known and early example of this was in free-ranging ducks and noted by Austrian ethologist Konrad Lorenz. Lorenz (1935) observed that an individual duck was more likely to escape from a pen through a hole in the fencing if that individual was in close proximity to another duck that happened to be passing through the hole as well. In social enhancement, it appears to be that other conspecifics in close proximity to some terminal goal increase an animal's attentiveness to that stimulus. You might imagine that this behaviour is very reminiscent of large flocks of waterfowl foraging or preening in bodies of water, and is especially clear when incoming birds land in the same general area as well.

Figure 2.3 Young animals learn a great deal from their parents, like this gorilla infant who is learning what should and shouldn't be eaten through observation. *Source: Sarel Kromer.* https://commons.wikimedia.org/wiki/File:Gorilla_mother_and_baby_at_Volcans_National_Park.jpg.

2.5.3 Observational Learning

Observational learning is a form of social learning that occurs through watching the behaviour of others. As a result, we can assume that learning occurs *during* observation. Observation of innovative behaviours may result in faster acquisition of adaptive novel behaviours in a group, but in doing so, may also enhance an individual member's chances of surviving and reproducing (Yeater and Kuczaj 2010). Observational learning has been reported in free-ranging individuals of many species, especially non-human primates, but also many other mammals as well as avian species.

One of the most well-known examples of observational learning is that wild and laboratory rats gain information about both the saliency and noxiousness of unfamiliar foods by observing others (Galef et al. 1984). Many rodents, both laboratory and free-roaming, avoid unfamiliar foods and rely instead on social experience to inform them of the relative safety of new food sources. Consider for a moment if animals needed to directly sample all unfamiliar foods themselves. In such a scenario, there would be a very high risk of consuming an item that could be highly toxic, which would have a fatal consequence for the individual. Opportunities to learn how food items should be most efficiently processed or obtained prior to eating can come from observing others as well. Through observational learning, great tits have also been able to successfully learn how to peck through the sealed caps of milk bottles to obtain the milk within.

Dolphins have also demonstrated the use of social learning. Bender et al. (2009) found that Atlantic spotted dolphin mothers (*Stenella frontalis*) used observational learning to teach their calves foraging techniques. Mothers chased prey for longer periods of time and made more referential body pointing movements while foraging when naive calves were present, suggesting that such behaviours provide extended opportunities for the calves to observe the mother's behaviour. In addition, when mothers were foraging with their attentive calves, the mothers sometimes would let the prey escape and burrow into the sand before recapturing the prey, and even allowed calves to chase the prey. By altering their foraging strategies, the mothers increased their calves' interest in the prey and provided a rich opportunity for the calves to learn foraging behaviours through the process of observational learning.

A major preoccupation for animals in the wild is to avoid predators, so we would expect natural selection to favour the ability to learn such things as recognising a predator, responding appropriately to different kinds of predators, and avoiding places where predators are likely to be. In turn, we would expect predators to learn how to recognise prey, how to respond to the prey animal's antipredator behaviours, and where to find the best places to encounter prey. Much research has been undertaken on alarm calls, which in a number of species such as vervet monkeys *C. aethiops* (Seyfarth and Cheney 1986) and Belding's ground squirrels *Spermophilus beldingi* (Mateo 1996; Mateo and Holmes 1997) are specific for different categories of predators, promote different responses when heard by others, and are learning from other members of the group (Griffin 2004; Hollén and Radford 2009). Some animals also learn and respond to the alarm calls of other species. For example fairy wrens (*Malurus cyaneus*) learn the acoustically similar alarm calls of the scrub wren (*Sericornis frontalis*), and also the acoustically dissimilar calls of the honeyeater (*Phylidonyris novaehollandiae*) (Magrath et al. 2009). Predators, of course, have to learn something about prey too. Meerkats (*Suricata suricatta*), for example, teach their pups prey-handling skills by providing them with opportunities to interact with live prey (Thornton and McAuliffe 2006). And prey species are likely to evolve antipredator strategies where the predator may need to learn a discrimination or a new behaviour. Some prey species evolve aposematic colouration, warning potential predators that they are distasteful, toxic, or in some way dangerous. Predators learn to discriminate and subsequently avoid prey with these warning colours (Lindström et al. 2001; Svádová et al. 2009).

It is, of course, not only predators who benefit from learning about how to acquire and process food. Some of the earliest evidence for social learning came from experiments showing how wild and laboratory rats gain information about the palatability and toxicity of unfamiliar foods by observing others (Galef et al. 1984). As opportunistic omnivores, rats can thus learn quickly to avoid dangerous novel foods without having to sample them all themselves. Opportunities to learn about what items can be safely eaten, and how food items should be processed prior to eating, can come from observing others, but also from being provided with samples of different foods to practice with, as for example in gorilla parenting (*G. gorilla*) (Nowell and Fletcher 2008; see Figure 2.3), and encountering food items previously processed by others, for example capuchins C. *apella* finding bamboo segments previously opened to extract beetle larvae by other members of the group (Gunst et al. 2008).

Apart from finding food and avoiding becoming food, the other main preoccupation for wild animals is to reproduce. Successful reproduction involves finding and selecting appropriate mates, engaging in copulation, and promoting the survival of offspring. A good example of the role of learning in this comes from studies of sexual imprinting, where young animals learn the characteristics of potential mates (such as species-membership), but also who to avoid mating with (because of genetic relatedness), and thus develop sexual preferences. Thus mate choice and subsequent species divergence can be influenced by learned preferences (Irwin and Price 2001; Witte and Nöbel 2011).

2.5.4 Imitation, Emulation, and Cultural Transmission

Imitation and emulation are also considered types of social learning, albeit with more stringent criteria. They both involve sharing information between conspecifics and sometimes also across generations within a species. Evidence of these types of social learning as it occurs in the wild may provide evidence for cultures in the animal kingdom.

2.5.5 Difficulties in Determining Learning Type

Cetaceans are also known to be highly imitative and capable of social learning, both in the wild and in captivity (Krützen et al. 2005). Killer whales (*Orcinus orca*) capture seal pups by intentionally stranding on breeding beaches off the coast of Argentina (Guinet and Bouvier 1995). Adult females modified their stranding behaviour in the presence of naive juvenile calves, suggesting that females were providing the calves with opportunities to observe various stranding techniques that could be used to capture seal pups. Guinet (1991) suggested that killer whale calves developed intentional stranding foraging skills through imitation of the successful hunting behaviours of their mothers (or other relatives). Dolphins have been shown to copy the actions of another dolphin for a food reward but can do so even when they are blindfolded (Jaakkola et al. 2010). Some have suggested that the observer dolphin may accomplish this by using the sound produced by the motions of the other dolphin or perhaps through vocal communication occurring between the dolphins (but outside the range of human hearing).

Horner and Whiten (2005) investigated emulation in wild-born chimpanzees from an African sanctuary and compared their behaviour to three- to four-year-old children who observed a human demonstrator use a tool to retrieve a reward from both a clear and an opaque puzzle-box. In the opaque condition, it was impossible to differentiate between the relevant and irrelevant parts of the demonstration, whereas, with the clear puzzle-box, it was possible to differentiate between the relevant and irrelevant responses made by the demonstrator necessary to obtain the reward. When chimpanzees were presented with the opaque box, they reproduced both the relevant and irrelevant actions, thus imitating the overall structure of the task. When the box was presented in the clear condition they instead ignored the irrelevant actions in favour of a more efficient, emulative technique. These results of Horner and Whiten (2005) suggest that emulation is a preferred strategy of chimpanzees when a demonstrator's necessary actions are observable, whereas this was not the case when the demonstrator's actions were not observable (i.e. with the opaque puzzle box). Interestingly, children employed imitation to solve the task in both conditions, even when it was not most efficient. The authors suggested that the difference in strategies between children and chimpanzees might be due to a greater susceptibility of children to cultural conventions, that is, where performing the actions of the demonstrator tends to be rewarded.

In wild populations, it can be difficult to identify the transmission of a behaviour based on social or observational learning; this can also be the case in captive settings

(see Figure 2.4). A behavioural trait is considered to vary culturally if it is acquired through social learning from conspecifics and transmitted repeatedly within or between generations. Although we discussed tool use previously, there is also scientific evidence that tool use is socially learned in the wild. This should come as no surprise because, as previously mentioned in the definition set forth by St Amant and Horton (2008), tool use can also be used to mediate information between the tool user and the surrounding environment, including other animals in the environment (St Amant and Horton 2008). Social learning of tool use in orangutans was demonstrated experimentally by Call and Tomasello (1994). Of 16 orangutans, 8 individuals observed a human demonstrator use a rake-like tool to extract an unattainable, high-value food. Whereas the remaining eight individuals observed the demonstrator use the tool in a different, non-functional way. Interestingly, there were no behavioural differences observed between the two groups of orangutans studied; instead, many individuals appeared to rely on idiosyncratic trial-and-error rather than

Figure 2.4 While the walruses here look like they're imitating their keeper, it is difficult to know how this behaviour originated without observing the animals more fully. *Source: Katharina Herrmann.*

mimicking their use of the tool in the same way as the demonstrator did. As you might have suspected based on your reading from the prior section, these findings suggest that these particular orangutans were displaying an excellent example of emulation learning, but not imitation learning. Nonetheless, it is a clear demonstration of how tool use can at least be facilitated by observing another social being engaging with the tool.

Another example is that different types of tool use are transmitted and practiced by individual free-ranging animals of different communities. However, individuals within the same community often demonstrate similar if not identical types of tool use. Although chimpanzees are perhaps the most well recognised species to exhibit community effects of tool use, cultural transmission of tool use has also been reported in wild bottlenose dolphins, specifically the transmission of sponging from mother to female offspring (Krutzen et al. 2005). Sponging, as discussed previously, is significantly sex biased to females, making it comparable with sex differences in learning tool use in chimpanzees.

2.6 Learning in Response to Human-induced Changes

Individuals may also learn how to behave in response to anthropogenic changes in their natural environments. Many species may alter their behaviour to avoid human-dense or human-disturbed areas. However, in some cases, species may change their dispersal or foraging patterns if a plentiful source of food is available in human-changed landscapes. For example, southern stingrays (*Dasyatis americana*) at Stingray City Sandbar (SCS) in the Grand Cayman Islands have been receiving supplemental feedings as a result of eco-tourism for nearly 30 years. Scientists, using tag-recapture data and acoustic telemetry field methods, were able to collect data on activity patterns of stingrays at this site and compare to stingrays at other control sites

where ecotourism and supplemental feedings were not occurring. The human-influenced stingrays' had significant alterations to their natural activity patterns and habitat use relative to stingrays at wild control sites (i.e. non-tourism sites). In contrast to nocturnal stingrays at control sites, supplemented stingrays were constantly active during the day with little movement at night, stayed in close proximity to the ecotourism site, and exhibited differently distributed social behaviour. Although these behavioural changes may be adaptive to this population of stingrays in the short-term (i.e. direct access to food with relatively few costs incurred), supplemental feeding has strikingly altered movement behaviour and spatial distribution of the stingrays, and generated an atypically high density of animals at SCS, which could have downstream fitness costs for individuals and potentially broader ecosystem effects (Corcoran et al. 2013). Given the popularity of stingray interaction exhibits at many zoos and aquariums, these may be viable alternatives to ecotourism interactions without the risk of influencing endemic populations of species and their ecosystems. Zoos and aquariums can also be useful in sending conservation messages about the impact of human-induced changes in native habitats, as well as what animals can learn as a result of changing environmental conditions (see Figure 2.5).

2.7 Limitations of Wild Animal Studies

Although this chapter covered a great deal of in situ examples of learning and cognitive abilities in wild animals, it should be noted that these studies are difficult to carry out compared to studies involving captive counterparts. Relative to research in artificial settings, cognitive research on wild animals is still lacking in some areas. This discrepancy between findings of cognitive abilities in wild and captive animals is due to a number of

Figure 2.5 Carefully thought out environmental enrichment provision (a plastic drum) provides this polar bear with an object to manipulate, whilst raising awareness in zoo visitors about the disaster that is 'plastics in our oceans'. *Source: Vicky Melfi.*

substantially more expensive and time-intensive than captive studies. For example, transporting and housing research staff at field stations, purchasing sophisticated equipment for tagging and tracking individuals over potentially large ranges as well as for monitoring behaviour are all likely costs that can be incurred when conducting research with wild animals. Considering the cost and time expenses one may need to account for when studying animals in the wild, we may expect that field research is limited by funding constraints than are studies that can be done in relatively artificial environments. We should keep in mind, however, that one important potential advantage of studying wild individuals is that we are more likely to observe behaviours that are the cumulative result of natural selection.

2.8 A Final Reminder: Using Different Approaches to Understand Learning and Behaviour

There is genetic variation in many behavioural traits, and this provides the raw material for the evolution of behaviour through natural selection, but whatever it is that the genes are doing to bring about that variability, they are doing it in an environment; both the external environment that the animal lives in and the internal environment within the animal's body. Taken together, all of this variability means that while closely related animals have similar behaviours, there may also be considerable individual differences in behaviour. How much they do this depends on what sort of environment it is, what sort of animal it is, and what sort of experiences that animal has. For this reason it makes no sense to ask how much of a particular behaviour an animal exhibits is due to its genes and how much to is due to its environment.

One of the many remarkable things about animals is their ability to use their experiences to modify their behaviour. This is such

important distinctions between captive and wild settings in our abilities to record observations and conduct experiments. First, wild animals are simply more difficult to see and to get to, to enable data collection. Second, the wild setting typically allows for a lower degree of experimental control. Rigorous experimental control and the ability to manipulate only certain variables while leaving others constant over time is often required to be able to demonstrate and provide convincing evidence of many of the cognitive abilities discussed above and to provide convincing evidence and replicate findings. These are often prerequisites to publishing findings as well as disseminating information to broader scientific audiences. A third (and related) difficulty of conducting research on cognitive abilities in the wild is that free-roaming individuals are often difficult to recognise individually. A final difficulty of conducting field research is that it can be a prominent feature of our own lives that it is easy to take learning for granted, but we can also often assume that much of what animals do is in some way innate, and does not require learning. The field of animal behaviour has long been influenced by two different traditions; on one hand, comparative psychology, promoted the view that most behaviour was learned, primarily through conditioning, and that these processes were best studied in the laboratory, whereas ethology (and more recently, behavioural ecology) promoted the view that most behaviour was innate, the result of evolutionary processes, and that these processes were best studied in the animals' natural habitat. More recently, the two approaches have started to come together, with more demonstrations of animals' learning abilities that have been observed in the wild, and how this influences their evolution (Dukas 2004; Shettleworth 2001).

We know today that certain sequences of behaviours are provided by genetic processes. These sometimes exhibit variability and the behaviours that are expressed, can be changed through developmental processes and learning during the lifetime of the animal. Hatchery-reared salmon (*Salmo salar*) show a greater response to predator odours at 10–15 weeks of age than at 26–36 weeks, and this recognition of the odour is innate. But they have a peak of learning about predator odours at an age of 16–20 weeks, when they would, if in the wild, change habitats, and this learning doesn't occur in the hatchery, leading to a decline in responsiveness of older fish (Hawkins et al. 2008). Squirrel monkeys (*Saimiri sciureus*) show an intense fear of snakes if they are wild-born or laboratory-born, but not if they are laboratory-born but not fed on insects, suggesting that experience of insects sensitises the monkeys to fear of snakes (Masataka 1993). Wild-caught and naive hand-reared blue tits and coal tits avoided and did not attack aposematic firebugs (which were novel to all the birds), indicating innate recognition, whereas in the related great tits and crested tits the wild-caught birds avoided the bugs but the hand-

reared birds had to learn to avoid them (Exnerová et al. 2007). Bumble bees (*Bombus terrestris*) have innate preferences for flowers of certain colours, but learn new preferences through reinforced exposure to different colours (Gumbert 2000).

All of this, of course, has important implications for the way we manage animals in zoos. The aim to maintain captive animals with all of their species-appropriate behaviours must clearly be not just to prevent the loss of behaviour through inbreeding or inadvertent genetic selection (i.e. 'domestication'), but also to ensure that they learn about as many as possible of the things that those species would learn about in the wild. These can range from learning about species-identity (e.g. ensuring that hand-reared birds do not imprint on the wrong species) to acquiring food-handling and predator-avoidance skills (particularly if the animals are destined for reintroduction to the wild). We know from our domesticated species that behaviours do change in comparison with the non-domesticated forms, and this can affect learning. For example, domesticated guinea pigs (*Cavia porcellus*) are less bold and aggressive than wild cavies, but are able to learn associations faster (Brust and Guenther 2014). Bengalese finches were domesticated from the white-rumped munia (*Lonchura striata*) on criteria related to good parenting ability, but their song learning has been affected too, with the munias showing a much more accurate learning of song from than the finches (Takahasi and Okanoya 2010). Hatchery-reared trout (*Salmo trutta*) show faster learning than wild trout when foraging on cryptic prey (Adriaenssens and Johnsson 2011). Long-term captivity can therefore influence what the animals learn, but also their ability to learn, sometimes in unexpected ways. Similar things appear to be true for animals that are not domesticated, living in long-term captivity. Captive spotted hyaenas (*Crocuta crocuta*), for example, were more successful at solving a novel problem and showed a greater diversity in their exploratory behaviours when first interacting with the problem than wild hyaenas. This appears to be because the captive animals were less neophobic and more exploratory than their wild counterparts (Benson-Amram et al. 2013). When they eat nettles, zoo gorillas process the plants in a different way from that of wild gorillas, suggesting that while the manual skills are innate, the precise techniques are acquired through enculturation (Byrne et al. 2011). More of these kinds of studies need to be done to help us understand how learning and its contribution to naturalistic behaviours are affected over long-term captivity.

2.9 Summary and Conclusions

This chapter provided an overview of the variety of cognitive abilities documented in wild animals. It is highly adaptive and important for individuals across species and habitats to learn in the wild. The physical and social environments that animals live in are dynamic, and it should follow that behaviours are able to be dynamic too. Differences in learning abilities are to have lifetime benefits as well as evolutionary consequences with respect to foraging, reproduction and survival success. Behaviour is the product of both natural selection as well as an individual animal's external and internal (biological) environments. Great variability in behaviour occurs both between and within species; even

closely related animals have similar behaviours, as well as considerable individual differences. Conditioning, both classical and operant, allows wild animals to learn about causal relationships among events in the world that are biologically meaningful to them by learning to associate environmental cues with the appearance of biologically important events. In addition, animals are either continuously or intermittently reinforced or punished for learned behaviours, the latter of which is more common in the wild. Wild animals, especially birds and primates exhibit an impressive array of cognitive abilities: tool use, spatial learning, memory, discrimination, observational and social learning, imitation, and cultural transmission. Animals of a wide range of taxa utilise a variety of sensory and perceptual modalities to navigate through their natural environments (see modalities boxes for more information). Furthermore, communication and acquisition of communicative repertoires is considered an essential form of social learning between conspecifics in the wild, and can positively influence attracting mates, locating food, and avoiding predators. All of these types of learning can be highly adaptive to animals living in the wild. Despite the challenges of studying animals in their natural habitats, doing so has important implications for how we ensure quality learning can occur in a wide variety of animal care facilities.

References

Adriaenssens, B. and Johnsson, J.I. (2011). Learning and context-specific exploration behaviour in hatchery and wild brown trout. *Applied Animal Behaviour Science* 132: 90–99.

Bender, C.E., Herzing, D.L., and Bjorklund, D.F. (2009). Evidence of teaching in atlantic spotted dolphins (*Stenella frontalis*) by mother dolphins foraging in the presence of their calves. *Animal Cognition* 12 (1): 43–53.

Benson-Amram, S., Weidele, M.L., and Holekamp, K.E. (2013). A comparison of innovative problem-solving abilities between wild and captive spotted hyaenas, *Crocuta crocuta*. *Animal Behaviour* 85: 349–356.

Biro, D., Inoue-Nakamura, N., Tonooka, R. et al. (2003). Cultural innovation and transmission of tool use in wild chimpanzees: evidence from field experiments. *Animal Cognition* 6: 213–223.

Bluff, L.A., Troscianko, J., Weir, A.A. et al. (2010). Tool use by wild New Caledonian crows *Corvus moneduloides* at natural foraging sites. *Proceedings of the Royal Society B: Biological Sciences* 277: 1377–1385.

Boesch, C. and Boesch, H. (1990). Tool use and tool making in wild Chimpanzees. *Folia Primatologica* 54 (1–2): 86–99.

Boinski, S. (1988). Use of a club by a wild white-faced capuchin (Cebus capucinus) to attack a venomous snake (*Bothrops asper*). *American Journal of Primatology* 14 (2): 177–179.

Breuer, T., Ndoundou-Hockemba, M., and Fishlock, V. (2005). First observation of tool use in wild gorillas. *PLoS Biology* 3 (11): e380.

Brooke, P.N., Alford, R.A., and Schwarzkopf, L. (2000). Environmental and social factors influence chorusing behaviour in a tropical frog: examining various temporal and spatial scales. *Behavioral Ecology and Sociobiology* 49 (1): 79–87.

Brust, V. and Guenther, A. (2014). Domestication effects on behavioural traits and learning performance: comparing wild cavies to guinea pigs. *Animal Cognition* 18 (1); 99–109. https://doi.org/10.1007/s10071-014-0781-9.

Byrne, R.W., Hobaiter, C., and Klailova, M. (2011). Local traditions in gorilla manual skill: evidence for observational learning of behavioral organization. *Animal Cognition* 11: 683–693.

Call, J. and Tomasello, M. (1994). The social learning of tool use by orangutans (*Pongo pygmaeus*). *Human Evolution* 9 (4): 297–313.

Caro, T. (2005). *Antipredator Defenses in Birds and Mammals*. University of Chicago Press.

Carr, A. and Hirth, H. (1961). Social facilitation in green turtle siblings. *Animal Behaviour* 9 (1–2): 68–70.

Chevalier-Skolnikoff, S. and Liska, J.O. (1993). Tool use by wild and captive elephants. *Animal Behaviour* 46 (2): 209–219.

Clarke, M.R. (1983). Brief report: Infant-killing and infant disappearance following male takeovers in a group of free-ranging howling monkeys (*Alouatta palliata*) in Costa Rica. *American Journal of Primatology* 5 (3): 241–247.

Connor, R.C., Heithaus, M.R., Berggren, P., and Miksis, J.L. (2000). 'kerplunking': surface Fluke-Splashes during shallow-water bottom foraging by Bottlenose Dolphins. *Marine Mammal Science* 16 (3): 646–653.

Corcoran, M.J., Wetherbee, B.M., Shivji, M.S. et al. (2013). Supplemental feeding for ecotourism reverses diel activity and alters movement patterns and spatial distribution of the southern stingray, *Dasyatis americana*. *PLoS One* 8 (3): e59235.

Cowie, R.J. (1977). Optimal foraging in great tits (Parus major). *Nature* 268 (5616): 137–139.

Doniol-Valcroze, T., Lesage, V., Giard, J., and Michaud, R. (2011). Optimal foraging theory predicts diving and feeding strategies of the largest marine predator. *Behavioral Ecology* 22 (4): 880–888.

Dugatkin, L.A. (2013). *Principles of Animal Behavior*. Third international student edition. New York, USA: WW Norton.

Dukas, R. (2004). Evolutionary biology of animal cognition. *Annual Review of Ecology, Evolution and Systematics* 35: 347–374.

Exnerová, A., Štys, P., Fučiková, E. et al. (2007). Avoidance of aposematic prey in European tits (Paridae): learned or innate? *Behavioural Ecology* 18: 148–156.

Fortin, D. (2003). Searching behavior and use of sampling information by free-ranging bison (Bos bison). *Behavioral Ecology and Sociobiology* 54 (2): 194–203.

Found, R., Kloppers, E.L., Hurd, T.E., and Clair, C.C.S. (2018). Intermediate frequency of aversive conditioning best restores wariness in habituated elk (*Cervus canadensis*). *PLoS One* 13 (6): e0199216.

Galef, B.G. (2012). *WIREs Cognitive Science* 3: 581–592. https://doi.org/10.1002/wcs.1196.

Galef, B.G., Kennett, D.J., and Wigmore, S.W. (1984). Transfer of information concerning distant foods in rats: a robust phenomenon. *Animal Learning and Behavior* 12: 292–296.

Gaulin, S.J.C. and Fitzgerald, R.W. (1989). Sexual selection for spatial-learning ability. *Animal Behaviour* 37 (2): 322–331.

Griffin, A.S. (2004). Social learning about predators: a review and prospectus. *Animal Learning and Behavior* 32: 131–140.

Griffin, A.S. (2008). Socially acquired predator avoidance: is it just classical conditioning? *Brain Research Bulletin* 76: 264–271.

Griffin, A.S. and Galef, B.G. (2005). Social learning about predators: does timing matter? *Animal Behaviour* 69: 669–678.

Gumbert, A. (2000). Color choices by bumble bees (*Bombus terrestris*): innate preferences and generalization after learning. *Behavioural Ecology and Sociobiology* 48: 36–43.

Guinet, C. (1991). Intentional stranding apprenticeship and social play in killer whales (*Orcinus orca*). *Canadian Journal of Zoology* 69 (11): 2712–2716.

Guinet, C. and Bouvier, J. (1995). Development of intentional stranding hunting techniques in killer whale (*Orcinus orca*) calves at Crozet Archipelago. *Canadian Journal of Zoology* 73 (1): 27–33.

Gunst, N., Boinski, S., and Fragaszy, D.M. (2008). Acquisition of foraging competence in wild brown capuchins (*Cebus apella*), with special reference to conspecifics' foraging artefacts asan indirect social influence. *Behaviour* 145: 145–229.

Hall, K.R.L. and Schaller, G.B. (1964). Tool-using behavior of the California Sea Otter. *Journal of Mammalogy* 45 (2): 287.

Hawkins, L.A., Magurran, A.E., and Armstrong, J.D. (2008). Ontogenetic learning of predator recognition in hatchery-reared Atlantic salmon *Salmo salar*. *Animal Behaviour* 75: 1663–1671.

Hayward, M.W., Hofmeyr, M., O'Brien, J., and Kerley, G.I.H. (2006). Prey preferences of the cheetah (*Acinonyx jubatus*) (Felidae: Carnivora): morphological limitations or the need to capture rapidly consumable prey before kleptoparasites arrive? *Journal of Zoology* 270 (4): 615–627.

Hollén, L.I. and Radford, A.N. (2009). The development of alarm call behaviour in mammals and birds. *Animal Behaviour* 78: 791–800.

Horner, V. and Whiten, A. (2005). Causal knowledge and imitation/emulation switching in chimpanzees (Pan troglodytes) and children (Homo sapiens). *Animal Cognition* 8 (3): 164–181.

Inoue-Nakamura, N. and Matsuzawa, T. (1997). Development of stone tool use by wild chimpanzees (*Pan troglodytes*). *Journal of Comparative Psychology* 111 (2): 159–173.

Irwin, D.E. and Price, T. (2001). Sexual imprinting, learning and speciation. *Heredity* 82: 347–354.

Jaakkola, K., Guarino, E., and Rodriguez, M. (2010). Blindfolded imitation in a Bottlenose Dolphin (*Tursiops truncatus*). *International Journal of Comparative Psychology* 23 (4). Retrieved from http://escholarship.org/uc/item/7d90k867.

Kie, J.G. (1999). Optimal foraging and risk of predation: effects on behavior and social structure in Ungulates. *Journal of Mammalogy* 80 (4): 1114.

Krützen, M., Mann, J., Heithaus, M.R. et al. (2005). Cultural transmission of tool use in bottlenose dolphins. *Proceedings of the National Academy of Sciences of the United States of America* 102 (25): 8939–8943.

Kuba, M.J., Byrne, R.A., and Burghardt, G.M. (2010). A new method for studying problem solving and tool use in stingrays (*Potamotrygon castexi*). *Animal Cognition* 13 (3): 507–513.

Lindshield, S.M. and Rodrigues, M.A. (2009). Tool use in wild spider monkeys (*Ateles geoffroyi*). *Primates* 50 (3): 269–272.

Lindström, L., Alatalo, R.V., Lyytinen, A., and Mappes, J. (2001). Predator experience on cryptic prey affects the survival of conspicuous aposematic prey. *Proceedings of the Royal Society of London, Series B: Biological Sciences* 268: 357–361.

Lorenz, K. (1935). Der Kumpan in der Umvelt des Vogels: Der Artgenosse als auslosendes Moment socialer Verhaltensweisen. *Journal für Ornithologie* 83 (137–213): 289–413.

Magrath, R.D., Pitcher, B.J., and Gardner, J.L. (2009). Recognition of other species' aerial alarm calls: speaking the same language or learning another? *Proceedings of the Royal Society of London, Series B: Biological Sciences* 276: 769–774.

Masataka, N. (1993). Effects of experience with live insects on the development of fear of snakes in squirrel monkeys, *Saimiri sciureus*. *Animal Behaviour* 46: 741–746.

Mateo, J.M. (1996). The development of alarm-call response behaviour in free-living juvenile Belding's ground squirrel. *Animal Behaviour* 52: 489–505.

Mateo, J.M. and Holmes, W.G. (1997). Development of alarm-call responses in Belding's ground squirrels: the role of dams. *Animal Behaviour* 54: 509–524.

Menzel, R. and Giurfa, M. (2001). Cognitive architecture of a mini-brain: the honeybee. *Trends in Cognitive Sciences* 5: 62–71.

Menzel, R. and Müller, U. (1996). Learning and memory in honeybees: from behaviour to neural substrates. *Annual Review of Neuroscience* 19: 379–404.

Much, R.M., Breck, S.W., Lance, N.J., and Callahan, P. (2018). An ounce of prevention: Quantifying the effects of non-lethal tools on wolf behavior. *Applied Animal Behaviour Science* 203: 73–80.

Nowell, A.A. and Fletcher, A.W. (2008). The development of feeding behaviour in wild western lowland gorillas (*Gorilla gorilla gorilla*). *Behaviour* 145: 171–193.

Ottoni, E.B. and Mannu, M. (2001). Semifree-ranging Tufted Capuchins (*Cebus apella*) spontaneously use tools to crack open nuts. *International Journal of Primatology* 22 (3): 347–358.

Owings, D.H. and Leger, D.W. (1980). Chatter vocalizations of California ground squirrels: predator-and social-role specificity. *Zeitschrift für Tierpsychologie* 54 (2): 163–184.

Ropert-Coudert, Y., Kato, A., Wilson, R.P., and Cannell, B. (2006). Foraging strategies and prey encounter rate of free-ranging Little Penguins. *Marine Biology* 149 (2): 139–148. https://doi.org/10.1007/s00227-005-0188-x.

Schneider, K.J. (1984). Dominance, predation, and optimal foraging in White-Throated sparrow flocks. *Ecology* 65 (6): 1820.

Seyfarth, R.M. and Cheney, D.L. (1986). Vocal development in vervet monkeys. *Animal Behaviour* 34: 1640–1658.

Shettleworth, S.J. (2001). Animal cognition and animal behaviour. *Animal Behaviour* 61: 277–286.

Sih, A. and Christensen, B. (2001). Optimal diet theory: when does it work, and when and why does it fail? *Animal Behaviour* 61 (2): 379–390.

Sousa, C., Biro, D., and Matsuzawa, T. (2009). Leaf-tool use for drinking water by wild chimpanzees (*Pan troglodytes*): acquisition patterns and handedness. *Animal Cognition* 12 (1): 115–125.

St Amant, R. and Horton, T.E. (2008). Revisiting the definition of animal tool use. *Animal Behaviour* 75 (4): 1199–1208.

Stryjek, R., Mioduszewska, B., Spaltabaka-Gędek, E., and Juszczak, G.R. (2018). Wild Norway rats do not avoid predator scents when collecting food in a familiar habitat: a field study. *Scientific Reports* 8 (1): 9475.

Suzuki, S., Kuroda, S., and Nishihara, T. (1995). Tool-set for termite-fishing by chimpanzees in the Ndoki Forest, Congo. *Behaviour* 132 (3–4): 219–235.

Svádová, K., Exnerová, A., Štys, P. et al. (2009). Role of different colours of aposematic insects in learning, memory and generalization of naïve bird predators. *Animal Behaviour* 77: 327–336.

Takahasi, M. and Okanoya, K. (2010). Song learning in wild and domesticated strains of white-rumped munia, *Lonchura striata*, compared by cross-fostering procedures: domestication increases song variability by decreasing strain-specific bias. *Ethology* 116: 396–405.

Thornton, A. and McAuliffe, K. (2006). Teaching in wild meerkats. *Science* 313: 227–229.

Tierney, A.J. (1986). The evolution of learned and innate behaviour: contributions from genetics and neurobiology to a theory of behavioral evolution. *Animal Learning and Behavior* 14: 339–348.

Vedder, A.L. (1984). Movement patterns of a group of free-ranging mountain gorillas (*Gorilla gorilla beringei*) and their relation to food availability. *American Journal of Primatology* 7 (2): 73–88.

van Lawick-Goodall, J. and Van Lawick-Goodall, H. (1966). Use of tools by the Egyptian Vulture, Neophron percnopterus. *Nature* 212 (5069): 1468–1469.

van Schaik, C.P., Fox, E.A., and Sitompul, A.F. (1996). Manufacture and use of tools in wild Sumatran orangutans. *Naturwissenschaften* 83 (4): 186–188.

Vince, M.A. (1964). Social facilitation of hatching in the bobwhite quail. *Animal Behaviour* 12 (4): 531–534.

Werner, E.E. and Hall, D.J. (1974). Optimal foraging and the size selection of prey by the Bluegill Sunfish (*Lepomis macrochirus*). *Ecology* 55 (5): 1042.

Witte, K. and Nöbel, S. (2011). Learning and mate choice. In: *Fish Cognition and Behaviour*, 2e (eds. C. Brown, K. Laland and J. Krause), 81–107. Chichester: Wiley Blackwell.

Yeater, D.B. and Kuczaj II, A.S. (2010). Observational learning in wild and Captive Dolphins. *International Journal of Comparative Psychology* 23 (3) Retrieved from http://escholarship.org/uc/item/3qf5v7mj.

Zentall, T.R. and Galef, B.G. (2013). *Social learning: Psychological and Biological Perspectives*. Hillsdale, NJ: Erlbaum.

3

The Ultimate Benefits of Learning

Kathy Baker and Vicky A. Melfi

3.1   Introduction

One of the first theories or concepts university undergraduates studying an animal behaviour degree come across in their academic careers is likely to be Tinbergen's four 'questions' relating to animal behaviour. Proposed in his (1963) article 'On aims and methods of ethology', the four fundamental questions, sometimes considered to be problems, and sometimes termed whys, of animal behaviour have had a lasting appeal and application within biology and changed little in over 50 years since their first inception (Bateson and Laland 2013). Tinbergen identified four fundamentally different questions that can be asked to explain animal behaviour which were: how does the behaviour contribute to the animals' survival, what is it for (survival/function); how has the behaviour changed over the animals' lifetime, how did it develop (ontogeny/development); how has the behaviour evolved over time within the species, how did it evolve (evolution/phylogeny); and how is the behaviour caused physiologically, how does it work (causation/mechanism). The four questions can be broadly split into two categories; proximate and ultimate explanations for behaviour (Mayr 1961). Ultimate explanations consider the fitness consequences of a trait and thus whether it is selected (*survival value* and *evolution*), whilst proximate explanations are concerned with the physiological mechanisms which enable a trait to be performed (*causation* and *ontogeny*).

Proximate explanations for learning are discussed to some degree in other chapters within this book, with an emphasis on ontology rather than causation, i.e. how learning develops within an animal's lifetime, which is explored in discussions about the processes and methods of learning (e.g. Chapters 1 and 4). In the current chapter we will focus on the ultimate explanations for learning; how learning can effect survival of an individual and evolution of a species. This focus will also explore why certain learned behaviours evolve in certain species and not others? We will begin by conducting a brief review of current literature on the survival value of different learned behaviours such as predator recognition and highlight case studies where behaviours with a learned component have determined the extent to which the species has succeeded, or failed, during reintroduction attempts. We will then explore the less obvious benefits of the learning process itself, such as enhanced brain development, which may not have immediate or obvious benefits to inclusive fitness but no doubt offers potential survival/reproduction advantages.

Zoo Animal Learning and Training, First Edition. Edited by Vicky A. Melfi, Nicole R. Dorey, and Samantha J. Ward.
© 2020 John Wiley & Sons Ltd. Published 2020 by John Wiley & Sons Ltd.

3.2 Survival Value of Learned Behaviours

As Tinbergen himself stated 'the ultimate test of survival value [of a behaviour] is survival itself, survival in the natural environment' (Tinbergen 1963, p. 423). Animals evolve cognitive abilities, as they would physical or behavioural abilities, to deal with challenges faced in their wild ecological niches (Meehan and Mench 2007). One concern regarding zoo populations is that animals will lose their natural survival behaviours (sometimes referred to as traits or skills [Snyder et al. 1996]), or their ability to cope with these challenges (Hill and Broom 2009), due to not being subjected to natural phenomenon such as predation (see Chapter 12). Some survival behaviours are instinctive and we observe these behaviours regardless of whether animals are wild or captive. For example in many primate species that conduct arboreal locomotion 'ventral clinging' behaviour in infants is essential; offspring must cling tightly to their mothers as falling would undoubtedly lead to death, and in many species we see this instinctive clinging behaviour within hours of birth (see Figure 3.1). However responding to predator alarm calls may require a much longer period of development, for example in wild vervet monkeys (*Chlorocebus pygeryth-rus*) the production of vocalisations in the correct social context, and the correct response to others' vocalisations, gradually develops across the first four years of life (Seyfarth and Cheney 1986). Within the zoological environment we are uniquely placed to evaluate the survival value of learned behaviours through evaluating the success, or indeed failure, of reintroduction attempts. A generalised benefit for zoo animals of learning appropriate behaviours is that it will aid reintroduction attempts as they require animals to demonstrate suitable 'survival behaviours' (Rabin 2003). We can therefore highlight the survival value of learning by focusing on which of these behaviours appear to have a learned component, i.e. are less reliant on instinct and instead the behaviour develops within an animal's lifetime.

Figure 3.1 An example of the innate behaviour displayed here by a newborn silvery gibbon (*Hylobates moloch*); ventral clinging displayed by newborn primates is essential to their survival. *Source:* reproduced with permission of Chester Zoo.

Historically the distinction between instinctive or learned (nature vs nurture) behaviours has been controversial. But it is generally considered that behavioural ontogeny is a complex interplay between genetics and the environment (Barlow 1991), where some behaviours arise from a greater contribution of instinct or learning.

In order to explore survival behaviours we conducted a brief review of recent (published since 2005–2018) studies on reintroduction attempts. The review was conducted using results from a 'Web of Science' database search, where the search terms were 'reintroduction*behaviour' and 'reintroduction*learning'. Studies included in the review were: reintroductions of captive born and raised individuals (from zoos or other captive environments); animal translocations, as they represented animals being moved from one habitat to another, where learned behaviours gained in the original habitat may not necessarily correspond to the behaviours required for survival in the new habitat; and studies which either directly demonstrated or formed a strong conclusion regarding the causes of reintroduction success and/or failure. The main factors affecting the 116 reintroduction studies, which met our criteria are illustrated in Figure 3.2. We have provided a discussion of these main factors, along with a case study, to better understand how learning within zoos is essential for the long term survival of animals if reintroduced; and likely results in animals which are better representatives of their wild counterparts in zoo education.

3.3 Survival Behaviours 'Lost' in Captivity

3.3.1 Antipredator Behaviour

'Antipredator behaviour can be viewed as falling along a continuum of innateness. At one extreme some defence behaviours are expressed fully on first encounter. Most other antipredator behaviours to some extent depend on experience' (Griffin et al. 2000, p. 1320). With this in mind it is unsurprising

that in our review, and indeed other reviews of reintroduction attempts (e.g. Moseby et al. 2011; Reading et al. 2013), the most important factor influencing the success or failure of reintroduction programmes is the exhibition of antipredator behaviour (Figure 3.2). Recognition of predators' visual and sensory cues and appropriate responses such as fleeing or increasing vigilance time have been shown to be affected by time in captivity and influence the success of reintroductions. For example during a study, captive peccaries (*Pecari tajacu*) did not respond to either canine or feline predator models by fleeing or showing threat displays, so would require periods of antipredator training should they be fit for reintroductions (de Faria et al. 2018).

What appear to be small differences in activity budgets, may belie larger issues; in this context small differences in grey partridge (*Perdix perdix*) behaviour appear to affect antipredator behaviour. In order to assess whether released grey partridges displayed appropriate survival behaviours, Rantanen et al. (2010) conducted a two year experiment in Gloucestershire, UK, during 2006–2007. The study birds originated from a game farm where they had been bred in captivity for seven generations, before being released at four different sites on arable farms. Each covey (family group) consisted

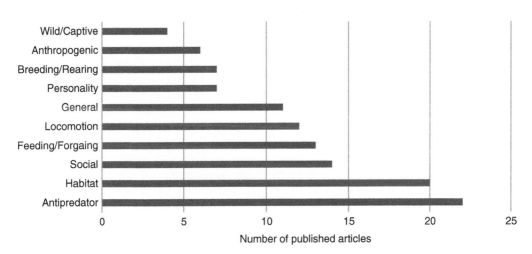

Figure 3.2 Factors affecting either success or failure of reintroduction attempts from the years 2005 to 2018. *Source:* reproduced with permission of the authors.

of 9–16 young and their foster parents (birds were brood reared artificially until three weeks and then fostered to an adult pair). Nineteen or twenty coveys were released on each of the four study sites when birds were four/five months old. The authors radio tagged 4/5 random birds from each covey resulting in 101 tagged birds in 2006 and 92 tagged birds in 2007. Behavioural observations of tagged birds were conducted after release, and mortality rates were also recorded. The released partridges showed poor vigilance behaviour compared to their wild counterparts; the mean percentage of individuals' activity budget included 3.8–5.4% of vigilance behaviour (2006–2007). Whereas comparable results from a wild study, across 20 sites in lowland UK farmland, observed wild birds spending 43% of their time in vigilance behaviour. Wild partridge also show a strong negative correlation between vigilance and group size, a relationship not seen in released animals. These studies demonstrate that the captive bred birds were not behaving adaptively to wild conditions, insofar as their performance of vigilance behaviour was too low and so they were at risk of predation (Rantanen et al. 2010).

Many antipredator behaviours that are considered 'lost' in captivity can be trained pre-release, i.e. through operant conditioning as discussed in Chapter 12; in brief, pairing a stimulus associated with a predator (e.g. visual, olfactory, auditory cues) with a stimulus perceived to be aversive (e.g. loud noise), has been used to train numerous different species to successfully avoid and or respond appropriately to predators. Allowing potential reintroduction candidates to experience a predator can confer behavioural benefits. For example in the case of burrowing bettongs (*Bettongia lesueur*), a small marsupial, captive individuals raised in conditions where they were exposed to feral cats on a regular basis displayed greater flight distance behaviour post-release than cat-naive individuals (West et al. 2018).

That these survival behaviours, which are seemingly 'lost', can be stimulated through the provision learning opportunities and training techniques, highlights the importance of providing adequate learning opportunities within zoo environments (see Chapter 5). Whether the zoo housed animals are to be used for reintroduction attempts themselves or not, survival behaviours and ensuring animals are developing a full behavioural repertoire is beneficial (Reading et al. 2013).

3.3.2 Habitat Use

Within our review making appropriate use of the habitat was the next most important factor in determining the success or failure of reintroduction attempts; as evidenced by it appearing in 20/116 studies within this review. With the studies reviewed, consideration of habitat use ranged from the ability to establish appropriate territories (Dunston et al. 2017) to dispersal patterns throughout the habitat (Richardson and Ewen 2016). Incorrect habitat use may also interact, and effect the successful display of other behaviours such as antipredator strategies. For example captive ratsnakes (*Pantherophis obsoletus*) held in captivity were less likely to survive reintroduction compared to translocated ratsnakes. This difference was thought to stem from the fact that captive individuals showed reduced concealment behaviour which made them visible to predators. Interestingly this study also showed a negative relationship between time spent in captivity and survival, indicating that short periods of time spent in captivity may have less of an effect on the display of these survival behaviours in this species (DeGregorio et al. 2017). The importance of habitat related behaviour on the success of reintroduction attempts becomes even clearer when exploring hard vs soft release techniques. Soft release, generally allows the animal to grow accustomed to the habitat, usually whilst it is still being provisioned and/or offered protection from predators whilst hard release does not offer any buffer to the reintroduced animals (e.g. de Milliano et al. 2016).

Soft release techniques may allow animals more time to learn about the new environment around them and behaviourally adapt accordingly. For example, in bird studies in particular, early dispersal from the release site is cited as a reason for reintroduction failure as the birds do not have the time to form pair bonds (Wang et al. 2017). Crested ibis (*Nipponia nippon*) under soft release conditions remained closer to the release site and flocked more rapidly, compared to those birds which were reintroduced using the hard release technique; it is suggested that both behaviours seen at soft release might offer a social advantage, in terms of facilitating the acquisition of mates (Wang et al. 2017). Interestingly in this example the survival rates of the two groups of birds remained similar, which poses the question of whether soft release programmes are important for encouraging habitat and social behaviours but may not result in a direct survival benefit for the released individuals.

3.3.3 Social Behaviour

The main issue relating to how social behaviour negatively effects reintroduction success seems to revolve around reintroduced animals inappropriately responding to social cohorts; responding to conspecifics with inadequate courtship behaviour which limits breeding opportunities, or engaging in aggression with superior adversaries potentially ending in fatality. The importance of captive rearing conditions on the exhibition of appropriate social behaviour has been demonstrated in the endangered Mexican fish (*Skiffia multipunctata*; Kelley et al. 2005). Fish raised in laboratory conditions (indoor aquaria), were observed to exhibit more courtship behaviours, and were generally more aggressive towards competitors, than their counterparts raised in semi-natural conditions (outdoor ponds), either in single taxa or mixed taxa groups. This difference in behaviour is potentially due to higher stocking densities in laboratory conditions, therefore males were more likely to encounter females and thus display more courtship behaviour. We may expect that fish that display higher levels of courtship behaviour are advantaged. Thus fish reared in higher stocking densities might be more successful if included in release programmes, as more courtship behaviours and greater aggression directed at rivals would potentially lead to more and/or better mates which would lead to greater inclusive fitness. However those fish which perform elaborate courtship displays and territorial fights, are also at risk of higher detection by predators. Therefore as the author concluded, captive-rearing conditions which promote the development of risk adverse behaviours, in this case fewer courtship and aggressive displays, would be favoured for reintroduction (Kelley et al. 2005).

The ultimate social indicator as to whether reintroductions have been successful is the ability of the reintroduced animals to reproduce. A review of artic fox (*Vulpes lagopus*) reintroductions since 2015 showed that the release of 385 individuals (over the course of 7 years) had resulted in 3 stable populations and most importantly, estimations that the number of wild-born pups descended from captive individuals had exceeded 600 (Landa et al. 2017). Artic fox introduction sites included artificial dens and food dispensers; methods of provisioning which appear to convey benefit in allowing the artic foxes time to learn about their new environment through soft release.

3.3.4 Feeding and Foraging

The development of inadequate foraging and feeding strategies is another key behavioural deficiency reported to lead to either success or failure in reintroduction programmes. For the same reason that antipredator behaviour is so essential for successful reintroduction, so too it is imperative that animals do learn how to obtain and process food resources; without these vital survival behaviours the ultimate negative consequence occurs, death. Released animals can differ from wild counterparts in relatively simple aspects of feeding and foraging such as the time of day they feed. Released grey partridges (*P. perdix*), for example, feed throughout the day rather than concentrating feeding bouts around dawn and dusk, as their wild counterparts do (Rantanen et al. 2010). This could be a result of the consistency of food availability in captivity, which leads to captive birds feeding throughout the day or it is possible that the birds are less efficient at feeding in the wild and therefore require more time to gather an adequate amount of energy from their food (Rantanen et al. 2010).

The ability to catch live prey is often reported as a reason for failed mammalian carnivore reintroductions (Jule et al. 2008), but it is likely to be as important in bird and reptile species destined for reintroduction too. A study by DeGregorio et al. (2013) demonstrated the importance that captivity can have on ratsnake foraging behaviour. Ratsnakes (*Elaphe obsolete*) that had been kept in captivity short term (under 2 weeks) reacted correctly to prey and at higher rates than counterparts who had been in captivity longer (between 1 and 60 months). Ratsnakes which had been in captivity short-term chose the correct arm (baited vs empty) of a three arm feed choice maze, and approached prey faster, than snakes kept in captivity longer term. Other variables such as time since last meal and body condition of snakes did not have an effect on feeding behaviour. There also appeared to be relationships between the type of prey cue provided and the snake reactions. Ratsnakes held in captivity short term reacted to prey at greater rates than expected by chance after chemical and visual cues, and when they were combined; they responded most quickly with a chemical cue alone. In contrast, ratsnakes held in captivity long term showed no trends, in either reaction to prey or latency to approach correct prey dependant on prey cue. The authors suggest that because ratsnakes in captivity are generally managed on a diurnal feeding pattern, that the longer they stay in captivity the less able they are to adapt to different prey items; which is problematic for a species which shifts from diurnal to nocturnal feeding strategies in the wild (DeGregorio et al. 2013).

As with antipredator training, operant conditioning can be used to enhance appropriate feeding/foraging behaviour. Conditioned taste aversion techniques have been used to encourage animals to avoid potentially toxic prey (e.g. Cremona et al. 2017), or undesirable food items (such as commercial crops), and exposing predators to live prey pre-release greatly increases their chances of successful hunting post-release (e.g. Houser et al. 2011).

3.3.5 Other Learned Factors Affecting Reintroduction

In many studies no one specific behaviour was cited as the reason for success and/or failure of a reintroduction attempt, but general differences in behaviour expressed by wild and captive conspecifics as a result of captivity were reported. For example northern water snakes (*Nerodia sipedon sipedon*) were the subject of an experimental test, of the feasibility of common reintroduction strategies, using assessments of post-release behaviour and physiological variables as indicators of success (Roe et al. 2010). Three groups of water snakes were compared, wild snakes at the study site, wild snakes translocated from their original range to the study site, and snakes reared in captivity at accelerated growth rates; a technique known as head-starting where the aim is to maximise growth rates and thus survivability upon release. All animals were captured and radio transmitters surgically inserted, after a 7–11 day recovery all snakes were released into a 500 ha nature reserve managed by the nature conservancy in northeast Indiana, USA. Snakes were located once a week during the active season (May–Sep), every two weeks during hibernation ingress and egress and once per month for the overwintering period. The captive reared snakes moved less, traversed smaller areas of land, and selected inadequate (not representative for the species) habitat; compared to the other two wild snake groups. The captive reared snakes were also rarely observed basking, foraging, or travelling and left their hibernation refuge one month earlier than resident snakes. As a result they failed to gain weight and had high rates of mortality in the overwintering period (Roe et al. 2010).

In summary, many behaviours have been shown to get 'lost' in captive populations and as such this can have substantial impacts on the likelihood that reintroduction attempts are successful. There are however, many techniques which employ learning theory such as operant conditioning (see Chapter 12) and species appropriate environmental enrichment (see Chapter 6), which are both discussed in more detail elsewhere in this text, which can bridge the learning gap between captive and wild environments (see Chapter 5).

3.4 Indirect Benefits of Learning

The review and case studies above demonstrate the immediate survival benefits of learned behaviours. To some extent they seem quite obvious to even the untrained observer; fail to develop survival behaviours and you will not survive. In addition to these obvious benefits, the process of learning itself may in fact offer benefits. 'Indirect benefits' of learning maybe conferred through changed physiological or psychological parameters, for example positive feedback loops that affect the animal's emotional state; sometimes referred to as secondary benefits. We do not propose that these indirect benefits of learning are any less important, than performing the learned behaviours themselves, rather that the benefits might be less obvious in terms of their immediate survival benefit to the animal.

Some authors have demonstrated the importance of learning in shaping naturally occurring phenomenon, such as brain size, through comparative studies (e.g. Krebs et al. 1996). Food storing birds have been shown in both experimental and field studies to have better spatial memory than species which do not regularly store food. This is also associated with a larger hippocampal region of the brain in food-storing birds (Krebs et al. 1996).

These 'natural' physiological variations in closely related species, which result from the degree of learning each species performs, offer interesting phylogenetic comparisons, but experimental studies can provide a further dimension to understanding the indirect benefits of learning. We conducted a brief review, again using recent literature (published 2005–2018) to assess the indirect benefits of providing captive animals with learning opportunities. The review was conducted on the results generated from a 'Web of Science' database search, where search terms were environmental enrichment*benefits and operant conditioning*benefits. The most common method of providing learning opportunities in captive settings was via environmental enrichment (see Figure 3.3); our review extended beyond zoo settings, to include other forms of captivity including laboratories. As such, some of the learning opportunities could be viewed as traditional forms of learning provision, such as maze tasks or operant conditioning.

The majority of experimental studies reviewed focused on laboratory animals and consequentially included common laboratory species such as rodents; which were the most documented species in our review (Figure 3.4). It is unsurprising that research focusing on physiological measures and/or recovery after brain injury are conducted in laboratories. It should be noted however, that there were also a number of studies in this review which were carried out on zoo housed animals as well as other domestic populations (Figure 3.4); note primates for example were housed in laboratories and zoos. These latter studies were less invasive in nature and focused on the benefits of learning, such as performance in cognitive testing after animals were given learning opportunities. We included all studies in our review which investigated environmental enrichment (i.e. not just environmental enrichment with a cognitive goal), as all enrichment, upon first presentation at least, can provide a cognitive challenge. Cognitive challenges or positive reinforcement training were also popular methods of providing learning opportunities.

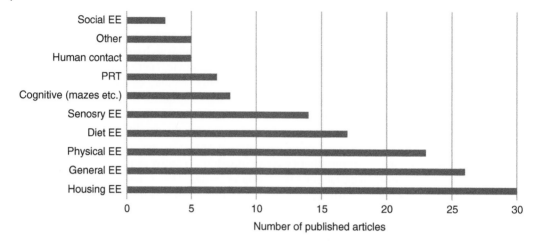

Figure 3.3 Type of learning opportunities provided in studies of indirect benefits of learning opportunities from the years 2005 to 2018 (*note:* some articles used more than one type of enrichment/learning opportunity so these have been recorded as separate learning opportunities where applicable). *Source:* reproduced with permission of the authors.

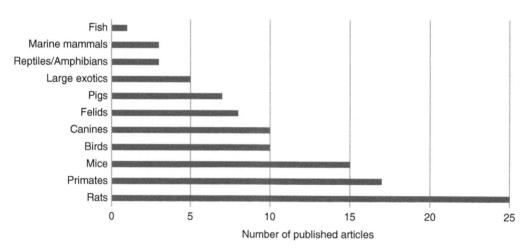

Figure 3.4 Species involved in studies of indirect benefits of learning opportunities from the years 2005 to 2018. *Source:* reproduced with permission of the authors.

The main benefits described associated with the provision of learning opportunities included, general behavioural benefits, enhanced cognitive function, and general improvements in physiological and/or brain development and/or recovery (Figure 3.5). From the studies collated in this review, we will now focus on some of the indirect benefits associated with providing learning opportunities and outline some case studies.

3.4.1 Learning Opportunities Lead to Generalised Behavioural Benefits

In the first section of this review, we demonstrated that there are behavioural benefits of learning survival behaviours, but there may be more generalised benefits to learning too. As humans we realise the importance of life-long learning, i.e. not just simply learning one task but giving ourselves a range of learning

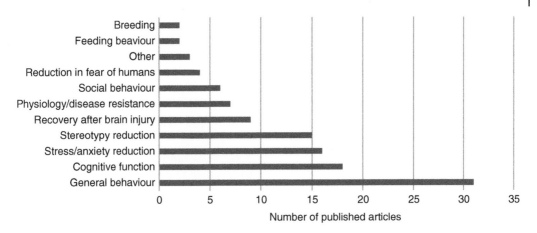

Figure 3.5 Indirect benefits of learning opportunities reported from the years 2005 to 2018 (*note:* some articles reported more than one benefit so these have been recorded as separate outcomes where applicable). *Source:* reproduced with permission of the authors.

opportunities (Duyff 1999). Throughout a lifetime both humans and animals must assimilate a wide range of information. The concept of lifelong learning in humans suggests that learning becomes easier when embedded in a lifelong learning context, learners can transfer knowledge across related learning tasks and become more experienced and generalise better (Thrun 1996); we learn to learn. Cognitive skills are 'the mechanisms by which animals acquire, process, store, and act upon information from the environment' (Shettleworth 2001). These mechanisms allow for the collection of representational information about the world, and knowledge can be exploited by animals even in the absence of the objects to which the knowledge relates (Meehan and Mench 2007). As such cognitive functions are 'higher level controls' of behaviour and it is difficult to distinguish them from behavioural controls that are part of a direct link between stimulus and response (Toates 2004). When providing environmental enrichment its evaluation often relies on the measurement of behavioural indicators or a priori 'goals' set out; measuring foraging time might be used to evaluate if a feeding device is successful and thus enriching. Whilst the relationship between feeding device and increased foraging behaviour may appear to be simple, the

expression of increased foraging involves the application of cognitive mechanisms, such as environmental perception, memory, and problem solving.

Perhaps the generalised benefits of learning opportunities provided to captive animals, can be best illustrated when considering learning opportunities that offer only intrinsic rewards, i.e. the performance of behaviour itself is rewarding unlike extrinsic learning opportunities such as feeding enrichment that offer an external reward. Sensory enrichment provides a learning opportunity for the animal but offers no immediate external reward, so could be considered an intrinsic learning opportunity. For example black-footed cats (*Felis nigripes*) spend longer investigating a cloth impregnated with novel scents (nutmeg, catnip, or prey scent) compared to control (no scent) cloths (Wells and Egli 2004). This behavioural response might be expected, as investigating a new scent could lead to an extrinsic reward, e.g. follow prey scent = find prey, but other general behavioural changes are also observed in the cats' activity budgets. Cats given scented cloths show an increase in active behaviours such as locomotion and exploring and a decrease in more sedentary behaviours during the scent conditions (Wells and Egli 2004); these additional behavioural changes might be viewed as a generalised effect of the enrichment on behaviour. There has been a recent rise in the number of published articles investigating sensory based enrichment such as music (e.g. chimpanzees, Wallace et al. 2017; gorillas, Brooker 2016; exotic song birds, Robbins and Margulis 2016), audiobooks (e.g. dogs, Brayley and Montrose 2016), naturalistic sounds (e.g. exotic songbirds, Robbins and Margulis 2016) and olfactory stimuli (e.g. lemurs, Baker et al. 2018; felids, Damasceno et al. 2017; Vidal et al. 2016; Martínez-Macipe et al. 2015). These studies cover numerous taxa and a wide range of different methodologies were used but as a general comment we can suggest that, where positive behavioural changes occur, these types of enrichments have enhanced animal welfare through offering intrinsic learning opportunities.

Providing learning opportunities may also infer a benefit to the animal in terms of shaping its personality; if you are raised in an enriched environment do you ultimately develop a personality type different to non-enriched counterparts? It seems that, in rats at least, this is the case. Using operant conditioning Brydges et al. (2011) trained rats to associate a stimulus, either rough or smooth sandpaper, with a high-value (chocolate) or standard (cheerio) reward. When given an intermediate stimulus, a sandpaper grade between the two trained scenarios, 'optimistic' rats chose to forage in the location where the high value reward was presented, whereas 'pessimistic' rats chose to forage in the location where the low value reward was presented. A comparison of the rats rearing environment showed that rats raised in enriched (physical enrichments such as tubes, sandpaper substrate) conditions showed more optimistic responses than their counterparts reared in standard cages.

These case studies demonstrate the importance of considering the 'whole animal's' response to learning opportunities. The process of learning itself may proffer benefits that we simply have not considered. For example when providing feeding puzzles we may gauge success by observing the time spent interacting with the puzzle and the ability of the animal to gain food, but has the process of gaining information from the environment itself given the animal choice and control, and as such are there other behavioural measures of enhanced welfare we should consider (see Chapter 11).

3.4.2 Learning Opportunities Lead to Enhanced Cognitive Function

Is the cliché of use it or lose it valid in the case of cognitive ability and if so why? This is the question posed in Milgram et al.'s (2006) review of cognitive enrichment in humans and animals. One of the main concerns in human neural research is the decline in cognitive function with age. Leading a physically and intellectually active life has been shown to provide a protective effect in retaining cognitive abilities. Many of gerontological studies are, by their nature, retrospective and often provide conflicting results with regards to the effect of lifestyle differences, such as education, job complexity, hobbies, and so forth on cognitive function (Milgram et al. 2006). For example, education level appears to have a positive effect on memory retention and a protective effect with respect to dementia. Benefits associated with education level in humans however, may only correspond to certain tests of cognition; education primarily affects performance on measures of crystallised intelligence (construct linked to acquired experience and knowledge) rather than fluid intelligence measures (reasoning ability, more dependent on biology rather than experience) (Kramer et al. 2004). Job complexity (regardless of age) appears to reduce the risk of developing cognitive impairment and the risk of developing Alzheimer's disease is reduced if hobbies are cognitively challenging rather than physically challenging.

Whilst we will not discuss in detail the mechanisms by which cognitive activity has been hypothesised to support indirect benefits, one suggested route is the cognitive reserve hypothesis. It is suggested that cognitive environmental enrichment is able to delay onset of dementia because the brain is able to utilise available neural structures which are active in those people who are still learning (Milgram et al. 2006; see Figure 3.6). The cognitive reserve hypothesis relies on the assumption that cognitive enrichment early in life affects brain organisation in later life, which is supported by MRIs of young and old human patients (Milgram et al. 2006). Similar findings have also been reported in rodent models, where it has been shown that cognitive enrichment leads to a number of structural changes in the brain, such as an increased number of neurons, synapses, and dendritic branches, especially in cortex and hippocampal formation (Würbel 2001).

Rodent models are often used to better understand the cellular mechanisms driving the neural effects of enrichment and as a result there is a wealth of literature investigating the effects of providing animals with both physical and cognitive challenges and the resulting effects on both cognitive function and brain physiology. For example Harati et al. (2013) subjected female rats to a series of behavioural tests (including the Morris water maze and plus maze tests) at various ages, 4, 13, and 25 months. Rats that were housed in enriched conditions showed better performance in the tasks at older ages compared to rats housed in standard laboratory conditions. Enriched conditions comprised of keeping groups of 10–12 animals in two adjoining wire-mesh cages, with objects such as tunnels and toys which were changed five times a week. Harati et al. (2013) concluded that their study demonstrated that enrichment delayed the onset of short term memory retention deficits. Whilst it can be assumed that the enrichment was providing animals with learning opportunities, those rats housed in enriched conditions will also likely be more active, so it is possible that the activity levels also contributed or were the cause of this positive effect. For example in humans, cognitive activity during mid-life is associated with a reduced risk of developing Alzheimer's disease but social and physical activity are also important (Milgram et al. 2006).

To address which component of environmental enrichment (social/physical/cogni-

Figure 3.6 An illustration of enriched laboratory mouse enclosure. *Source:* reproduced with permission of the Institute of Animal Technology/NC3Rs.

tive) can have the largest impact on the onset of Alzheimer's disease, Cracchiolo et al. (2007) carried out an experiment using Alzheimer's disease transgenic mice. At six weeks of age, mice were transferred from standard social cages to one of the following conditions: (i) impoverished, animals were housed individually in a standard Plexiglas© mouse cage; (ii) social housing, animals were housed in standard cages with other mice of the same gender; (iii) physical enrichment, animals were socially housed, and had access to running wheels; and (iv) complete environmental enrichment, animals were socially housed, and had access to tubes, tunnels and toys, etc. (within all housing conditions all items were changed weekly, and mice were placed in a novel complex environment three times a week). At 6 months of age all mice were tested, over a period of 5 weeks, on a range of behavioural tests: Y maze; Morris water maze; circular platform task (1 escape box, 16 choices/holes); and platform recognition, radial arm water maze. The authors found that during certain tasks mice raised with complete environmental enrichment outperformed mice raised in all other conditions. These data suggest that enhanced cognitive activity, beyond social and/or physical environmental enrichment, is required to protect against cognitive impairment as a result of Alzheimer's disease. In their discussion of the topic Cracchiolo et al. (2007) highlight that whilst physical activity may protect against 'normal' cognitive decline associated with ageing, or give enriched rodents an advantage in cognitive tasks, these are not the same as protecting against the effects of a neurodegenerative disease such as Alzheimer's disease.

A highly reported physiological benefit of providing learning opportunities is enhanced recovery after brain injury. In these studies, animals (traditionally rodent models) are subjected to some form of brain injury to mimic naturally occurring incidents such as brain lesions (e.g. Will et al. 2004). Animals with brain injuries raised in enriched conditions generally had better recovery than those housed in standard conditions. The benefits of enrichment do seem to vary depending on the type of brain injury and the type of enrichment provided; activities requiring learning whether environmental enrichment and/or training were better than physical activities at enhancing motor performance after brain injury (Will et al. 2004).

3.4.3 Learning Opportunities Lead to Reduced Stress and/or Stereotypy Reduction

One of the behavioural goals of providing environmental enrichment can often be a reduction in stress levels. Stereotypies and self-directed behaviours (SDBs) can develop when animals are housed in conditions which do not meet their physical or psychological needs (e.g. Lutz et al. 2003). In many cases a reduction in stereotypies and SDBs are used as indicators of reduced stress and thus the goal of many enrichment programmes may be to reduce the occurrence of these behaviours (Swaisgood and Shepherdson 2005). For example 14 captive sloth bears (*Mehursus ursinus*) provided with 'honey logs' as a type of feeding enrichment, showed a significant reduction in the percentage of time they spent performing various stereotypic behaviours (Anderson et al. 2010); the enrichment consisted of logs with drilled holes in them, which were filled with honey and then closed with wooden plugs. The effect of the 'honey logs' was seen regardless of whether the enrichment was provided on a continuous (every day for five days) or intermittent (every other day) schedule (Anderson et al. 2010).

Stereotypies may not be the most reliable indicator of stress levels. Once stereotypies have become established, they can become emancipated from their original causal factors and difficult to reduce and also be indicative of previous rather than current stress levels (Mason 1991). In order to explore whether providing learning opportunities can reduce stress we can directly measure the physiological indicators associated with stress such as cortisol and heart rate (HR) (Fraser 2008). Langbein et al. (2004) tested visual discrimination in 12 young (15–22 weeks of age) Nigerian dwarf goats *Capra hircus* using a computer controlled learning device; buttons, corresponding to stimuli on screen, had to be pressed by the goats to gain drinking water. The experiment consisted of four phases, the shaping phase where goats were taught how to use the apparatus and three testing phases. The testing phases all involved simultaneous presentation of four visual stimuli on the screen and the goats had to pick the correct stimuli to get rewarded. In testing phase 1 all three negative stimuli were identical, but in test phase 2 and 3 they were not. Throughout the experimental phases the researchers recorded HR of all individuals, their hypothesis being that unpredictable or uncontrollable situations will activate the hypothalamic–pituitary–adrenal axis, leading to depression of behaviour combined with reduced or stable HR (HR, while the ability to cope with a stressor is thought to be under the control of the sympathetic nervous system and is generally accompanied by a raise in HR). In the first training tasks, when goats were naive to the apparatus and procedure, they showed low HR suggesting a level of frustration and/or stress in response to the new stimuli. In the second and third training stages, when the training procedure and apparatus was well established, the opposite relationship was true. These data indicated that whilst the goats found the learning task challenging, it was a task they could cope with. The authors suggest that the learning task presented the goats with a 'positive stress' once the animals understood the task and learned to recognise the positive stimulus (Langbein et al. 2004).

Another interesting topic of research is whether reduction in stress can effect the impact of learning opportunities and so lead to better performances during cognitive testing such as maze trials. For example rats provided with enrichment, in the form of novel nesting material, tunnels, hanging tubes, and novel objects, were observed to outperform their non-enriched counterparts in a Morris water maze; non-enriched rats took significantly longer to find the hidden platform (Harris et al. 2009). Thigmotaxis, tendency to 'hug' the side of the apparatus, was also measured and

considered an indication of stress or anxiety; whereby those animals which are less stressed will show low levels of thigmotaxis. Enriched rats were observed to be significantly less thigmotactic than their non-enriched counterparts. The authors concluded that the enriched rats performed well in the cognitive tests, not due to elevated cognitive skills, but instead because they showed less thigmotaxis and thus were able to engage in the cognitive tests more readily (Harris et al. 2009).

3.5 Summary – How Learning Supports Zoo Animals

Many of the examples provided throughout this chapter have referred to laboratory animal studies, which might make you wonder how does this apply to zoo animal learning? Maybe you're wondering whether benefits which arise from being given learning opportunities, result because baseline levels of housing and husbandry are impoverished and therefore represent a different situation to those which zoo animals are maintained in. Notwithstanding that laboratory animal housing and husbandry is highly variable, so too is that provided within the zoo profession. Furthermore, numerous case studies have been identified based in zoos which clearly demonstrate the benefits of learning; paralleling the findings published on laboratory animals. It is true, that studies of physiological function and certainly brain development are limited within the zoo profession, and thus our understanding of how the zoo environment, including the provision of learning opportunities affects these metrics in the varied species housed in zoos is limited. It would be foolish however, to consider that the animal models used in laboratory animal studies are sufficiently different from the animals we care for in zoos, to negate the potential benefits that learning opportunities can have on mortality, morbidity, reproduction, psychological well-being, and physical welfare. Certainly is seems likely that just as learning opportunities are observed to have behavioural benefits in zoo animals, i.e. with the reduction of behaviours associated with high levels of stress, similar processes, which lead to positive brain development and function, are also likely to occur in zoo settings too.

Finally, and probably most importantly we arrive at a concept that is particularly difficult to demonstrate but central to all that we hope to understand, to ensure the zoo animals in our care have a good life – their emotions – and more specifically their emotional reaction to learning opportunities. It has been suggested that as humans, we gain an emotional benefit from engaging in learning opportunities which sufficiently challenge our abilities but are within our skill set. When the challenge and skill set are appropriately matched (see Figure 3.7) people are considered to enter a state of 'flow', a term coined by Csikszentmihalyi (1990) which has been considered one of the biggest influences on human happiness (Myers and Diener 1995). When challenges exceed our skill set we can feel overwhelmed and stressed, but if the challenges are too easily accomplished we can feel apathetic and bored. By contrast, the state of flow which can be achieved through engaging in a variety of different activities, has been associated with being described as 'in the zone' and people profess to losing track of time because they are so enthralled and absorbed by the activity they are in. For different people, different activities open up 'flow', it might be a crossword puzzle, a computer game, painting, or a musical instrument. Experiencing 'flow' seems to represent the height of positive emotional engagement and whether animals can also reach this state was initially explored by Meehan and Mench (2007). Meehan and Mench (2007) suggested that if appropriately chosen, forms of environmental enrichment could enable captive ani-

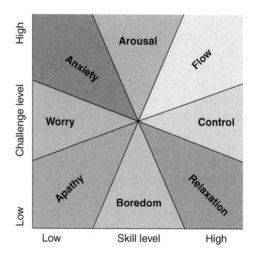

Figure 3.7 According to the flow phenomenon, if the challenge experienced can be met by an appropriate level of skill, it is possible to enter a state of flow. *Source: after Csikszentmihalyi (1997).* Reproduced with permission of creative commons licence.

mals to experience a state of 'flow'; the environmental enrichment would need to provide a challenge which the animals' species-specific behaviours were capable of fulfilling. The incorporation of 'flow' into the concept of cognitive enrichment has provided a framework, to explore the relationship between providing learning opportunities (sometimes termed cognitive challenge) and the animals' emotional experience and is being applied in zoos (Clark and Smith 2013; Clark 2017; Hopper 2017; see also Chapter 6).

Incorporating the animals' emotional response to learning opportunities, has taken puzzle feeders, a potentially overly used and under monitored object, to new sophisticated heights. There is much we can learn about animals' responses to learning from studies of ourselves, and potentially, with the right studies, we could learn about ourselves from studies of animal learning.

References

Anderson, C., Arun, A.S., and Jensen, P. (2010). Habituation to environmental enrichment in captive sloth bears – effect on stereotypies. *Zoo Biology* 29 (6): 705–714.

Baker, B., Taylor, S., and Montrose, V.T. (2018). The effects of olfactory stimulation on the behavior of captive ring-tailed lemurs (*Lemur catta*). *Zoo biology* 37 (1): 16–22.

Barlow, G.W. (1991). Nature-nurture and the debates surrounding ethology and sociobiology. *American Zoologist* 31 (2): 286–296.

Bateson, P. and Laland, K.N. (2013). Tinbergen's four questions: an appreciation and an update. *Trends in Ecology & Evolution* 28: 712–718.

Brayley, C. and Montrose, V.T. (2016). The effects of audiobooks on the behaviour of dogs at a rehoming kennels. *Applied Animal Behaviour Science* 174: 111–115.

Brooker, J.S. (2016). An investigation of the auditory perception of western lowland gorillas in an enrichment study. *Zoo Biology* 35 (5): 398–408.

Brydges, N.M., Leach, M., Nicol, K. et al. (2011). Environmental enrichment induces optimistic cognitive bias in rats. *Animal Behaviour* 81 (1): 169–175.

Clark, F.E. (2017). Cognitive enrichment and welfare: current approaches and future directions. *Animal Behavior and Cognition* 4 (1): 52–71.

Clark, F.E. and Smith, L.J. (2013). Effect of a cognitive challenge device containing food and non-food rewards on chimpanzee well-being. *American Journal of Primatology* 75 (8): 807–816.

Cracchiolo, J.R., Mori, T., Nazian, S.J. et al. (2007). Enhanced cognitive activity – over and above social or physical activity – is required to protect Alzheimer's mice against cognitive impairment, reduce Aβ deposition, and increase synaptic immunoreactivity. *Neurobiology of Learning and Memory* 88 (3): 277–294.

Cremona, T., Crowther, M.S., and Webb, J.K. (2017). High mortality and small population size prevent population recovery of a reintroduced mesopredator. *Animal Conservation* 20 (6): 555–563.

Csikszentmihalyi, M. (1990). *Flow: The Psychology of Optimal Experience*. New York: Harper and Row.

Csikszentmihalyi, M. (1997). *Flow and the Psychology of Discovery and Invention*, 39. New York: Harper Perennial.

de Faria, C.M., de Souza Sá, F., Costa, D.D.L. et al. (2018). Captive-born collared peccary

(*Pecari tajacu*, Tayassuidae) fails to discriminate between predator and non-predator models. *Acta Ethologica* 21 (3): 175–184.

de Milliano, J., Di Stefano, J., Courtney, P. et al. (2016). Soft-release versus hard-release for reintroduction of an endangered species: an experimental comparison using eastern barred bandicoots (*Perameles gunnii*). *Wildlife Research* 43 (1): 1–12.

Damasceno, J., Genaro, G., Quirke, T. et al. (2017). The effects of intrinsic enrichment on captive felids. *Zoo Biology* 36 (3): 186–192.

DeGregorio, B., Weatherhead, P., Tuberville, T., and Sperry, J. (2013). Time in captivity affects foraging behavior of ratsnakes: implications for translocation. *Herpetological Conservation and Biology* 8 (3): 581–590.

DeGregorio, B.A., Sperry, J.H., Tuberville, T.D., and Weatherhead, P.J. (2017). Translocating ratsnakes: does enrichment offset negative effects of time in captivity? *Wildlife Research* 44 (5): 438–448.

Dunston, E.J., Abell, J., Doyle, R.E. et al. (2017). Investigating the impacts of captive origin, time and vegetation on the daily activity of African lion prides. *Journal of ethology* 35 (2): 187–195.

Duyff, R.L. (1999). The value of lifelong learning: key element in professional career development. *Journal of the American Dietetic Association* 99 (5): 538–543.

Fraser, D. (2008). *Understanding Animal Welfare: The Science in Its Cultural Context*. Chichester, UK: Wiley.

Griffin, A.S., Blumstein, D.T., and Evans, C.S. (2000). Training captive-bred or translocated animals to avoid predators. *Conservation Biology* 14 (5): 1317–1326.

Harati, H., Barbelivien, A., Herbeaux, K. et al. (2013). Lifelong environmental enrichment in rats: impact on emotional behavior, spatial memory vividness, and cholinergic neurons over the lifespan. *Age* 35 (4): 1027–1043.

Harris, A.P., D'Eath, R.B., and Healy, S.D. (2009). Environmental enrichment enhances spatial cognition in rats by reducing thigmotaxis (wall hugging) during testing. *Animal Behaviour* 77 (6): 1459–1464.

Hill, S.P. and Broom, D.M. (2009). Measuring zoo animal welfare: theory and practice. *Zoo Biology: Published in affiliation with the American Zoo and Aquarium Association* 28 (6): 531–544.

Hopper, L.M. (2017). Cognitive research in zoos. *Current Opinion in Behavioral Sciences* 16: 100–110.

Houser, A., Gusset, M., Bragg, C.J. et al. (2011). Pre-release hunting training and post-release monitoring are key components in the rehabilitation of orphaned large felids. *South African Journal of Wildlife Research* 41 (1): 11–20.

Jule, K.R., Leaver, L.A., and Lea, S.E. (2008). The effects of captive experience on reintroduction survival in carnivores: a review and analysis. *Biological Conservation* 141 (2): 355–363.

Kelley, J.L., Magurran, A.E., and Macías-Garcia, C. (2005). The influence of rearing experience on the behaviour of an endangered Mexican fish, *Skiffia multipunctata*. *Biological Conservation* 122 (2): 223–230.

Kramer, A.F., Bherer, L., Colcombe, S.J. et al. (2004). Environmental influences on cognitive and brain plasticity during aging. *The Journals of Gerontology. Series A, Biological Sciences and Medical Sciences* 59 (9): 940–957.

Krebs, J.R., Clayton, N.S., Healy, S.D. et al. (1996). The ecology of the avian brain: food-storing memory and the hippocampus. *Ibis* 138 (1): 34–46.

Landa, A., Flagstad, Ø., Areskoug, V. et al. (2017). The endangered Arctic fox in Norway – the failure and success of captive breeding and reintroduction. *Polar Research* 36 (sup1): 9.

Langbein, J., Nürnberg, G., and Manteuffel, G. (2004). Visual discrimination learning in dwarf goats and associated changes in heart rate and heart rate variability. *Physiology & Behavior* 82 (4): 601–609.

Lutz, C., Well, A., and Novak, M. (2003). Stereotypic and self-injurious behavior in rhesus macaques: a survey and retrospective analysis of environment and early experience. *American Journal of Primatology* 60 (1): 1–15.

Martínez-Macipe, M., Lafont-Lecuelle, C., Manteca, X. et al. (2015). Evaluation of an innovative approach for sensory enrichment in zoos: semiochemical stimulation for captive lions (Panthera leo). *Animal Welfare* 24 (4): 455–461.

Mason, G.J. (1991). Stereotypies: a critical review. *Animal Behaviour* 41 (6): 1015–1037.

Mayr, E. (1961). Cause and effect in biology. *Science* 134 (3489): 1501–1506.

Meehan, C.L. and Mench, J.A. (2007). The challenge of challenge: can problem solving opportunities enhance animal welfare? *Applied Animal Behaviour Science* 102 (3–4): 246–261.

Milgram, N.W., Siwak-Tapp, C.T., Araujo, J., and Head, E. (2006). Neuroprotective effects of cognitive enrichment. *Ageing Research Reviews* 5 (3): 354–369.

Moseby, K.E., Read, J.L., Paton, D.C. et al. (2011). Predation determines the outcome of 10 reintroduction attempts in arid South Australia. *Biological Conservation* 144 (12): 2863–2872.

Myers, D.G. and Diener, E. (1995). Who is happy? *Psychological Science* 6 (1): 10–19.

Rabin, L.A. (2003). Maintaining behavioural diversity in captivity for conservation: natural behaviour management. *Animal Welfare* 12 (1): 85–94.

Rantanen, E.M., Buner, F., Riordan, P. et al. (2010). Vigilance, time budgets and predation risk in reintroduced captive-bred grey partridges *Perdix perdix*. *Applied Animal Behaviour Science* 127 (1–2): 43–50.

Reading, R.P., Miller, B., and Shepherdson, D. (2013). The value of enrichment to reintroduction success. *Zoo Biology* 32 (3): 332–341.

Richardson, K.M. and Ewen, J.G. (2016). Habitat selection in a reintroduced population: social effects differ between natal and post-release dispersal. *Animal Conservation* 19 (5): 413–421.

Robbins, L. and Margulis, S.W. (2016). Music for the birds: effects of auditory enrichment on captive bird species. *Zoo biology* 35 (1): 29–34.

Roe, J.H., Frank, M.R., Gibson, S.E. et al. (2010). No place like home: an experimental comparison of reintroduction strategies using snakes. *Journal of Applied Ecology* 47 (6): 1253–1261.

Seyfarth, R.M. and Cheney, D.L. (1986). Vocal development in vervet monkeys. *Animal Behaviour* 34 (6): 1640–1658.

Shettleworth, S.J. (2001). Animal cognition and animal behaviour. *Animal Behaviour* 61 (2): 277–286.

Snyder, N.F., Derrickson, S.R., Beissinger, S.R. et al. (1996). Limitations of captive breeding in endangered species recovery. *Conservation Biology* 10 (2): 338–348.

Swaisgood, R.R. and Shepherdson, D.J. (2005). Scientific approaches to enrichment and stereotypies in zoo animals: what's been done and where should we go next? *Zoo Biology* 24 (6): 499–518.

Thrun, S. (1996). Is learning the n-th thing any easier than learning the first? In: *Advances in Neural Information Processing Systems* (ed. S. Thrun), 640–646.

Tinbergen, N. (1963). On aims and methods of ethology. *Zeitschrift für Tierpsychologie* 20 (4): 410–433.

Toates, F. (2004). 'In two minds'–consideration of evolutionary precursors permits a more integrative theory. *Trends in Cognitive Sciences* 8 (2): 57.

Vidal, L.S., Guilherme, F.R., Silva, V.F. et al. (2016). The effect of visitor number and spice provisioning in pacing expression by jaguars evaluated through a case study. *Brazilian Journal of Biology* 76 (2): 506–510.

Wallace, E.K., Altschul, D., Körfer, K. et al. (2017). Is music enriching for group-housed captive chimpanzees (*Pan troglodytes*)? *PloS one* 12 (3): e0172672.

Wang, M., Ye, X.P., Li, Y.F. et al. (2017). On the sustainability of a reintroduced Crested Ibis population in Qinling Mountains, Shaanxi, Central China. *Restoration Ecology* 25 (2): 261–268.

Wells, D.L. and Egli, J.M. (2004). The influence of olfactory enrichment on the behaviour of captive black-footed cats, *Felis nigripes*. *Applied Animal Behaviour Science* 85 (1–2): 107–119.

West, R., Letnic, M., Blumstein, D.T., and Moseby, K.E. (2018). Predator exposure improves anti-predator responses in a threatened mammal. *Journal of Applied Ecology* 55 (1): 147–156.

Will, B., Galani, R., Kelche, C., and Rosenzweig, M.R. (2004). Recovery from brain injury in animals: relative efficacy of environmental enrichment, physical exercise or formal training (1990–2002). *Progress in Neurobiology* 72 (3): 167–182.

Würbel, H. (2001). Ideal homes? Housing effects on rodent brain and behaviour. *Trends in Neurosciences* 24 (4): 207–211.

4

Choosing the Right Method

Reinforcement vs Punishment

Ken Ramirez

4.1 Introduction

The professional who wishes to train faces a dilemma when making decisions about what methods to follow when training animals. The science, which underpins animal learning, clearly states that when punishment is applied after a behaviour it decreases the frequency of that behaviour, whereas a reinforcement will increase the frequency of the behaviour (Chance 2009; Kazdin 2001; Pierce and Cheney 2008). If we consider that fact alone, it would seem to successfully train we should be prepared to use both reinforcement and punishment in equal measures. Current popular trends, however, lean towards avoiding punishment when possible and focusing on reinforcement (more specifically positive reinforcement). This can be seen in the mission statements of leading zoo training organisations such as the International Marine Animal Trainers Association (IMATA 2019) and the Animal Behavior Management Alliance (ABMA 2019). Similarly, it is part of the mission statement or values of leading domestic animal training organisations such as the Association of Professional Dog Trainers (APDT 2019), the International Association of Animal Behavior Consultants (IAABC 2019), and Karen Pryor Academy (KPA 2019). When choosing a training method you may be inclined to ask whether this trend is based on perception and public relations or is there a solid scientific basis for the popular support for reinforcement over punishment?

4.2 Consequences

When using operant conditioning to train an animal to perform a behaviour, the key factor that needs to be remembered is that behaviour is modified by its consequences (Thorndike 1898). Understanding consequences forms the foundation of operant conditioning and our ability to effectively train any behaviour (Schneider 2012), for example, when a gate is opened and an animal is requested to move from its outdoor exhibit to an indoor enclosure, the animal may be given a piece of food. If that piece of food increases the animal's behaviour of moving from outdoor to indoor then the food is considered a reinforcer because it increases the likelihood that the animal will move into the enclosure the next time they are asked. By contrast a punisher decreases the frequency of the behaviour it follows. If an animal tries to climb out of an exhibit and receives a mild shock when it encounters a live electric wire at the top of the wall, the animal is likely to retreat into its enclosure. The shock would be considered a punisher because it decreases the likelihood that the animal would attempt to climb out of its exhibit in the future.

Zoo Animal Learning and Training, First Edition. Edited by Vicky A. Melfi, Nicole R. Dorey, and Samantha J. Ward.
© 2020 John Wiley & Sons Ltd. Published 2020 by John Wiley & Sons Ltd.

The basic explanation of reinforcement and punishment is relatively straightforward: most zoo professionals who train do not have difficulty understanding the concepts as described above. When faced with the many types of reinforcers and punishers, however, even an experienced trainer can get confused and feel uncertain about how to apply them.

Consequences can be either unconditioned or conditioned. Unconditioned reinforcers include food, water, social interactions, and anything that serves a biological need. These types of reinforcers are often referred to as primary reinforcers. But animals can learn to accept other things including a clicker, a whistle, or a word like 'good' as reinforcers. If the trainer pairs those sounds with food on a regular basis, they may become conditioned reinforcers, also referred to as secondary reinforcers. When trainers work closely with animals, other conditioned reinforcers may be established; animals may learn to appreciate a rub down or the opportunity to play with a toy as these are also referred to as conditioned reinforcers that can be used to increase the frequency of behaviour.

Punishers also can be unconditioned or conditioned. Unconditioned punishers can be anything the animal perceives as aversive that decreases behaviour. There are also many things in an animal's world that may not be initially aversive but the animal learns to dislike them, for example, a crack of a tree limb indicating the approach of a predator and certain people who are aversive, all of which would be considered conditioned punishers. Primary reinforcers and punishers are generally considered more effective when training behaviour because no previous learning history is needed to make them effective (Chance 2009; Kazdin 2001; Pierce and Cheney 2008). However, when used properly, both conditioned reinforcers and conditioned punishers can be effective and useful in training behaviour (Chance 2009; Ramirez 2010; Pryor and Ramirez 2014; see Figure 4.1).

4.3 Choices, Choices, Choices

Most behaviours can be trained using any of the consequence options outlined in the scientific literature (see Chapter 1). It is precisely because

Figure 4.1 In this example of a zoo animal training programme, the bear has been reinforced when a keeper touches a part of its body; this type of behaviour facilitates different husbandry goals including health cheques. *Source:* Steve Martin.

there are so many options available, that the young or inexperienced keeper can be uncertain about how to train an animal's behaviour. Let's examine a typical behaviour needed in a zoo: gating or shifting; moving from one location in the enclosure to another on cue. For a variety of reasons shifting is beneficial, for example when animals move from an old enclosure to a new enclosure. This behaviour can be accomplished by focusing on either the desired behaviour (going through the gate when cued) or the undesired behaviour (staying in enclosure A and not moving).

If we focus on getting the desired behaviour, science informs us that we must use reinforcement. We can use either positive or negative reinforcement. To use positive reinforcement, we can train the animal's behaviour to approach the zoo professional when called, by giving the animal food when it arrives at the desired location. If this behaviour is first trained within the enclosure, it can later be cued (calling the animal) when the zoo professional is in a different position, i.e. on the other side of the gate, in a different enclosure. With appropriate use of positive reinforcement animals can learn to move to a new enclosure when called.

Sometimes an animal might be nervous about crossing a threshold or moving to a location where they have never been before, in which case negative reinforcement might be necessary. If the presence of the trainer or the food reinforcer isn't rewarding enough for them and the need to move them is urgent, trainers might choose to use a net or a board etc. to push the animal through the gate or door. If this equipment has been used previously and the animal associates them with the negative experience (of being caught, pushed, or pulled), the presence of them or of them moving into the animal's flight distance can be enough to cause the animal to move, in an attempt to escape from the equipment and move into the enclosure that does not have the aversive stimulus present. The movement into the new enclosure will have been negatively reinforced, particularly if the aversive stimulus (net, board, etc.) is immediately removed after the animal moves through the gate.

The other behavioural option is to focus on the unwanted behaviour of not moving when cued. Science indicates that if our focus is on decreasing a behaviour, we must use punishment. Continuing with the example above, moving an animal into a new enclosure, the addition of a net, a board, or other aversive equipment into the animal's habitat, the behaviour of staying in the original enclosure is positively punished. Or, another option available, is to simply end the training session if the animal does not move to the new enclosure when cued. By removing the potential for food reinforcement, sometimes termed a timeout, this would be considered negative punishment. By removing something the animal presumably wants, food, right after not responding to the cue to move, it is hoped that the animal will learn that not moving caused the food to disappear. If this timeout is a successful positive punishment, there would be a decrease in the likelihood that the animal would refuse to move following a cue next time the request was made.

If the trainer focuses on getting a desired behaviour, reinforcement must be used. If the focus is reducing a behaviour, punishment must be used. The examples above illustrate how a behaviour can be approached and managed from multiple perspectives. Just because the science seems straightforward, and all methods will achieve the desired outcome, it does not mean that the proper choices are always that clear. For example, one might think a good punishment would be a timeout where the food and the attention of a keeper is taken away. However, in some cases this could reinforce the animal if it doesn't enjoy the keepers attention or the food wasn't of high value. The sections below outline how to choose between reinforcement and punishment and what the zoo professional must consider when choosing the right tool.

4.4 Uses in Training

Professionals skilled in animal training have been using consequences to effectively train an animal's behaviour long before the study of animal learning. The science is clear and the practical applications are evident everywhere that behaviour is trained. Examples of this are provided below:

- Falconry has been popular for centuries. The bird's natural hunting behaviour is used in training; the birds will fly, search for, and catch prey for the reinforcement derived from catching and eating the prey. This behaviour occurs in the wild and can be harnessed in human care, where the prey functions as a positive reinforcer increasing the likelihood of hunting behaviour occurring again in the future.
- In the search and rescue community, dogs are trained to use their nose to find victims trapped or lost in a building. On completion of the task, some search and rescue dogs' behaviour is reinforced after finding a lost person with the opportunity to play with a squeaky toy, a tennis ball, or the chance to play a game of tug. For these dogs these are examples of toys or play being used as positive reinforcement because the likelihood of expressing searching behaviour will increase in the future.
- In typical horse riding, a horse is taught to take riders from place to place. The rider gives the horse direction through the use of reins (straps affixed to a halter around the horse's face and body); these reins are pulled in one direction or another to guide the horse in a desired direction. The horse will feel the pressure of the reins on one side of his face and will move in the opposite direction to relieve the pressure. This is an example of a mild aversive, depending on the force being used by the rider, which is used to change the animal's behaviour and is an example of negative reinforcement because the likelihood of moving in the direction indicated is likely to increase in the future.

- In puppy training, to assist in teaching a young dog not to bite people, new puppy owners might be given two sets of instructions: (i) each time the puppy bites, thump it sharply on the nose to teach the puppy that biting causes it discomfort. This is an example of positive punishment, because it decreases the likelihood that the puppy will bite in the future; (ii) an alternative approach that is sometimes taught is, each time the puppy bites, the owner should stop playing and leave the puppy alone in the room. The removal of play and attention serves as a negative punishment because it reduces the likelihood that the puppy will bite in the future (this technique is sometimes called a timeout).

4.5 Misuses and Challenges

The fact that it is possible to train animals to perform the desired behaviour in so many different ways adds to the dilemma of which methods to choose and which will be most effective? Part of the challenge in using any method is in recognising that its effectiveness is based on using it properly and understanding that misuse can make even the best technique ineffective. A few of the key challenges of applying consequences properly include the following:

4.5.1 Timing

Consequences have proven to be most effective when they are delivered immediately after a behaviour. The longer the gap between the completion of a behaviour and delivery of the consequence, the more likely that the animal learns something other than what is intended. This is true of both reinforcements and punishments, as the following examples illustrate (Chance 2009; Kazdin 2001; Pierce and Cheney 2008).

Poorly timed punishment: a dog owner returns home after a long day at work. He comes into the house to find that his dog has ripped up the sofa, broken several lamps, and made a huge mess. The dog happily rushes up to the owner when he arrives home. But having seen the mess made by the dog, the owner yells 'bad dog' and perhaps in his frustration even smacks the dog on the nose. The smack and yelling 'bad dog' are certainly aversive and could be perceived as positive punishers; but the timing was very poor. The dog made the mess in the house several hours earlier. By applying the punishment when he did, the owner actually punished the behaviour of happily greeting the owner at the door, not the behaviour of making the house a mess. This misuse of punishment creates confusion for the dog and does not teach the dog what is desired (and in extreme cases can even teach the dog to be fearful of the owner).

Poorly timed reinforcement: an animal has been taught to hold still and present its feet for nail trims. Once the nail has been cut, the animal makes a cooing noise and prances around after which it is given a treat for allowing the nails to be trimmed. Even though the treat followed the nail trim, so did the noises and prancing. The reinforcement could result in an animal that becomes more vocal and more boisterous rather than calm behaviour that allows nails to be trimmed.

Good timing requires good observation and mechanical skills that can only be acquired through practice. In my opinion to assist with timing, it is helpful to use some type of signal to indicate to the animal when a behaviour is performed correctly; this marker may be in the form of a whistle, a clicker, or a word such as 'good'.

4.5.2 Value and Intensity

Consequences can be described in terms of intensity and perceived value to an animal, both of which will vary. In my experience, understanding intensity and value has proven to be one of the hardest things for inexperienced trainers to understand and apply.

Reinforcement value: everything we offer to an animal may be perceived of value, though this value may vary. A morsel of grain, a carrot, an apple, a toy, a scratch behind the ears (for animals where contact is permissible), and the opportunity to play a game may all serve as reinforcers for one animal. But it is likely that each of these items would rank differently in terms of how the animal perceives their value. The morsel of grain might be of low value compared to the apple. The scratch behind the ear might be of low value compared to the toy. Every animal is different and even with the same animal the value of the reinforcer might change across environmental situations. It is important to evaluate which reinforcers are going to be the most effective in any given situation. A morsel of grain may be all that is needed to train a behaviour with an animal that is alone in a habitat with no distractions. But that same morsel of grain might prove inconsequential in a training session that includes several other animals and lots of distractions. It is important to take the time to understand the value of each reinforcer being used whilst training.

Punishment intensity: the value of a punisher is often measured by its intensity. An animal that is afraid of people might find the mere presence of a person in their space as sufficiently aversive to cause that animal to move or change its behaviour. Whilst for another animal the presence of a person may not be aversive at all, yet any type of touch, no matter how gentle, might be unpleasant and thus aversive to that animal. Meanwhile, an animal that enjoys human contact, might find neither of these situations aversive. To use an aversive (so that a punisher or negative reinforcer could be applied) with an animal already comfortable around people might require that a 'scary' object also be present (i.e. an object which is associated with previous situations that were aversive) or that the trainer hit the animal with a certain amount of force. Because of the nature of aversive stimuli, i.e. they need to be unpleasant, it is possible to inadvertently escalate the intensity of aversive stimuli used. It can be difficult to recognise if the intensity is too low to have any impact or too intense and cause unwanted side effects. There are many possible side effects that may develop from the use of aversive tools, but the most problematic is aggression.

4.5.3 The 'Sneaky' Animal

Another major challenge associated with punishment is that the learner begins to associate punishment as coming from the trainer. When this happens, the animal learns to behave appropriately in the presence of the trainer, because that is the only time that punishment occurs. But in the absence of the trainer, the animal will continue to exhibit the unwanted behaviour, creating the appearance of being 'sneaky', when in reality this happens because punishment is seldom a consequence in the absence of the trainer.

4.5.4 Emotional Responses

Training is successful because of the proper application of scientifically proven learning principles. However, when training animals we can become quite emotionally attached when an animal does something extremely well or extremely poorly. These emotions often cloud our judgement and make it difficult for us to reason through what is happening or adhere as precisely to proven principles. It is often the frustration and anger that comes when an animal is perceived to misbehave that causes the use of punishment to get out of hand. Emotional responses can often be less precise and in many cases too severe; there is a fine line between a well-timed punishment and inhumane abuse.

4.5.5 Misuse of Common Definitions

Too often when training an animal there is too much reliance on common uses of the terms 'reinforce' and 'punish'. In our society we tend to reinforce or punish people and animals, as a result of what we consider good or bad actions. However, that is not how the science of animal learning works. We should never reinforce or punish people or animals. We should only aim to reinforce or punish

specific behaviours performed. This difference may seem semantic in nature but there are key pragmatic differences. Two examples in the human world will illustrate the difference:

A child gets a bad report card. The parents choose to punish the child for the grades they received by not allowing them computer privileges for a week. The behaviour of getting bad grades was not punished, that behaviour occurred weeks if not months earlier. Instead the behaviour that was punished was the behaviour of showing the report card and the punishment is more likely to teach the child to not show the report card or to forge their parent's signature on the card in the future. The child was certainly punished, but not for behaviour the parents were targeting!

A criminal commits a serious crime. He eventually goes to jail but sentencing and jail time happen many months (in some cases years) after the crime was committed. A large percentage of criminals go on to commit additional crimes once they are released (Flora 2004). Jail punishes the person (which may be a necessary aspect in modern society) but it does not punish the behaviour.

When punishment or reinforcement is directed at the learner instead of the behaviour, at its best the animal is receiving vague guidance, at worst it is meaningless or teaches the wrong thing. Only when training is focused on behaviour and uses good timing and consequences of appropriate value will training be successful and an animal learn desired behaviour.

4.6 Making an Informed Choice

Choosing the appropriate method requires knowledge and skill. But most importantly it requires setting goals and knowing what you ultimately want to accomplish. By having a clear vision of the behaviour you want from an animal and a set of expectations for that animal's behaviour, you can then create a training plan to achieve those goals and vision (MacPhee 2008). Training new behaviour and increasing the reliability of the animal to perform on cue is most ably accomplished through proper application of reinforcement.

The scientific study of human–animal relationships (HAR) is an emerging discipline (Hosey and Melfi 2014), but they still remain difficult to explain or quantify scientifically. It is recognised by most zoo professionals that a good relationship with an animal is helpful in training. A well-established relationship appears to create trust between both human and animal, which seems to allow much more to be accomplished, than if no HAR exists. Strong HAR will often open the door to new reinforcement opportunities that otherwise might not be available when training (Ramirez 2010). The use of punishments by contrast, seems to cause trust to break down and HAR to deteriorate. From my experience, a one-time use of a single punisher or the one-time use of an aversive can break down the trust, which has been hard to establish. I have also found that it often takes many reinforcers to offset the damage caused by a single punisher (Ramirez 2013; see Figure 4.2).

Frequently, if we are dealing with unwanted behaviour, it is usually a sign that something happened to punish the desired behaviour that had previously been present. This is not to suggest that the behaviour was purposely punished; we must remember that punishers and reinforcers abound in the natural world. The heat of the day, the physical demands of a behaviour, the aggression displayed by other animals in the environment are all examples of potentially aversive events that could punish the original desirable behaviour. If we try to counter this unwanted behaviour with more punishment, then a situation of competing punishers may be created, and this will often require that the new punishers be of greater intensity to outweigh the already present punishers. Instead of taking such an aversive approach a creative animal care professional will seek out the already present punishers that may be occurring in the environment and attempt to remove or block them. If the aversive stimuli that blocked the

Figure 4.2 When an animal extends part of its body through the enclosure barrier, it can be viewed as an example of trust between keeper and animal, which has likely been formed through repeated interactions between both parties. *Source:* Steve Martin.

desired behaviour and prompted the undesirable behaviour to occur are gone, it opens up the possibility of using reinforcement to get the desired behaviour back.

The biggest drawback to using punishment is that no information is provided to the animal about what behaviour is desired. Simply punishing behaviour does not help an animal to understand what it should do in that situation instead. Most behaviours that humans find unacceptable in animals are very natural behaviours: animals bark, scream, roar, or bellow when they are nervous or scared. We often find these behaviours unacceptable, despite the fact that they are naturally occurring and fulfil a function for the animal. Punishing the behaviour of making too much noise does not help the animal understand what would be acceptable in those circumstances. It will learn, not to perform these vocalisations, much faster if provided with an alternative behaviour to perform under those conditions; one that will keep it safe and earn it reinforcement. So many unwanted behaviours that animals exhibit (fighting, urinating, defecating, digging, climbing, etc.) are behaviours that serve a purpose in the natural world. The fact that we find them undesirable or unacceptable in our world is not something we can expect an animal to understand. Punishing these undesirable behaviours often just creates greater confusion for most animals. We often can mitigate these problems by training the animal to perform an alternative behaviour (O'Heare 2010; Ramirez 1999). Specific examples include:

- Animals that fight and compete with each other at feeding time are behaving as they might in the wild. If we teach the animals that they are only fed if they are sitting on separate rocks, tree branches, platforms, etc., they learn an alternative behaviour to fighting for their food and an acceptable

behaviour (sitting in the desired location) that will earn reinforcement.

- A primate that grabs for the trainer (or throws things at the veterinarian) every time a medical behaviour is attempted, can be taught that both hands and both feet must hold on to an object when humans are present to gain reinforcement; specific bars or targets in the enclosure can also be provided. The animal cannot grab the trainer or throw things at the veterinarian if its hands are occupied (see Figure 4.3).
- Animals in free contact that jump all over zoo staff when they enter the enclosure, either to gain attention or to drive staff out of the enclosure, do so because it produces results. An alternative behaviour which can be trained is to provide the animals with a station/location to go to when zoo staff enter the enclosure. For the animal that is attention seeking, reinforcing the behaviour of going to their station and provide appropriate attention to them at the station. For the animal that is trying to

Figure 4.3 It is important to consider keeper safety when training and this might mean training animals so that they are not in a position to grab you; this is especially important if the person training 'has their hands full'. *Source:* Steve Martin.

drive zoo staff out of the enclosure, reinforce them after the zoo staff have left the enclosure. Either way you have taught them an alternative behaviour.

When working with zoo animals, many have reported on the great success they have had focusing on redirection and teaching alternative behaviours to undesirable behaviour through positive reinforcement as opposed to using punishment. The marine mammal community has dealt with aggressive sea lions and focused on positive reinforcement solutions using alternative behaviours and redirection (Graff 2013; Keaton 2014; Streeter et al. 2013). In 1990, Turner and Tompkins wrote about a positive approach to aggression reduction which has become a must-read for the training community and although written about marine mammals, has applications to all zoo animals (Turner and Tompkins 1990). Some zoos have traditionally used punishment and making the transition from forced based training can be a challenging journey and looking at the success achieved with camels should be inspirational for anyone facing the need to make a transition (Urbina et al. 2014).

To aid in the transition and/or adoption of positive reinforcement methods of training, I've outlined a variety of published case studies below, including a wide range of different species being trained for many different purposes, which provide further insight into this journey.

- Using redirection and alternative behaviours to resolve sea lion aggression (Graff 2013; Keaton 2014; Streeter et al. 2013).
- General positive approach to aggression reduction (Turner and Tompkins 1990).
- Camel training transition from force based methods to positive reinforcement (Urbina et al. 2014).
- Resolution of challenges for improved giraffe care with positive reinforcement (Mueller 2003; Stevens 2002).
- Wolf recall behaviour to assist in improved safety (McKeel 2005).
- Improved primate care through positive reinforcement (Hickman and Stein 2009;

Russell and Gregory 2003; Russell and Varsik 2002).
- Removal of aversive equipment when training birds of prey (Anderson 2009).
- Gaining trust with skittish birds (Tresz and Murphy 2008).
- Distraction training with positive reinforcement with a variety of species (Leeson 2006).
- Aggression reduction with spotted eagle rays (McDowell et al. 2003).
- Transition from free-contact to protected-contact with elephants (Andrews et al. 2005; Priest et al. 1998).
- General approach to training multiple species and eradicating unwanted behaviour by focusing on positive reinforcement (Joseph and Belting 2002; Lacinak 2010; Ramirez 2012; Scarpuzzi et al. 1991; Seymour 2002).

4.7 Ethical Considerations

The science behind the use of reinforcers and punishers is very clear as described above. By their very definition, both types of consequences work when applied with vigilance and good timing. The debate over which method to use cannot be won by arguing that one is more effective than the other. Additionally, there are very few trainers that rely entirely on one side of the equation. It would be nearly impossible to use nothing but reinforcement or nothing but punishment. Because even if you attempted to only use one of these methods, the environment is full of reinforcers and punishers which impact the learning progress of your animal continuously. So, to be effective you are constantly having to adjust your training decisions to compensate for consequences which exist in the environment, your animals past experience and their natural behavioural tendencies.

Many who train animals are bound by rules and guidelines put in place by their respective organisations. These guidelines may be based on scientific principles, but in many cases they are influenced by other outside factors:

- Speed of behavioural acquisition: in some training environments getting results fast may lead to job promotions and be perceived as a better trainer. The choices about which methods were used to get the desired behaviour/results may not matter to those organisations. Or, whilst it may matter, they choose not to ask how the behaviour was acquired.
- Behavioural reliability: in many training circles the key to success is whether or not the behaviour is just as reliable weeks, months, or even years after its initial acquisition. The animal trained to give a blood sample to monitor its diabetic condition is not very helpful if the behaviour breaks down after only one or two samples are drawn. The animal trained to cooperate in a research study will fail to be useful if they are not reliable for repeated trials. An animal trained for educational programming is not very useful if it cannot be counted on week after week for those programmes. In the domestic animal world the importance of reliability can be the difference between life and death. The animal that is used to detect bombs in a stadium needs to be as reliable in its ability to detect explosives two years after completing training as it was the week it completed training. The guide dog that assists its blind handler to navigate the world needs to be as skilled three years after training as it was one month after training. Behavioural reliability is often a key indicator of success in many organisations.
- Professional organisations: many people who train are bound by rules and guidelines put forth by professional associations and groups designed to manage a species or a breed of animal. These guidelines are often derived through compromises and discussions by many professionals with a wide range of skills, knowledge, and agendas.
- Public relations: sometimes organisations make choices based on appearances, and public perception. These are not usually scientifically based decisions, but they are important to most organisations and can have a huge impact on what choices are available to those who train animals.

Generally, we as humans are a compassionate species. We train animals in an attempt to give them better care and help them live in our world safely. The methods that we choose to use must be governed not only by efficacy but also by ethics. Just because we can train something doesn't mean that we should. One of the biggest factors that guides us in our decision about which training methods to use is our personal ethics. There are ethical guidelines laid down by our employers, by our profession, and by our peers, but as individuals we are also bound by our own ethical considerations and beliefs. Many wise trainers and scientists have used ethics and written about the importance of having an ethical framework to guide animal training in their decision making. Three of the most significant include:

- Least intrusive and minimally aversive (LIMA) principle (Lindsay 2005): Stephen Lindsay describes what he refers to as a cynopraxic (dog friendly) approach to training that is reinforcement based, but recognises that the need for aversive tools may at times be necessary. The LIMA principle advocates for a 'least intrusive and minimally aversive' approach to training. He emphasises that any ethical approach must be competency based, because it requires skill and experience to know when to use a more aversive approach. He also describes the dangerous effects of unnecessary escalation of utilising aversive tools.
- Hierarchy of effective procedures (Friedman 2009): Susan Friedman asks the question 'Is effectiveness enough?' Just because a tool gets the job done is that sufficient reason to use it? She concludes that it is not, nor should it be the only criteria in determining which methods to use. She proposes a hierarchy in which the use of punishers are a last resort, used only when

all other techniques have been tried and proven ineffective.

- LIEBI algorithm (O'Heare 2013): James O'Heare proposed a model that he labelled as a 'least intrusive effective behaviour intervention' algorithm. He refers to it as a best practices model that includes a decision-making algorithm with a 'levels of intrusion table' designed to help professionals work through the decision-making process of when to use aversive intervention. He describes a 'red zone' that involves a high degree of invasiveness and the goal of the procedure is to help professionals avoid ever getting to the red zone.

All three of these frameworks are similar but approach the problem from a different perspective.

Each of them acknowledges the science, but they also make a compelling argument for using the least intrusive methods first. They don't suggest that good trainers never use punishment, just that they use it wisely and avoid it whenever possible. These types of guidelines have been adopted by many leading training certification bodies for example: the Association of Animal Behaviour Professionals (O'Heare 2013), the International Association of Animal Behaviour Consultants (IAABC 2019), and the Certification Council for Professional Dog Trainers (CCPDT 2019).

4.8 A Personal Note, the Author's Approach: Balancing Ethics, Efficacy, and Best Practices

As my training style has evolved with experience, I have transitioned from a traditional approach to training, using corrections to teach impulse control during my early years as a guide dog trainer. Later, when I entered the zoo community, I was introduced to positive reinforcement training. There were certainly punishers and aversive stimuli in the environment, but they were not methods we regularly employed to shape behaviour. It

became clear to me that punishment did not have to be a significant part of the training equation to be successful. That is not to say that minor aversive stimuli were not implemented from time to time to assist in shaping behaviour faster or making a concept clearer. But the use of those methods were rare and their use was restricted to more experienced trainers who had the skill to understand when and how to apply them.

We were positive reinforcement trainers who used mild aversive stimuli on very rare occasions. As I became a supervisor and was responsible for teaching new zoo professionals how to train, I was challenged with the question, 'Does being a positive reinforcement trainer mean that we never use punishment, ever?' This then led to the follow up question, 'and if we find ourselves needing to use an aversive tool or apply a punisher, does that mean we can no longer call ourselves positive reinforcement trainers?' These questions perplexed me until I read Friedman (2009). In this article, Friedman (2009) states that it is necessary that the method used to train aligns with your ethical beliefs. It showed me a clear path to determine when and why I might need to use something beyond positive reinforcement. As with most procedures, each trainer adapts them to fit their particular training style. My interpretation of the hierarchy as I teach it to young trainers is as follows, I always start at number 1 and move down the hierarchy only when needed:

1) *Animal needs come first*: animal welfare must always be a top priority. Therefore, before taking any training steps you should always assure that the animal is physically and mentally healthy and getting appropriate nutrition, housing, and care daily (see Figure 4.4).

2) *Include primary reasons for training in all decision making:* if it is determined that training is needed, always put the primary reasons for training before all others. The training must benefit the individual animal being trained by assuring that the

Figure 4.4 An example of how training animals in zoos can facilitate preventative and proactive veterinary care; here a bear has been trained for venepuncture whilst it stays in its enclosure. *Source:* Steve Martin.

training meets one of these goals and does not compromise the animal's welfare:

a) physical exercise – gives the animal appropriate physical exercise;

b) mental stimulation – provides appropriate mental stimulation;

c) leads to cooperative behaviour – contributes to the safe management and care of the animal (medical behaviours, taking medication or vitamins, shifting, etc.).

3) *Set the environment up for success:* before embarking on a complex training plan, ensure the environment has been set up to make it easy for the animal to succeed in meeting desired behavioural goals.

4) *Use positive reinforcement:* once it is determined that training is desired, search for the best way to accomplish the goals using positive reinforcement. Remembering that the best reinforcement varies from individual to individual and from situation to situation.

5) *Use redirection:* if the animal is exhibiting unwanted behaviour, teach it something to do that is acceptable and will earn it reinforcement.

6) *Extinction may be used in conjunction with other tools:* if there is unacceptable behaviour, look for reinforcers that are strengthening or maintaining that unwanted behaviour and try to remove or withhold them.

7) *Negative reinforcement or negative punishment may be employed if absolutely needed:* when unwanted behaviour persists, review previous steps in the hierarchy and ensure that a positive alternative hasn't been overlooked, then use the method (negative reinforcement or negative punishment) that will likely achieve the desired result with the least fallout.

8) *Use positive punishment as a last resort:* if all else has failed, is impractical or is impossible and the undesirable behaviour must be stopped, apply a carefully thought-out positive punishment.

This hierarchy is not absolute, it is a guide. The wise application of these rules must be accompanied with skill and knowledge to assess when moving down the ladder is required.

4.9 Concluding Thoughts and Considerations

No matter the rhetoric or discussion regarding training methods used, it is unlikely that many professional trainers are purely all punishment based or all reinforcement based. Trainers must use a mix of methods to be effective and successful trainers. But not every method is equal nor is any method always effective!

Positive reinforcement training has become widespread because it has shown great effectiveness and success in achieving the needed goals in the modern zoo as well as in the modern domestic animal training world. Much of the challenge when choosing the right methods is mired in the confusion of terminology where the scientific meaning of reinforcement and punishment differs so greatly from the general public's understanding of these terms. Additionally, those who want to train animals are challenged with developing the mechanical skills to apply the methods properly and with the correct intensity. Finally, in animal training it is not possible to separate ethics and animal welfare from the equation when choosing the appropriate tools to use.

These considerations are required to make good choices and highlight the importance for all training programmes to have skilled leadership in the areas of behaviour, training, and enrichment and for programmes to have clear goals and guidelines to assist staff in making an informed decision.

No one training animals should make any training decision without being fully informed, and knowledgeable of the mechanisms that are at work when animals learn (animal learning theory, see Chapter 1). Only through a thorough knowledge of learning theory as underpinned by scientific research and an awareness of what practical applica-

tions have demonstrated, can the available options be compared and the appropriate choices for their animals and their programme be made. It is clear, starting a training programme using positive reinforcement is the most practical and effective approach. Applying other types of training methods, requires skill and a deeper understanding of the science, practical considerations, and ethics in each situation. The skilled trainer should find that the need to purposely use punishment in training is rare, but having full knowledge about the use of punishment and reinforcement will always be critically important in making an informed decision.

References

ABMA (2019). The Animal Behavior Management Alliance Mission Statement, Vision, And Values. https://theabma.org/abma.

Anderson, T. (2009). Why are we using equipment on birds of prey? *Proceedings of the 2009 Annual Conference of the Animal Behavior Management Alliance*. Providence, RI, USA: Animal Behavior Management Alliance.

Andrews, J., Boos, M., Young, G., and Fad, O. (2005). Elephant management: making the change. *Proceedings of the 2005 Annual Conference of the Animal Behavior Management Alliance*. Houston, TX, USA: Animal Behavior Management Alliance.

APDT (2019). The Association of Professional Dog Trainers Position Statements. https://apdt.com/about/position-statements.

CCPDT (2019). Certification Council for Professional Dog Trainers Humane Hierarchy. http://www.ccpdt.org.

Chance, P. (2009). *Learning and Behavior*, 6e. Thomas Wadsworth: Belmont, CA.

Flora, S.R. (2004). *The Power of Reinforcement*, 195–197. Alnamy, NY: State University of New York Press.

Friedman, S.G. (2009). What's wrong with this picture? Effectiveness is not enough. *Journal of Applied Companion Animal Behavior* 3 (1): 41–45.

Graff, S. (2013). Training fundamental behaviors in California Sea Lions (*Zalophus californianus*) to decrease aggression. *Proceedings of the 41st Annual Conference of the International Marine Animal Trainers Association*. Las Vegas, NV: International Marine Animal Trainers Association.

Hickman, J. and Stein, J. (2009). Who's training whom? Implementation of a positive reinforcement program for shifting a larger group of hooded capuchin monkeys. *Proceedings of the 2009 Annual Conference of the Animal Behavior Management Alliance*. Providence, RI: Animal Behavior Management Alliance.

Hosey, G. and Melfi, V. (2014). Human-animal interactions, relationships and bonds: a review and analysis of the literature. *International Journal of Comparative Psychology* 27 (1): 117–142.

IAABC (2019). The International Association of Animal Behavior Consultants Mission Statement. https://iaabc.org/about.

IMATA (2019). The International Marine Animal Trainers Association Mission Statement and Values. http://www.imata.org/mission_values.

Joseph, B. and Belting, T. (2002). Operant conditioning as a tool for the medical management of non-domestic animals. *Proceedings of the 2002 Annual Conference* of the Animal Behavior Management Alliance. San Diego, CA: Animal Behavior Management Alliance.

Kazdin, A.E. (2001). *Behavior Modification in Applied Settings*, 6e. Wadsworth/Thomas Learning: Belmont, CA.

Keaton, L. (2014). Houston we have a biter … but we need an x-ray! Training a male California Sea Lion with a history of aggressive behavior for voluntary protected contact radiographs. *Proceedings of the 42nd Annual Conference of the International Marine Animal Trainers Association.* Orlando, FL: International Marine Animal Trainers Association.

KPA (2019). Karen Pryor Academy for Animal Training and Behavior Who We Are. https://www.karenpryoracademy.com/about.

Lacinak, T. (2010). Safety and responsibility in zoological environments. *Proceedings of the 38th Annual Conference of the International Marine Animal Trainers Association.* Boston, MA: International Marine Animal Trainers Association.

Leeson, H. (2006). Training with distractions: a proactive approach for success. *Proceedings of the 2006 Annual Conference of the Animal Behavior Management Alliance.* San Diego, CA: Animal Behavior Management Alliance.

Lindsay, S.R. (2005). *Handbook of Applied Dog Behavior and Training, Volume Three: Procedures and Protocols*. Blackwell: Ames, IA.

MacPhee, M. (2008). Techniques to expand views on animal training. *Proceedings of the 2008 Annual Conference of the Animal Behavior Management Alliance.* Phoenix, AZ: Animal Behavior Management Alliance.

McDowell, A., Muraco, H.S., and Stamper, A. (2003). Training spotted eagle rays (*Aetobatus narinari*) to decrease aggressive behavior toward divers. *Journal of Aquariculture and Aquatic Sciences* VIII (4): 88–98.

McKeel, B. (2005). Total Recall: training a recall for safety with free contact gray wolves. *Proceedings of the 2005 Annual Conference of the Animal Behavior Management Alliance.* Houston, TX: Animal Behavior Management Alliance.

Mueller, T.A. (2003). Target training with a male giraffe: Management considerations in a free contact environment. *Proceedings of the 2003 Annual Conference of the Animal Behavior Management Alliance.* Tampa, FL: Animal Behavior Management Alliance.

O'Heare, J. (2010). *Changing Problem Behavior: A Systematic and Comprehensive Approach to Behavior Change Management*, 109–120. Ottawa, Canada: BehaveTech Publishing.

O'Heare, J. (2013). The least intrusive effective behavior intervention (LIEBI) algorithm and levels of intrusiveness table: a proposed best-practices model. *Journal of Applied Companion Animal Behavior* 3 (1): 7–25. Retrieved from http://www.associationof animalbehaviorprofessionals.com/vol3no1oheare.pdf.

Pierce, W.D. and Cheney, C.D. (2008). *Behavior Analysis and Learning*, 4e. New York, NY: Taylor & Francis Group.

Priest, G., Antrim, J., Gilbert, J., and Hare, V. (1998). Managing multiple elephants using protected contact at San Diego's wild animal park. *Soundings* 23 (1): 20–24.

Pryor, K. and Ramirez, K. (2014). Modern animal training: a transformative technology. In: *The Wiley Blackwell Handbook of Operant and Classical Conditioning* (eds. F.K. McSweeney and E.S. Murphy), 455–482. Oxford: Wiley.

Ramirez, K. (1999). Problem solving. In: *Animal Training: Successful Animal Management Through Positive Reinforcement*. Chicago, IL: Shedd Aquarium Publishing.

Ramirez, K. (2010). Smart reinforcement: a systematic look at reinforcement strategies. In: *Curriculum for Karen Pryor Clicker Expo 2010*. Waltham, MA: KPCT.

Ramirez, K. (2012). Oops! What to do when mistakes happen. In: *Curriculum for Karen Pryor Clicker Expo 2012*. Waltham, MA: KPCT.

Ramirez, K. (2013). Husbandry training. In: *Zookeeping: An Introduction to the Science and Technology* (eds. M.D. Irwin, J.B. Stoner and A.M. Cobaugh), 424–434. Chicago, IL, Ch. 43: The University of Chicago Press.

Russell, C.K. and Gregory, D.M. (2003). Evaluation of qualitative research studies. *Evidence-based nursing* 6 (2): 36–40.

Russell, I.A. and Varsik, A. (2002). To the Max: Addressing behavioral and health challenges with a 32-year-old male gorilla (*Gorilla gorilla gorilla*). *Proceedings of the 2002 Annual Conference of the Animal Behavior Management Alliance*. San Diego, CA.

Scarpuzzi, M.R., Lacinak, C.T., Turner, T.N. et al. (1991). Decreasing the frequency of behavior through extinction. In: *Animal Training: Successful Animal Management through Positive Reinforcement* (ed. K. Ramirez). Chicago, IL: Shedd Aquarium Press.

Schneider, S.M. (2012). *The Science of Consequences: How They Affect Genes, Change the Brain, and Impact Our World*. Amherst, NY: Prometheus Books.

Seymour, H. (2002). Training, not restraining. *Proceedings of the 2002 Annual Conference of the Animal Behavior Management Alliance*. San Diego, CA.

Stevens, B.N. (2002). Use of operant conditioning for unrestrained husbandry procedures of giraffes. *Proceedings of the 2002 Annual Conference of the Animal Behavior Management Alliance*. San Diego, CA.

Streeter, K., Montague, J., Brackett, B., and Schilling, P. (2013). Meeting in the middle. Giving an aggressive sea lion the power of choice. *Proceedings of the 2013 Annual Conference of the Animal Behavior Management Alliance*. Toronto, Canada: Animal Behavior Management Alliance.

Thorndike, E.L. (1898). Animal intelligence: an experimental study of the associative processes in animals. *Psychological Monographs: General and Applied* 2 (4): i–109.

Tresz, H. and Murphy, L. (2008). Techniques for training apprehensive animals. *Proceedings of the 2008 Annual Conference of the Animal Behavior Management Alliance*. Phoenix, AZ: Animal Behavior Management Alliance.

Turner, T.N. and Tompkins, C. (1990). Aggression: exploring the causes and possible reduction techniques. *Soundings* 15 (2): 11–15.

Urbina, E., Garduno, M., Mata, A., et al. (2014). The use of positive reinforcement techniques to retrain dromedary camels (*Camelus dromedaries*) that had been previously trained using aversive conditioning techniques for interactive programs. *Proceedings of the 42nd Annual Conference of the International Marine Animal Trainers Association*. Orlando, FL.

Box A1

Animal Vision

Andrew Smith

Vision is a key sense for humans, yet for other species the world isn't always quite as we see it. Species may differ in their field of view, depth perception, visual acuity (ability to discriminate objects from a distance), colour vision and the perception of time. For example, to distinguish colour an animal's eye must typically have more than one type of cone cell, since colour perception is based on comparing the outputs of the different classes of cone. As such, those animals with rod-only retinas such deep sea fish (Hunt et al. 2001) and deep diving cetaceans, e.g. the sperm whale (*Physeter macrocephalus*) (Meredith et al. 2013) see the world in monochrome (shades of grey), perceiving no colour.

The physical area that an animal can see is referred to as its field of view. Humans, like other primates, have forward facing eyes, and thus have a relatively narrow field of view, ca. 180°, compared to an animal with eyes on the sides of its head, such as the scalloped hammerhead shark (*Sphyrna lewini*), which has a full 360° field of vision (McComb et al. 2009). The area seen by both eyes simultaneously is the binocular field of vision, and that seen by one eye, the monocular field. Whilst monocular cues can be used to determine depth (Timney and Keil 1995; Martin 2009), experiments show that performance, for mammals at least, may be better when an animal uses its binocular field of vision to judge depth (Timney and Keil 1999). The relative size of an animal's binocular field depends on where its eyes are located. Our forward facing eyes provide us with 140° of binocular vision, whereas dogs (*Canis lupus familiaris*) with eyes set slightly to the side of their heads have a larger 240° field of view of which 30–60% is binocular, and woodcock (*Scolopax* spp.) and other birds with eyes on the very sides of their heads can have 360° visual fields with only 5° of binocularity in the horizontal plane (Miller and Murphy 1995; Martin 2009). Animals differ in their vertical fields of view too, thus those birds with eyes on the sides of their heads may overlap both behind and above them, thus providing them with total panoramic vision of the space around their head.

Animals not only differ in their fields of view, but also show variation of vision within these fields. A key aspect of vision is the ability to resolve detail (visual acuity). Some animals, such as birds of prey, have a much greater visual acuity than humans. Whereas a person might just be able to discern a 2 mm object from 6 m, it's estimated an eagle can detect it from 35 m away (Hodos 2012). In contrast, the visual acuity of other species is poorer than in humans, that of dogs for example has been estimated at 20–40%. Thus what we can distinguish from 9 m away, a dog can only see at 2 m (Miller and Murphy 1995).

Zoo Animal Learning and Training, First Edition. Edited by Vicky A. Melfi, Nicole R. Dorey, and Samantha J. Ward.
© 2020 John Wiley & Sons Ltd. Published 2020 by John Wiley & Sons Ltd.

Similar to visual acuity is accommodation, this is the eye's ability to change its focal length so that objects at varying distances are brought into focus. This is done by changing the shape of the lens, and also to some extent the shape of the eye itself. Thus as a general rule, larger eyes have a greater range of accommodation than smaller ones. Large eyes provide long focal lengths and space for a larger lens which in combination gives an ability for higher resolution over a greater range of distances (Land and Nilsson 2012). Differences in accommodation mean differences in the shortest distance at which an object can be brought into clear focus. Thus humans with their greater accommodation can focus on an object just 7 cm away, whereas for a dog anything closer than 33–50 cm will be blurred (Miller and Murphy 1995).

Colour is perhaps one of the most obvious aspects of the world around us, yet few animals see it in the way we do. Colour vision typically requires cone cells, a special type of receptor cell in the retina. Cones normally require higher levels of light to function which is why we cannot see colour well in dim light. However there are some exceptions, for example it has recently been shown that some animals, such as geckos and hawk-moths, have special highly efficient cones to allow colour vision at night (Kelber and Roth 2006). In order to perceive colour the brain compares the outputs of different classes of cones, each maximally responsive to different wavelengths of light. Humans have three cone types, and are thus referred to as trichromats. As a general rule, animals with fewer cone types can distinguish fewer colours. Our colour vision, and that of Old World monkeys and apes, is amongst the best found in mammals, since most mammals are dichromats and have two classes of cones and as such cannot distinguish as many colours as we can (Jacobs 1993). However birds, reptiles and amphibians typically have four types of cone which allows them to perceive a wider array of colours than we can. Often the more colourful a species is, the better its colour vision. However there are some exceptions. Cephalopods, well known for their colour-based displays and camouflage, are completely colour blind (Marshall and Messenger 1996) (Figure A1.1). Some species can use properties of light that we cannot detect, such as ultraviolet wavelengths, the near infrared, or polarisation.

Figure A1.1 This broadclub cuttlefish, *Sepia latimanis*, like other cuttlefish has the ability to change colour yet is colour blind, so body colour itself is unlikely to be the visual signal which this species uses to communicate with one another. *Source:* Jeroen Stevens.

These abilities allow them to make use of signals that we cannot see. It is important to remember that whilst birds, insects, and some mammals such as rodents (Jacobs et al. 1991), reindeer (*Rangifer tarandus*) (Hogg et al. 2011), and a phyllostomid flower bat (*Glossophaga soricina*) (Winter et al. 2003), can detect ultraviolet light, they do not see solely in ultraviolet. The images that they see will depend on how their brain combines the information from the ultraviolet portion of the spectrum with that from the receptors tuned to what we refer to as the visible spectrum. The same will be true of the butterflies and fish that can perceive near infrared (Land and Nilsson 2012).

Animals also differ in the way that they perceive movement. Like colour, movement is such a key aspect of our own visual world it is similarly hard to comprehend it is not always seen in the same way by other species. Motion perception is generated by the brain comparing the series of successive snapshots or frames that the eyes send to it. The rate at which these are generated influences an animal's visual perception of time. Humans generate and process information at 60 images per second, whereas smaller species with higher metabolic rates typically process images at a higher frequency (Healy et al. 2013). To them the world will appear as if in slow motion, whilst larger animals which process images at a lower rate will experience the world in a manner akin to a time-lapse film – for them perceiving slow moving objects as moving may be difficult. The refresh rate of televisions and monitor screens is set to just above the human threshold for seeing the individual images, and thus we perceive motion rather than a flickering series of images. However species with higher rates of visual processing, such as dogs and small animals, would see a flickering screen. This is important to remember if using a monitor screen as a stimulus, either for enrichment or research, with other species.

References

Healy, K., McNally, L., Ruxton, G.D. et al. (2013). Metabolic rate and body size are linked with perception of temporal information. *Animal Behaviour* 86 (4): 685–696.

Hodos, W. (2012). What birds see and what they don't: luminance, contrast, and spatial and temporal resolution. In: *How Animals See the World: Comparative Behavior, Biology, and Evolution of Vision* (eds. O.F. Lazareva, T. Shimizu and E.A. Wasserman), 5–25. Oxford University Press.

Hogg, C., Neveu, M., Stokkan, K.A. et al. (2011). Arctic reindeer extend their visual range into the ultraviolet. *Journal of Experimental Biology* 214 (12): 2014–2019.

Hunt, D.M., Dulai, K.S., Partridge, J.C. et al. (2001). The molecular basis for spectral tuning of rod visual pigments in deep-sea fish. *Journal of Experimental Biology* 204 (19): 3333–3344.

Jacobs, G.H. (1993). The distribution and nature of color vision among the mammals. *Biological Review* 68: 413–471.

Jacobs, G.H., Neitz, J., and Deegan, J.F. II (1991). Retinal receptors in rodents maximally sensitive to ultraviolet light. *Nature* 353 (6345): 655.

Kelber, A. and Roth, L.S. (2006). Nocturnal colour vision – not as rare as we might think. *Journal of Experimental Biology* 209: 781–788.

Land, M.F. and Nilsson, D.E. (2012). *Animal Eyes*. Oxford University Press.

Marshall, N.J. and Messenger, J.B. (1996). Colour-blind camouflage. *Nature* 382: 408–409.

Martin, G.R. (2009). What is binocular vision for? A birds' eye view. *Journal of Vision* 9 (11): 14.

McComb, D.M., Tricas, T.C., and Kajiura, S.M. (2009). Enhanced visual fields in hammerhead sharks. *Journal of Experimental Biology* 212 (24): 4010–4018.

Meredith, R.W., Gatesy, J., Emerling, C.A. et al. (2013). Rod monochromacy and the coevolution of cetacean retinal opsins. *PLoS Genetics* 9 (4): e1003432.

Miller, P.E. and Murphy, C.J. (1995). Vision in dogs. *Journal of the American Veterinary Medical Association* 207: 1623–1634.

Timney, B. and Keil, K. (1995). Horses are sensitive to pictorial depth cues. *Perception* 25 (9): 1121–1128.

Timney, B. and Keil, K. (1999). Local and global stereopsis in the horse. *Vision Research* 39 (10): 1861–1867.

Winter, Y., López, J., and von Helversen, O. (2003). Ultraviolet vision in a bat. *Nature* 425 (6958): 612–614.

Box A2

Do You Hear What I Hear? Hearing and Sound in Animals

Erik Miller-Klein

The capacity to hear sounds provides animals with the ability to effectively monitor the activities and dangers in their environment usually outside the range of the other senses. This hypersensitivity to the sounds around them means that zoo professionals should be acutely aware of the animal's listening abilities, and potential issues within their captive environment. Hearing sensitivity is measured with respect to sound's range in amplitude and frequency, and can be recorded through audiometric testing for a representative sample of each species. An audiogram is a visual depiction of the sounds which can be perceived, for example a human audiogram has been compared to a few example animals, shown in Figure A2.1.

The audiograms in Figure A2.1 present the sensitivity of the ear to sounds with respect to frequency. The lowest points on the graph are where the hearing of that species is the most sensitive, and where the line ends on the *x*-axis is the frequency range limit for that species. Research has shown that species are most sensitive to the frequencies associated with communication and survival. For example, elephants can hear low frequency airborne sounds more efficiently than most species (Herbst et al. 2012), as their inner ear lets them hear and communicate through low frequency ground-borne surface vibrations (Reuter et al. 1998). Whilst the mouse has a limited frequency range, it is most sensitive to the high frequency sounds associated with their vocalisations and physical size. In general, most medium to large land animals have hearing that is equivalent to or greater than humans', so animals' should be able to perceive the human voice. The human vocal range should not be a hurdle for communicating and training zoo animals.

Animals with mobile pinna can significantly change their sensitivity to the direction of sound by adjusting the direction of their ear. When loudspeakers were located in front of an animal, those animals with mobile pinnae were able to optimally position their pinnae for detecting sound. For example reindeer could change their threshold of sensitivity by as much as 21 dB depending on whether the reindeer's pinnae were pointing towards or away from the sound source, which is a 105% reduction in perceived sound (Flydal et al. 2001). Therefore it is important to watch the ears of the animal you are working with, and training, to ensure their pinnae are faced towards you, otherwise your voice may not be perceptible above the other sounds that are distracting them.

The function of the ear and pinna is only one part of the challenging acoustic conditions of working with, and training, animals in captivity. One of the greatest challenges that is rarely discussed, but has the most significant impact on the zoo professionals' ability to communicate with the animal is

Zoo Animal Learning and Training, First Edition. Edited by Vicky A. Melfi, Nicole R. Dorey, and Samantha J. Ward.
© 2020 John Wiley & Sons Ltd. Published 2020 by John Wiley & Sons Ltd.

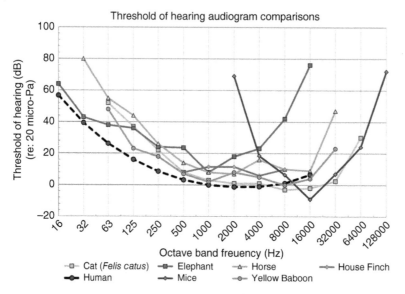

Figure A2.1 An Illustration of several audiograms for different mammal species, illustrating their hearing threshold (dB) and frequency (Hz). *Source:* created by Erik Miller-Klein based on information available in Fay (1988).

described as speech intelligibility for human communication. This is the evaluation of the sound level of the human's voice (signal) compared to the background noise level in the training environment (noise); defined as the signal-to-noise ratio. Research in classroom environments has found that the minimum signal-to-noise ratio must be 15 dB which means the sound from the zoo professional's voice must be at least 15 dB louder than the other potentially distracting noises in the animal's environment (Klatte et al. 2010). The average human speaks at a sound level of 65 dB at 1 m, and sound decays about 3 dB per doubling of distance within an enclosed space and about 6 dB per doubling of distance in an outdoor space. Based on these average levels in an indoor environment the zoo professional must be less than 5 m from the animal for a space that is quiet enough to hear your shoes land as you walk on a dry concrete floor, which would be a background noise level of about 40 dB(A). Whereas with that same vocal level the trainer would need to be no more than 3 m from the animal in an outdoor training environment assuming there is no additional background noise from air traffic, zoo patrons, or nearby roadways. These potential noisy distractions coupled with the mobile pinna of most animals can make communication more challenging for the trainer and difficult for the animal.

Most healthy adult humans, which is assumed to be the pool of zoo professionals, have trained their brains to be acutely aware of human speech patterns and effective at filtering out distracting noises. But most animals, much like juvenile humans, are still learning the patterns of human speech and will have a much more difficult time understanding instructions when there is competing noise and stimuli. Consider evaluating where you communicate with your animals, and/or train them, with respect to noise through the following simple experiment: get two zoo professionals to stand the distance expected for training exercises and have one zoo professional wear a blindfold, and have the other read a paragraph of a newspaper or magazine. Could the blindfolded zoo professional clearly understand all of the words? Were there noises or sounds that were distracting? Consider evaluating your facilities through the ears of your animals.

References

Fay, R.R. (1988). *Hearing in Vertebrates: A Psychophysics Databook*. Hill-Fay Associates.

Flydal, K., Hermansen, A., Enger, P.S., and Reimers, E. (2001). Hearing in reindeer *(Ragnifer tarandus)*. *Journal of Comparative Physiology A* 187: 265–269.

Herbst, C.T., Stoeger, A.S., Frey, R. et al. (2012). How low can you go? Physical production mechanism of elephant infrasonic vocalizations. *Science* 337 (6094): 595. https://doi.org/10.1126/science.1219712.

Klatte, M., Lachmann, T., and Meis, M. (2010). Effects of noise and reverberation on speech perception and listening comprehension of children and adults in a classroom-like setting. *Noise & Health* 12 (39): 270–282.

Reuter, T., Nummela, S., and Hmilä, S. (1998). Elephant hearing. *Journal of Acoustical Society of America*. 104: 1122.

Box A3

Making Sense of Scents: Olfactory Perception in Animals

Neil Jordan

Olfaction is the action or capacity of smelling and is arguably the oldest form of communication in nature. Olfaction underpins critical interactions as diverse as chemotaxis in single celled organisms (Vickers 2000), the pollination of flowering plants (Wright and Schiestl 2009), and the nipple-searching behaviour exhibited by newborn mammals (Hudson and Distel 1983). Whilst all forms of communication involve the provision of information by a sender and the receipt and response to that information by a receiver (Bradbury and Vehrencamp 1998), olfaction contrasts with other forms of communication in that it does not require close proximity between senders and receivers (Bradbury and Vehrencamp 1998). This increases the breadth of contexts in which olfaction may be employed and has profound effects on the design, distribution, and perception of olfactory signals and the anatomical equipment used in this process.

Scent Production and Distribution

Animals from ants to elephants are covered with an array of exocrine glands. These congregations of sebaceous and apocrine tissues are often under the control of hormones, which control the production and release of chemical signals (Ebling 1977). In addition to the secretions of such specialised glands, body orifices associated with digestion and reproduction are also sources of scent signals used in olfactory communication (e.g. Jordan et al. 2013). Whilst chemical information may be excreted or secreted passively from the bodies of animals as they move through their environment, animals often actively deposit olfactory signals into the environment using a diverse range of behaviours collectively called scent marking.

As scent marks may be costly to produce and supplies are limited (Gosling et al. 2000), scent marks are adapted and released so as to maximise their longevity in the environment and the likelihood of their detection by target receivers (e.g. Roberts and Lowen 1997). Both chemical and behavioural adaptions may be employed to ensure that signals remain in the environment for as long as possible. For example, fatty acids are commonly found in scent secretions and are thought to slow the release of signalling components (Alberts 1992), a goal which may be further promoted by the observation that animals commonly place their scents in sites protected from the elements (Eisenberg and Kleiman 1972). Additionally species often scent mark in prominent places, such as the crossroads of trails (Barja et al. 2004), presumably to promote the discovery of signals by intended receivers. The intended recipients of these olfactory messages vary

Zoo Animal Learning and Training, First Edition. Edited by Vicky A. Melfi, Nicole R. Dorey, and Samantha J. Ward.
© 2020 John Wiley & Sons Ltd. Published 2020 by John Wiley & Sons Ltd.

considerably depending on the particular message being sent, and olfactory signals have been shown to convey a staggering amount of information on factors such as sex, age, condition, reproductive status, group membership and even individual identity (e.g. see Epple et al. 1987). But how are these olfactory signals perceived?

Scent Perception

Successful olfactory communication relies on the receipt and accurate discrimination of scent signals, all of which is handled by specialised olfactory processing equipment, which is divided into two discrete systems. The first, the main olfactory system, is the olfactory bulb, and this receives information passed along millions of olfactory receptor neurons travelling from the nose to the brain. Neurons expressing the same olfactory receptor are clustered together in individual glomeruli in the olfactory bulb, forming a kind of layer that has been described as a spatial odour map (Uchida et al. 2000). Importantly, in the context of learning and perception, the olfactory bulb appears to exhibit enormous plasticity in response to experience with scents, and exposure to odours can modify the perceptions of those odours enormously (e.g. Wilson and Stevenson 2003). In the second system, the accessory olfactory system, the vomeronasal organ or Jacobson's organ (Keverne 1999), is responsible for scent detection and is usually found in the main nasal chamber of many but not all amphibians, reptiles, and mammals. Neurons from the vomeronasal organ connect it to the accessory olfactory bulb in the brain, which ultimately feeds the hypothalamus via the amygdala. As in the olfactory bulb, odours are detected following binding to receptors, but in this case the signals are non-volatile and so are received in the liquid phase. Additionally, animals may exhibit various specialist behaviours in order to transfer signals to the vomeronasal organ. For example, snakes touch their tongue to the vomeronasal organ after 'tasting' the air for

signals from potential prey (Gillingham and Clark 1981), and elephants transfer potential signals there using the tip of their trunk (Rasmussen and Schulte 1998). In fact, the typical 'flehmen' response of cats and many ungulates, whereby they lift and curl the nose whilst raising and contorting their lips, relates to the stimulation of the vomeronasal organ with urine from conspecifics. Overall, the vomeronasal organ appears to be particularly important in modulating reproduction and social behaviour, and many neurons are activated by urinary-borne chemicals signalling physiological state.

Perception and Learning

Using this olfactory equipment, receivers must be able to extract and categorise the relevant olfactory information from encountered odours, all within environments infused with irrelevant and potentially confusing smells. Correct categorisation, or recognition, of different odours may be important in many aspects of social life, including the modulation of competitive relationships and the maintenance of stable social groups, navigation and reproduction. Although innate responses to some chemical signals are known, novel odours are generally more difficult to discriminate than familiar ones (Rabin 1988). In fact, learning is perhaps the most important component of olfactory perception. Unsurprisingly, human adults and older children perform better at odour discrimination tasks than young children (Lehrner et al. 1999), but importantly improvement in the discrimination often improves after exposure to that odour, whilst discrimination of other odours remains poor (Rabin 1988). This suggests that maturity of the receptor system itself is not driving the observed odour improvement in discrimination with age. Rather it is specific experience or learning that enhances olfactory perception. In fact, under conditions where a person's capability for memory is reduced, such as is the case in Alzheimer's disease, the ability to perceive odour quality is itself lost or impaired, despite the olfactory apparatus presumably remaining intact (Koss et al. 1988). In animals, including humans, the olfactory system is a flexible processing network that is shaped and honed by learning. In a world of changing olfactory conditions and pressures where the importance and relevance of specific odours varies along with the appropriate responses to them, such plasticity is essential.

References

Alberts, A.C. (1992). Pheromonal self-recognition in desert iguanas. *Copeia* 1992 (1): 229–232.

Barja, I., de Miguel, F.J., and Bárcena, F. (2004). The importance of crossroads in faecal marking behaviour of the wolves (*Canis lupus*). *Naturwissenschaften* 91 (10): 489–492.

Bradbury, J.W. and Vehrencamp, S.L. (1998). *Principles of Animal Communication.* Sunderland, Massachusetts: Sinauer.

Ebling, F.J. (1977). Hormonal control of mammalian skin glands. In: *Chemical Signals in Vertebrates* (eds. D. Müller-Schwarze and M.M. Mozell), 17–33. Boston, MA: Springer.

Eisenberg, J.F. and Kleiman, D.G. (1972). Olfactory communication in mammals. *Annual Review of Ecology and Systematics* 3 (1): 1–32.

Epple, G., Belcher, A.M., Greenfield, K.L. et al. (1987). Making sense out of scents-species-differences in scent glands, scent marking behavior and scent mark composition in the Callitrichidae. *International Journal of Primatology.* 8 (5): 434–434.

Gillingham, J.C. and Clark, D.L. (1981). Snake tongue-flicking: transfer mechanics to Jacobson's organ. *Canadian Journal of Zoology* 59 (9): 1651–1657.

Gosling, L.M., Roberts, S.C., Thornton, E.A., and Andrew, M.J. (2000). Life history costs of olfactory status signalling in mice. *Behavioral Ecology and Sociobiology* 48 (4): 328–332.

Hudson, R. and Distel, H. (1983). Nipple location by newborn rabbits: behavioural evidence for pheromonal guidance. *Behaviour* 85 (3): 260–274.

Jordan, N.R., Golabek, K.A., Apps, P.J. et al. (2013). Scent-mark identification and scent-marking behaviour in African wild dogs (*Lycaon pictus*). *Ethology* 119 (8): 644–652.

Keverne, E.B. (1999). The vomeronasal organ. *Science* 286 (5440): 716–720.

Koss, E., Weiffenbach, J.M., Haxby, J.V., and Friedland, R.P. (1988). Olfactory detection and identification performance are dissociated in early Alzheimer's disease. *Neurology* 38 (8): 1228–1228.

Lehrner, J.P., Walla, P., Laska, M., and Deecke, L. (1999). Different forms of human odor memory: a developmental study. *Neuroscience Letters* 272 (1): 17–20.

Rabin, M.D. (1988). Experience facilitates olfactory quality discrimination. *Perception & Psychophysics* 44 (6): 532–540.

Rasmussen, L.E.L. and Schulte, B.A. (1998). Chemical signals in the reproduction of Asian (*Elephas maximus*) and African (*Loxodonta africana*) elephants. *Animal reproduction science* 53 (1–4): 19–34.

Roberts, S.C. and Lowen, C. (1997). Optimal patterns of scent marks in klipspringer (*Oreotragus oreotragus*) territories. *Journal of Zoology* 243 (3): 565–578.

Uchida, N., Takahashi, Y.K., Tanifuji, M., and Mori, K. (2000). Odor maps in the mammalian olfactory bulb: domain organization and odorant structural features. *Nature Neuroscience* 3 (10): 1035.

Vickers, N.J. (2000). Mechanisms of animal navigation in odor plumes. *The Biological Bulletin* 198 (2): 203–212.

Wilson, D.A. and Stevenson, R.J. (2003). The fundamental role of memory in olfactory perception. *Trends in Neurosciences* 26 (5): 243–247.

Wright, G.A. and Schiestl, F.P. (2009). The evolution of floral scent: the influence of olfactory learning by insect pollinators on the honest signalling of floral rewards. *Functional Ecology* 23 (5): 841–851.

Part B

Types of Learning That Can Be Achieved in a Zoo Environment

We learned in the previous section that animals learn from their surroundings. In this section, chapters will discuss how learning opportunities can be afforded in the zoo environment, enabling animals to learn throughout their lifetime. Specifically, we will dive into how training regimes based on the expression of wild-type behaviours can lead to the creation of enrichment and improved management overall. We will also discuss how training, which certainly has its foundations in science, can benefit from good intuition; based on personal experience of training and animal behaviour, as well as derived from being well trained themselves. Thus we celebrate that training as an art and consider how this art can be integrated into the animal's daily management routine. Finally we will explore the human–animal interactions that occur in a zoo and how these might result in learning opportunities for the animals; and people.

5

What Is There to Learn in a Zoo Setting?

Fay Clark

5.1 Introduction

Contrary to the belief of many, animals can have a rich learning experience in a zoo setting. Learning can be defined as the process of adaptive change in a behaviour as a result of experience (Thorpe 1963). At face value, it may seem that animals housed in zoos have restricted learning opportunities; space restrictions and routine husbandry procedures remove environmental variation, choice, and control (Watters 2014). Whilst it is known that highly predictable husbandry routines can have detrimental effects on welfare (Bassett and Buchanan-Smith 2007), zoos can still be highly variable environments in which animals are learning frequently. Environmental variability comes in the form of staff, volunteer, and researcher turnover; visitor presence; changing climate; and animals moving between exhibits for breeding programmes, exhibit developments or through the natural cycle of births and deaths. And in terms of fostering agency, in other words allowing animals to act independently and to make their own decisions (Clark 2018), modern zoos purposely provide animals with more choices and control over their daily lives through training and enrichment programmes (Westlund 2014; Young 2013).

Learning is a very broad topic intimately linked with memory and cognition (Shettleworth 2010); for that reason I have chosen to focus in on the three overarching questions posed by Shettleworth (2010): (i) what conditions/circumstances stimulate learning? (ii) what is being learned? and (iii) how does learning affect behaviour? This chapter will be a broad overview of the types of learning an animal may experience during its lifetime in the zoo, and the implications of these for their captive management. Although some learning opportunities occur at particular life stages, for example shortly after birth, at weaning, or at sexual maturation (Shettleworth 2010), other learning occurs on a daily basis, relating to the timing of food and cues associated with other events such as exhibit cleaning and veterinary checks. Animals are constantly being exposed to different stimuli and their motivations change too; and learning is, by definition, impacted by previous experience (Thorpe 1963).

5.2 Early Life

5.2.1 Embryonic Learning

We tend to think of an animal's learning journey beginning at birth, but in fact many mammals, birds, amphibians, and fish learn from the tastes, smells, and sounds that surround them during their prenatal development (Hepper 1996; Hepper and Waldman

Zoo Animal Learning and Training, First Edition. Edited by Vicky A. Melfi, Nicole R. Dorey, and Samantha J. Ward.
© 2020 John Wiley & Sons Ltd. Published 2020 by John Wiley & Sons Ltd.

1992; Sneddon et al. 1998). For example, superb fairy wren (*Malurus cyaneus*) mothers call to their eggs, and when an egg hatches, the nestling produces calls containing important characteristics of the mother's call. This embryonically learned 'password' bonds parent and nestling, and helps parents detect foreign cuckoo nestlings (Colombelli-Négrel et al. 2012). Cuttlefish (*Sepia officinalis*) can 'see' (i.e. their visual system is active) several weeks before hatching; when researchers provided embryonic cuttlefish with images of crabs (a prey species) they had a significantly higher preference for preying on this species once hatched (Darmaillacq et al. 2008). Domestic dogs, amongst other mammalian species as well as birds, can learn food preferences from prenatal exposure to certain flavours in the mother's diet (Wells and Hepper 2006).

Even though embryonic learning studies have been undertaken under highly controlled laboratory conditions, the results are relevant to the same or similar species living in zoos. We should be mindful of providing zoo animals with appropriate prenatal signals; for example paying special attention to minimising the stressors perceived by gravid females, and providing meaningful conspecific, predator, or diet-related cues they may learn from before birth. This could be particularly challenging for endangered birds cross-fostered by other species (Conway 1988). The finding that young frogs and salamanders show adaptive behaviours such as shelter-seeking cues learned before hatching (Mathis et al. 2008) has great implications for zoo endangered species breeding programmes where individuals may eventually be reintroduced to the wild (Crane and Mathis 2010; also refer to Chapter 12 on training and reintroduction).

5.2.2 Recognising Parents and Mates

Once an animal is born, it may need to recognise its parent(s) in order to receive care and begin to learn survival skills. This poses a challenge for zoos in rare cases where animals have to be removed from their mothers, shortly after birth for veterinary care or because the mother rejects them. Firstly, let me acknowledge that some behaviours are partially innate, dictated by an animal's genes, and partially learned from experience, either from interacting with the world or by being taught (Shettleworth 2010). Imprinting is a type of time-sensitive learning which generally occurs within hours or days after birth with some genetic input, in which an animal gains a sense of identity. Filial imprinting refers to when a young animal acquires several of its behavioural characteristics from its parent. However, in the absence of a parent the animal will imprint on any moving stimulus (Sluckin 2017). The behavioural development of precocial birds such as geese and ducks under human care is particularly delicate. However, Horwich (1989) found that sandhill crane (*Antigone canadensis*) chicks could be successfully hand-reared by humans, if the human form is disguised, and chicks imprinted on realistic models of their parents such as hand puppets with accompanying crane brooding calls. Other birds which have been hand-reared by zoos using similar methods include California condors *Gymnogyps californianus* (Utt et al. 2007) and kakapo *Strigops habroptilus* (Sibley 1994).

A young animal also has to learn its kin and who to mate with, and there is evidence that the rearing environment is important for the development of mating preferences (Slagsvold et al. 2002). Sexual imprinting is the process by which a young animal learns the characteristics of a desirable mate, which has very important implications for animals as part of zoo endangered species breeding programmes. Kendrick et al. (2001) found that domestic sheep and goats cross-fostered at birth, then reared in mixed-species groups, had social behaviour and mate choice more closely resembling their foster species than genetic species. Infant chimpanzees (*Pan troglodytes*) reared by human caregivers, for the purposes of language acquisition and other cognitive research, have demonstrated

Figure 5.1 Though there are reasons that might be found to hand-rear apes, it is common practice in zoos to reintroduce offspring to their mother or a conspecific surrogate as soon as possible. *Source:* Jeroen Stevens.

how they can learn a rich suite of skills and characteristics from a 'parent' figure (e.g. Gardner and Gardner 1998). This explains why, when hand-rearing great apes in zoos nowadays, it is considered best practice (if possible) to return the infant to its mother or a conspecific surrogate as soon as possible (Porton and Niebruegge 2006; see Figure 5.1).

A third type of imprinting less relevant to zoo species is habitat imprinting, where a young animal learns the characteristics of a suitable breeding habitat (Davis and Stamps 2004). This is most important to migratory species travelling long distances over divergent environments, and needing to learn how to return to a particular type of environment for breeding. In a zoo setting, natal homing (the process by which animals return to their birthplace to reproduce) is largely redundant since animals cannot move large distances, but animals may still associate conditions learned in early life with conditions suitable for breeding. Natal homing has been demonstrated in freshwater turtles *Graptemys kohnii* (Freedberg et al. 2005) and marine turtles (Lohmann et al. 2013) for example; caregivers of these species in zoos should make sure environmental conditions are kept consistent from early life to breeding age.

5.3 Maturation

5.3.1 Weaning and Independence

The weaning period which takes place in many animals broadly refers to the period of time where a young animal will gradually gain independence from its parents, learning to be one of its own species; this may coincide with the gradual replacement of mother's milk with adult diet. Early social development in long-lived social mammals has important implications for adult behaviour, particularly in taxa which exhibit stable, long-term bonds such as great apes, elephants, and dolphins (de Waal and Tyack 2009). Non-human great apes wean relatively late compared to other animals; on average weaning occurs at around six years of age (Galdikas and Wood 1990). Like the time-sensitive imprinting period described above, the weaning period can also be a sensitive time and crucial for certain learning experiences. For example, domestic piglets socialised prior to being weaned were scored by their caregivers as being more 'relaxed/ contented', indicating that key social skills could be learned before they gain independence from the mother (Morgan et al. 2014). An interesting zoo-based study on plains zebra (*Equus burchelli*) indicated that the sex of a foetus determined the time of weaning of the previous offspring; weaning current offspring was faster when the dam was carrying a male foetus than a female foetus (Pluháček et al. 2007). The study of Pluháček et al. (2007) requires replication in other zebra populations and would also be interesting to replicate in other equids, so the implications for zoo-housed zebra are conservative for now. If the sex of the foetus is known, caregivers may consider adjusting the time period allotted for weaning care accordingly, and the time-sensitive learning in the weaning period of the current offspring (if the foetus is male) will be more pressured because it will be shorter.

5.3.2 Learning Through Play

Play behaviour has widely been cited as having no clear function (Burghardt 2005), but conversely, young animals can learn much about their environment through play. Given that play behaviour in young animals often resembles foraging or social sequences, play may serve as a form of 'practice' for adult survival (Smith 1982). For example, play helps to prepare animals for serious fighting later in their lives (Pellis and Pellis 2017). Spinka et al. (2001) propose that play is a form of 'training for the unexpected', learning how to escape threats in a relatively relaxed context before those skills are needed in a real-world, dangerous situation. The fact that some animals also play throughout adult life indicates that continued practice of survival skills is important (Smith 1982; see Figure 5.2).

Zoos can promote play in their animals through the provision of enrichment (Chapter 6), but it will also arise sporadically in a social group. Zoo-housed gorillas (*Gorilla gorilla gorilla*) have shown that actions performed during play-fighting are functional; one gorilla hitting another during a play-fight was seen as an 'unfair' action, to which the 'hitee' was sensitive (Van Leeuwen et al. 2011). Furthermore, juvenile zoo gorillas have been observed to make careful, context-dependent decisions about who to engage in social play, as well as how 'rough' to make their actions (Palagi et al. 2007).

Figure 5.2 In a zoo setting play can also occur between different species, as illustrated here between a juvenile gorilla *Gorilla gorilla gorilla* and adult black mangabey *Lophocebus aterrimus*. *Source:* Tjerk van Meulen.

5.4 Adult Life

5.4.1 Being Social

Learning to be social encompasses how and when to interact with conspecifics and understanding social cues and rules. In a zoo setting, social group membership can change regularly with births, deaths, and between-zoo transfers, thus requiring animals to learn how to regularly recognise individuals and establish and maintain social bonds. The myriad of ways that animals learn to recognise group members is beyond the scope of this chapter, but briefly this can be via scent, sound and vision (Shettleworth 2010). Scent-based learning can be difficult in zoo exhibits where hygiene standards are paramount; frequent cleaning with disinfectant could mark or remove natural scents (Clark and King 2008). Songbirds are amongst the best-studied vocal learners, with research showing that individuals need to hear themselves sing to develop song normally (Brainard and Doupe 2000). Elephants are also vocal learners; Poole et al. (2005) showed that captive African elephants (*Loxodonta africana*) can modify their vocalisations in response to vocalisations they have previously heard. In the zoo, anthropogenic noise coming from visitors and amplified sound systems could disrupt the learning and maintenance of animal song and other vocalisations. Anthropogenic sound measurement and analysis in relation to zoo animal well-being is in its infancy, but we can expect to see more of this research emerging in the next few years (Orban et al. 2017). A number of mammals and birds are also capable of visually discriminating between familiar and unfamiliar individuals, and in some species this is based on facial patterns (e.g. Brown and Dooling 1992; Kendrick et al. 1995; Rosenfeld and Van Hoesen 1979). Recent research also suggests that cichlid fish are capable of conspecific face recognition (Hotta et al. 2017; Satoh et al. 2016).

The ability of an animal to be a bystander and 'eavesdrop' on the social status of others has been studied in primates, birds, and fish (e.g. Crockford et al. 2007; Oliveira et al. 1998; Peake 2005), and the ability of an animal to learn its relative position within a group has been studied in primates (Tomasello and Call 1997). Furthermore, being able to use what has been learned to exploit others in a social group, known as 'Machiavellian intelligence' (Gavrilets and Vose 2006; Whiten and Byrne 1997), is seen as the highest level of social intelligence. In captive settings, primates have learned to deceive their groupmates and hide knowledge from others in elaborately competitive strategies (Byrne and Whiten 1989). The 'social politics' of zoo chimpanzees was popularised by de Waal and de Waal (2007), but is not only confined to great apes; there is also perhaps some evidence for Machiavellian intelligence in fish (Bshary 2011).

Zoo exhibits that house more than one species, referred to as 'mixed-species exhibits' are commonplace in modern zoos, allowing species with the same ecological niche or from the same geographical area to be housed together (Clark and Melfi 2012; see Figure 5.3). These exhibits provide an additional layer of social opportunity and learning the meaning of both homo- and heterospecific cues. Different species can sometimes learn from each other; Krebs (1973) found that two species of chickadees (*Parus* spp.) learned from one another about the location and nature of potential feeding locations when they were housed together in a large, mixed species aviary. Heterospecific social learning (i.e. social learning from one species to another) was observed in a mixed-species zoo exhibit of tufted capuchins (*Sapajus* sp.) and common squirrel monkeys (*Saimiri sciureus*); capuchins were influenced by squirrel monkeys when foraging for food in mixed species groups (Messer 2013).

For many species in a zoo setting, being social also encompasses interactions and relationships with caregivers, known more formally as 'human–animal interactions' and 'human–animal relationships' (Hosey and Melfi 2014; see Chapter 9 and Figure 5.4).

Figure 5.3 Mixed species exhibits can offer many learning opportunities; the example here comprises of the North African barbary sheep *Ammotragus lervia* and gelada baboons *Theropithecus gelada*. *Source:* Jeroen Stevens.

Learning to attend to human gaze and read body language has been well studied in mammals, for example in primates (Hare and Tomasello 2005), domestic horses (Dorey et al. 2014; Proops and McComb 2010), domestic dogs (Shepherd 2010), and goats (Nawroth and McElligott 2017). Begging behaviour towards humans is also worthy of mention here. Carlstead et al. (1991) state that bears (Ursidae) can easily develop begging habits in zoos as a frustrated appetitive activity, and has similarly been observed in Asian short-clawed otters *Aonyx cinereus* (Gothard 2007). Begging behaviour has also been observed in zoo-housed orangutans *Pongo* spp. (Choo et al. 2011) and lion-tailed macaques *Macaca silenus* (Mallapur et al. 2005) in response to visitor presence.

5.4.2 Finding Food

Even though zoo animals are fed routinely by their caregivers, sometimes multiple times a day and often predictably over the course of a year (i.e. without significant seasonal variation), zoo animals still have to learn what is good to eat within the provided diet, and where and how to get it. The feeding of zoo

Figure 5.4 Social interactions for zoo animals should include those which occur between the animals and their caregivers; illustrated here as a positive interaction between an okapi *Okapia johnstoni* and their keeper. *Source:* Jeroen Stevens.

animals can broadly be split into three categories: (i) routine feeding in a focal location (e.g. from a bowl in a specific exhibit area); (ii) feeding within a large, naturalistic exhibit in multiple locations; and (iii) feeding the majority of the diet as part of a positive reinforcement training programme. These will now be discussed in turn.

It could be argued than routine focal feeding requires the least from animals in terms of learning skills. However, anyone who has worked with domestic or wild animals will be familiar with their ability to learn to recognise the sound of keys rattling, buckets clanging, or any other noise and movement associated with the impending arrival of feeding time. This type of learning is through classical conditioning, where an animal learns to associate one stimulus (such as rattling keys) with another (such as food arriving), until the stimulus alone may invoke a response (such as rattling keys making the animal run to the exhibit door, even if the food does not arrive). Acoustically sensitive animals can learn to become attuned to the sound of automatic feeders (food containers set to automatically release food randomly or at specific times) and therefore caregiver attempts to make feeding times less predictable are thwarted. Feeding animals via environmental enrichment may help decrease feeding predictability and thus reduce conditioned responses to unintentional feeding cues, to be discussed further in Chapter 6.

The second approach is feeding animals in large, naturalistic exhibits at several locations across time and space. In some instances, animals may be able to forage from scattered and hidden food items, and even from natural vegetation. Spatial learning is of relevance here, and several learning mechanisms have been identified across the animal kingdom. Juvenile chimpanzees have been shown to learn the locations of up to 18 hidden foods in a large outdoor enclosure, and use 'optimum routing' to visit these locations with minimal backtracking or revisiting empty locations (Menzel 1973). Caching animals, including but not limited to squirrels and corvid birds, can learn the locations of tens or hundreds of hidden food items, recalling these locations

over many weeks or months (Clayton et al. 2003; Kamil and Gould 2008). Some species are also able to take into account their previous actions; for example, zoo-housed short-beaked echidnas (*Tachyglossus aculeatus*) learned to avoid a location which had previously contained food (a 'win-shift' strategy) as long as the period of time to memorise this location was less than 90 minutes (Burke et al. 2002; see Figure 5.5). In contrast, a study on managed black-tailed deer *Odocoileus hemionus columbianus* (Gillingham and Bunnell 1989) showed that deer used the same search path repeatedly if it was previously successful, and were not adept at taking into account changes in food availability. This has great implications in a zoo setting, where different species will have different spatial learning abilities and therefore their food needs to be arranged in space accordingly.

Figure 5.5 A common method of providing learning opportunities for zoo animals is the provision of puzzle feeders or manipulating food provision in some other way, as seen here with a short-beaked echidna *Tachyglossus aculeatus. Source:* Ray Wiltshire.

Finally, some zoo animals, particularly marine mammals such as sea lions and dolphins, are fed the vast majority of their daily diet through participation in positive reinforcement training sessions (Ramirez 2012). Animals will learn through operant conditioning to associate a cue (such as a caregiver hand signal) with a behaviour they must perform (or indeed a sequence of behaviours) in order to receive a food reward (Laule 1999).

In terms of *how* to feed, many feeding behaviours are learned at a young age through social learning. A zoo-based feeding experiment found that young Japanese squirrels (*Sciurus lis*) learned an optimal method of processing and eating nuts after watching a demonstrator squirrel, but this learning period was fairly rigid and took place before the squirrels reached three years old (Tamura 2011). Young koala bears (*Phascolarctos cinereus*) will consume digested eucalyptus in their mother's faeces; undigested eucalyptus is toxic to their immature digestive systems and so coprophagy is an important feeding behaviour to learn (Martin and Handasyde 1999). Experiments in a range of species including primates and elephants show the importance of obtaining novel foraging skills through social learning, especially when trial-and-error learning may be inefficient. A suite of captive studies have used the paradigm of a food (puzzle) box with two possible actions to open (for example a lid which lifts and slides); a demonstrator (human or animal) can be observed performing one action, and researchers measure if the rest of the group learn that action socially and the behaviour spreads (Tomasello and Call 1997). For example, in the case of captive black-and-white ruffed lemurs *Varecia variegata* (Stoinski et al. 2011) and African elephants *Loxodonta africana africana* (Greco et al. 2012), viewing a demonstrator's interactions facilitated the learning of a novel task. Greco et al. (2012) provided a zoo herd of six adult female elephants with food-acquisition tasks that could be solved using two possible methods. A 'demonstrator' elephant (the dominant female) solved the tasks in the presence of the other elephants. Although the method used by the demonstrator did not predict the methods used by the 'observer' elephants, subjects spent a greater percentage of their time interacting with the apparatus if they had observed the demonstrator doing so first.

5.4.3 Learning Through Exploration

Spatial learning is important to all animals because at some point they must navigate their environment to find food as well as other resources such as mates and shelter. Amongst vertebrates, rats and mice are known for their ability to navigate complex 3D mazes (Vorhees and Williams 2014). The spatial learning skills of honeybees (*Apis* spp.) are also impressive; they can learn to fly through a complex maze using coloured markers as 'sign-posts'. Furthermore they can use these sign-posts to navigate a novel maze efficiently (Zhang et al. 1996). In guppies (*Poecilia reticulata*), a complex maze can be learned rapidly, over around 5 days (30 trials of exposure). These types of studies are highly controlled and therefore are not sited in zoos; however, as spoken about before, the results of highly controlled laboratory experiments can still have implications for zoo animals. Since many animals 'know where they are going' far better than we give them credit for, this strengthens the case for larger, more elaborate zoo enclosures moving forwards. Clark (2013) remarked on how most captive cetacean exhibits are smooth-sided concrete pools, and how this does not cater for their innate echolocative abilities; a substitute for a complete exhibit rebuild could be to add underwater obstacles to navigate into the current pool. A recent development for captive bottlenose dolphins (*Tursiops truncatus*) has been an underwater touchscreen, which dolphins can use to 'hunt' moving images of fish (Fenz and Kaplan 2017).

Exploration is known to be a highly motivated behaviour in both wild and captive animals (reviewed by Clark 2017). If exploration is split into two broad types, inspective and inquisitive, the purpose of the latter type of exploration is to gather information for the future (Berlyne 1960; Russell 1983). Learning about future opportunities in the environment can be particularly useful for animals living in changeable environments and therefore inquisitive exploration could be thwarted by relatively static zoo environments (Clark 2018). I touch on this more below, in the 'solving complex problems' section.

5.4.4 Mating and Parenting

I have already discussed the young animal's need to recognise and bond to a parent, and how mate choice can be learned at a young age (sexual imprinting). But when animals are sexually mature, how do they learn to find a mate and eventually care for their own offspring? In zoos, where mate choice is significantly restricted (as humans are responsible for creating and maintaining social assemblages), is mate recognition and mating behaviour really that important? I would argue affirmatively, since animals must learn how to mate effectively, even if their choice of potential mates is few. Birds are frequently cited as having elaborate courtship displays which may include sequenced head bobbing, wing flapping, and crouching movements (Rogers and Kaplan 2002). The male golden-collared manakin (*Manacus vitellinus*), for example, performs an elaborate courtship display including a sequence of acrobatic jumps unique to the individual (Fusani et al. 2007). When the mating season begins, males repeatedly practise their display to establish a choreographed routine (Coccon et al. 2012). This represents one animal's learning experience over time, whereas other animals may learn courtship behaviours off each other. In brown-headed cowbirds (*Molothrus ater*), courtship behaviours can be socially transmitted across generations, and different populations of birds have slightly different courtship behaviours as a result.

This brings us to parenting skills. Many primates must learn infant care from parents and siblings in order to be successful. When managing the breeding of highly social zoo primates, housing females together maximises females learning infant care skills from each other (Bard 1995). The orangutan has the longest duration of single parental care of any mammal, weaning young at around eight years of age (Galdikas and Wood 1990). During this period of one-on-one learning, the young orangutan will be learning what foods are safe to eat, and how to build a nest; when orangutans are orphaned and find their way into zoos or sanctuaries, the rehabilitation process involves humans or surrogate apes attempting to mimic natural parenting in what have been dubbed 'forest schools' (Russon et al. 2016). In contrast to direct teaching from parent to offspring, meerkats (*Suricata suricatta*) have a social system characterised by alloparental care; offspring are reared by their parents as well as additional group members called 'helpers' or 'carers' (Thornton and McAuliffe 2006).

5.4.5 Avoiding Danger

What dangers do zoo animals face? Although good zoos provide their inhabitants with shelter, warmth, comfort, and food, no environment can ever be 100% safe. Zoo animals may still face a threat of predation (e.g. in the UK, native fox predation on zoo birds), wounding aggression from conspecifics (e.g. Alford et al. 1995; Hosey et al. 2016; Ruehlmann et al. 1988), and the growing impact of climate change on ex situ conservation in zoos (Mawdsley et al. 2009).

Some responses to salient danger cues will be instinctive whilst others will be shaped by experience. Animals that have lived their whole lives in zoos may still respond to wild predator alarm calls, showing that antipredator responses are not necessarily lost in species living under human care. As an example, Hollén and Manser (2007) found that meerkats living in zoos demonstrated the same suite of alarm calls documented in wild meerkats, under broadly similar contexts. Furthermore, the zoo meerkats could discriminate between faeces of carnivores (potential predators) and herbivores (nonpredators), even though they did not have prior experience of these scent cues. Similarly, it has been found that the calls of birds of prey elicit alarm calls in captive-born Geoffroy's marmosets (*Callithrix geoffroyi*). The fact that prey animals have these types of responses to potential predators, which require little learning experience (Hollén and Manser 2007), means that we should be cautious when designing zoo exhibits where predators and prey are living close to each other (Stanley and Aspey 1984; Wielebnowski et al. 2002).

Other responses to danger require more active learning through teaching. Meerkat adults teach pups how to handle live prey such as scorpions through observational learning; furthermore, teaching methods are responses to changes in pup begging calls, making the learning process very efficient (Thornton and McAuliffe 2006). Knowing which species have sophisticated teaching strategies should be taken into account for zoo reintroduction programmes (see Chapter 12).

Finally, other fear responses may be learned more gradually, as and when required. Naive vervet monkeys (*Chlorocebus pygerythrus*) learned to avoid the electric fence of a new enclosure in a matter of days, which the authors believe was due to trial and error learning (Weingrill et al. 2005). Conditioned taste aversion refers to when an animal learns not to consume a food after experiencing an adverse effect such as illness or pain; ultimately this helps animals avoid foods which are poisonous. The animal forms an association between eating the food (e.g. berries) and the negative experience (e.g. vomiting). Forthman and Ogden (1992) proposed that conditioned taste aversion could be used to control pest species in the zoo, in other words by laying down bad-tasting food to warn them away from other food supplies. Whilst I am on the topic of 'pest' species in zoos, it is worthwhile taking a moment to consider what non-zoo animals can learn in the zoo. By this I am referring to the many thousands of cockroaches, rats, mice, pigeons, seagulls, coyotes, and foxes which

may enter the zoo grounds to take advantage of plentiful food supplies, warmth and shelter. Aside from conditioned taste aversion, they may learn to recognise the same or similar 'good food' cues as zoo animals do (for example rattling keys), how to beg or steal food from zoo visitors or animals, and to avoid dangers such as electric fences and traps.

To end this section, I discuss the zoo animal learning mechanism used to deal with stimuli which are *not* dangerous, such as regular visitor presence and noises made by caregivers that do not signal the impending arrival of food. Habituation is defined as a decrease in responsiveness to a stimulus that is presented repeatedly over time (Blumstein 2016). It is known to occur across the entire animal kingdom, and saves animals the energy of responding to harmless stimuli. Habituation is a relatively simple form of learning and can readily occur in response to enrichment which lacks novelty. For example, octopuses (*Octopus dofleini*) habituated within the first trial of exposure to a plastic object (Mather and Anderson 1999), and captive great apes are particularly well known for rapid habituation to enrichment (Clark 2011). The opposite of habituation is sensitisation; this is an increased response to a repeated stimulus. For example, the job of the zoo vet can become incredibly difficult if an animal becomes sensitised to their mere presence, manifesting in behaviour as running away or becoming stressed. To combat sensitisation towards aversive but vital veterinary and husbandry procedures, positive reinforcement training can be used to habituate the animal (Young and Cipreste 2004).

5.4.6 Solving Complex Problems

So far, I have considered how animals learn skills for their survival, but also for more routine needs such as learning regular food provision cues. It could be argued that selection pressures imposed on zoo animals are significantly weaker than those imposed on their wild counterparts; even though animals in managed care experience environmental changes these are not at the same frequency or magnitude as the wild. Environmental enrichment is an established management tool which stimulates captive animal learning, whether learning is a conscious goal of enrichment or not (refer to Chapter 6).

At the upper end of the enrichment scale is what I refer to as 'cognitive enrichment', which requires an animal to use its evolved cognitive skills to solve a complex (yet species-appropriate and skill-level-relevant) problem (Clark 2011, 2017). Solving the problem should be connected to some sort of outcome or reward; the animal should realise that the problem was solved, perhaps by receiving a food reward (Clark 2017). There is research to suggest that, in addition to the feel-good experience of the 'aha' moment when the problem is solved (Hagen and Broom 2004), the learning process in itself can have welfare benefits (Langbein et al. 2004). In contrast to the unstructured learning contexts of play and exploration (discussed earlier), learning through complex problem solving requires caregivers to actively provide animals with that problem, unless their normal exhibit

already provides it. Animal participation in pure cognitive research trials (i.e. research set up to investigate whether an animal has a particular cognitive skill) will actively stimulate learning, and could be a form of cognitive enrichment for many cognitively advanced species (Hopper 2017).

Some of the learning mechanisms involved in complex problem solving are trial-and-error learning, and insight learning. Trial and error learning was discussed earlier, where animals make repeated responses to 'see what works'. Insight learning, on the other hand, occurs when an animal uses its past experiences and reasoning to solve a novel problem (see Figure 5.6). Unlike operant conditioning, insight learning does not involve trial and error. Rooks (*Corvus frugilegus*) are not known to habitually use tools in the wild, but in captivity birds were found to use insightful learning to use tools to acquire a piece of food from a tube (Bird and Emery 2009). Insight was also identified in an Asian elephant *Elephas maximus* (Foerder et al. 2011) who, similar to the rooks, was able to acquire hidden food using tools by piecing together several prior experiences. Armed

Figure 5.6 It is likely that different learning styles are being used by this Goeldi monkey as it learns how to get food from the puzzle feeder, which compromises of a simple log with food hidden inside, but to access the log a film canister has to be pushed aside. *Source:* Nicky Needham.

with evidence of these 'extremely clever' forms of animal learning (with the obvious disclaimer of a species bias, judging animal skills relative to human abilities, Rowe and Healy 2014), zoos need to ask themselves the following questions: 'What *have* our animals learned? But more importantly what *could* they learn?'

5.4.7 Learning Across a Lifetime

Recent research has found a positive relationship between animal 'personality' traits such as boldness and aggressiveness, and learning ability (Carere and Locurto 2011). For example, in cavies (*Cavia aperea*) there was a strong positive correlation between the speed of learning a food acquisition task, and three different personality measures: boldness, activity level, and aggressiveness (Guenther et al. 2014). This suggests that individual differences in learning may persist fairly predictably throughout life; some individuals will generally have a higher aptitude for learning than others. From a management perspective, it is useful for zoo staff to identify learning differences in their animals, if this knowledge can help customise enrichment or training programmes.

Some learning outcomes may be predicted by 'personality', but not all learning stays consistent over an animal's lifetime. I wrap up this chapter with a brief consideration of what happens to learning as zoo animals inevitably age, especially considering that many animals in human care exceed their natural lifespans due to a lack of predators and excellent veterinary care (Krebs et al. 2018). The bulk of research on the topic of ageing and learning comes from laboratory animals and humans; from this we know that learning ability generally decreases with age in vertebrates (Kausler 1994; Riddle 2007). In their review of caring for aged zoo animals, Krebs et al. (2018) state the importance of giving geriatric animals extra human assistance to learn new things, whereas long-term memory may be impacted less.

5.5 Conclusions

- This chapter is a whistle-stop tour of some of the learning opportunities available to zoo animals, and their practical implication for the management of those animals. It is not intended to be an all-inclusive review, and readers are encouraged to read fundamental animal learning texts for greater detail on the mechanisms involved (e.g. Byrne 2017; Mackintosh 1994; Pearce 2013; Shettleworth 2010).

- Zoo environments are neither static nor devoid of learning opportunities. Learning can begin before birth and continues until death; zoo managers therefore need to provide optimum environments at every life stage and pay particular attention to the importance of the maternal environment and time-sensitive imprinting phases in early life.

- Learning through play and exploration allow animals to practice important skills and seek new information; these unstructured learning opportunities should be fostered in a zoo setting in order to place an animal in charge of its own learning processes.

- In addition to unstructured learning opportunities (play and exploration), structured learning opportunities should be provided to animals in zoos. These take the form of positive reinforcement training sessions, and cognitive enrichment involving complex problem solving.

- The vast majority of formal animal learning studies have been undertaken under highly controlled laboratory conditions, but there is no reason why the findings should not be applied to animals living in zoos. For example, if we discover a particular species possesses a particular learning skill, we should endeavour to facilitate the expression of this skill in captivity. We still have much to learn about how most zoo animals learn, but cognitive studies are becoming more commonplace under the remit of modern zoos being viable sites for research.

Acknowledgements

I would like to thank the book editors for asking me to contribute this chapter. I also thank Lucy Mason and Christa Emmett for editorial assistance, and Lewis Dean for helpful discussions regarding animal learning theories.

References

Alford, P.L., Bloomsmith, M.A., Keeling, M.E., and Beck, T.F. (1995). Wounding aggression during the formation and maintenance of captive, multimale chimpanzee groups. *Zoo Biology* 14 (4): 347–359.

Bard, K.A. (1995). Parenting in primates. In: *Handbook of Parenting, Vol. 2. Biology and Ecology of Parenting* (ed. M.H. Bornstein). New Jersey: Lawrence Erlbaum Associates, Inc.

Bassett, L. and Buchanan-Smith, H.M. (2007). Effects of predictability on the welfare of captive animals. *Applied Animal Behaviour Science* 102 (3–4): 223–245.

Berlyne, D.E. (1960). *Conflict, Arousal and Curiosity*. New York: McGraw Publishing.

Bird, C.D. and Emery, N.J. (2009). Insightful problem solving and creative tool modification by captive nontool-using rooks. *Proceedings of the National Academy of Sciences* 106 (25): 10370–10375.

Blumstein, D.T. (2016). Habituation and sensitization: new thoughts about old ideas. *Animal Behaviour* 120: 255–262.

Brainard, M.S. and Doupe, A.J. (2000). Auditory feedback in learning and maintenance of vocal behaviour. *Nature Reviews Neuroscience* 1 (1): 31.

Brown, S.D. and Dooling, R.J. (1992). Perception of conspecific faces by budgerigars (*Melopsittacus undulatus*): I. Natural faces. *Journal of Comparative Psychology* 106 (3): 203.

Bshary, R. (2011). Machiavellian intelligence in fishes. In: *Fish Cognition and Behaviour*, 2e (eds. C. Brown, K. Laland and J. Krause), 277–297. London: Wiley-Blackwell.

Burghardt, G.M. (2005). *The Genesis of Animal Play: Testing the Limits*. New York: MIT Press.

Burke, D., Cieplucha, C., Cass, J. et al. (2002). Win-shift and win-stay learning in the short-beaked echidna (*Tachyglossus aculeatus*). *Animal Cognition* 5 (2): 79–84.

Byrne, J.H. (2017). *Learning and Memory: A Comprehensive Reference*, 2e. London: Academic Press.

Byrne, R. and Whiten, A. (1989). *Machiavellian Intelligence: Social Expertise and the Evolution of Intellect in Monkeys, Apes, and Humans*. Oxford Science Publications.

Carere, C. and Locurto, C. (2011). Interaction between animal personality and animal cognition. *Current Zoology* 57 (4): 491–498.

Carlstead, K., Seidensticker, J., and Baldwin, R. (1991). Environmental enrichment for zoo bears. *Zoo Biology* 10 (1): 3–16.

Choo, Y., Todd, P.A., and Li, D. (2011). Visitor effects on zoo orangutans in two novel, naturalistic enclosures. *Applied Animal Behaviour Science* 133 (1–2): 78–86.

Clark, F.E. (2011). Great ape cognition and captive care: can cognitive challenges enhance well-being? *Applied Animal Behaviour Science* 135 (1–2): 1–12.

Clark, F.E. (2013). Marine mammal cognition and captive care: a proposal for cognitive enrichment in zoos and aquariums. *Journal of Zoo and Aquarium Research* 1 (1): 1–6.

Clark, F.E. (2017). Cognitive enrichment and welfare: current approaches and future directions. *Animal Behavior and Cognition* 4 (1): 52–71.

Clark, F.E. (2018). Competence and agency as novel measures of zoo animal welfare. *Measuring Behaviour* 11: 171–172.

Clark, F.E. and King, A.J. (2008). A critical review of zoo-based olfactory enrichment. In: *Chemical Signals in Vertebrates 11* (eds. J. Hurst, R.J. Beynon, S.C. Roberts and T. Wyatt), 391–398. New York: Springer.

Clark, F.E. and Melfi, V.A. (2012). Environmental enrichment for a mixed-species nocturnal mammal exhibit. *Zoo Biology* 31 (4): 397–413.

Clayton, N.S., Yu, K.S., and Dickinson, A. (2003). Interacting cache memories: evidence for flexible memory use by Western Scrub-Jays (*Aphelocoma californica*). *Journal of Experimental Psychology: Animal Behavior Processes* 29 (1): 14.

Coccon, F., Schlinger, B.A., and Fusani, L. (2012). Male Golden-collared Manakins *Manacus vitellinus* do not adapt their courtship display to spatial alteration of their court. *Ibis* 154 (1): 173–176.

Colombelli-Négrel, D., Hauber, M.E., Robertson, J. et al. (2012). Embryonic learning of vocal passwords in superb fairy-wrens reveals intruder cuckoo nestlings. *Current Biology* 22 (22): 2155–2160.

Conway, W. (1988). Can technology aid species preservation? In: *Biodiversity* (eds. E.O. Wilson and F.M. Peter), 263–268. Washington: National Academy Press.

Crane, A.L. and Mathis, A. (2010). Predator-recognition training: a conservation strategy to increase postrelease survival of hellbenders in head-starting programs. *Zoo Biology* 30 (6): 611–622.

Crockford, C., Wittig, R.M., Seyfarth, R.M., and Cheney, D.L. (2007). Baboons eavesdrop to deduce mating opportunities. *Animal Behaviour* 73 (5): 885–890.

Darmaillacq, A.S., Lesimple, C., and Dickel, L. (2008). Embryonic visual learning in the cuttlefish, *Sepia officinalis*. *Animal Behaviour* 76 (1): 131–134.

Davis, J.M. and Stamps, J.A. (2004). The effect of natal experience on habitat preferences. *Trends in Ecology and Evolution* 19 (8): 411–416.

De Waal, F.B. and Tyack, P.L. (2009). *Animal Social Complexity: Intelligence, Culture, and Individualized Societies*. Harvard University Press.

De Waal, F. and Waal, F.B. (2007). *Chimpanzee Politics: Power and Sex Among Apes*. Johns Hopkins University Press.

Dorey, N.R., Conover, A.M., and Udell, M.A.R. (2014). Interspecific communication from people to horses (*Equus ferus caballus*) is influenced by different horsemanship training styles. *Journal of Comparative Psychology* 128 (4): 337–342. https://doi.org/10.1037/a0037255

Fenz, K. and Kaplan, H. (2017). Scientists to probe dolphin intelligence using an interactive touchpad. Press release by Hunter College and Rockefeller University. http://www.m2c2.net/wp-content/uploads/2017/05/Official-Press-Release.pdf.

Foerder, P., Galloway, M., Barthel, T. et al. (2011). Insightful problem solving in an Asian elephant. *PLoS One* 6 (8): e23251.

Forthman, D.L. and Ogden, J.J. (1992). The role of applied behavior analysis in zoo management: today and tomorrow. *Journal of Applied Behavior Analysis* 25 (3): 647–652.

Freedberg, S., Ewert, M.A., Ridenhour, B.J. et al. (2005). Nesting fidelity and molecular evidence for natal homing in the freshwater turtle, *Graptemys kohnii*. *Proceedings of the Royal Society of London B: Biological Sciences* 272 (1570): 1345–1350.

Fusani, L., Giordano, M., Day, L.B., and Schlinger, B.A. (2007). High-speed video analysis reveals individual variability in the courtship displays of male golden-collared manakins. *Ethology* 113 (10): 964–972.

Galdikas, B.M. and Wood, J.W. (1990). Birth spacing patterns in humans and apes. *American Journal of Physical Anthropology* 83 (2): 185–191.

Gardner, B.T. and Gardner, R.A. (1998). Development of phrases in the early utterances of children and cross-fostered chimpanzees. *Human Evolution* 13 (3–4): 161–188.

Gavrilets, S. and Vose, A. (2006). The dynamics of Machiavellian intelligence. *Proceedings of the National Academy of Sciences of the United States of America* 103 (45): 16823–16828.

Gillingham, M.P. and Bunnell, F.L. (1989). Effects of learning on food selection and searching behaviour of deer. *Canadian Journal of Zoology* 67 (1): 24–32.

Gothard, N. (2007). What is the proximate cause of begging behaviour in a group of captive Asian short-clawed otters. *IUCN/SSC Otter Specialist Group Bulletin* 24 (1): 14–35.

Greco, B.J., Brown, T.K., Andrews, J.R. et al. (2012). Social learning in captive African elephants (*Loxodonta africana africana*). *Animal Cognition* 16 (3): 459–469.

Guenther, A., Brust, V., Dersen, M., and Trillmich, F. (2014). Learning and personality types are related in cavies (Cavia aperea). *Journal of Comparative Psychology* 128 (1): 74.

Hagen, K. and Broom, D.M. (2004). Emotional reactions to learning in cattle. *Applied Animal Behaviour Science* 85: 203–213.

Hare, B. and Tomasello, M. (2005). Human-like social skills in dogs? *Trends in Cognitive Sciences* 9 (9): 439–444.

Hepper, P.G. (1996). Fetal memory: does it exist? What does it do? *Acta Paediatrica* 85: 16–20.

Hepper, P.G. and Waldman, B. (1992). Embryonic olfactory learning in frogs. *The Quarterly Journal of Experimental Psychology Section B, Comparative and Physiological Psychology* 44 (3-4b): 179–197.

Hollén, L.I. and Manser, M.B. (2007). Motivation before meaning: motivational information encoded in meerkat alarm calls develops earlier than referential information. *The American Naturalist* 169 (6): 758–767.

Hopper, L.M. (2017). Cognitive research in zoos. *Current Opinion in Behavioral Sciences* 16: 100–110.

Horwich, R.H. (1989). Use of surrogate parental models and age periods in a successful release of hand-reared sandhill cranes. *Zoo Biology* 8 (4): 379–390.

Hosey, G. and Melfi, V. (2014). Human-animal interactions, relationships and bonds: a review and analysis of the literature. *International Journal of Comparative Psychology* 27 (1): 117–142.

Hosey, G., Melfi, V., Formella, I. et al. (2016). Is wounding aggression in zoo-housed chimpanzees and ring-tailed lemurs related to zoo visitor numbers? *Zoo Biology* 35 (3): 205–209.

Hotta, T., Satoh, S., Kosaka, N., and Kohda, M. (2017). Face recognition in the Tanganyikan cichlid *Julidochromis transcriptus*. *Animal Behaviour* 127: 1–5.

Kamil, A.C. and Gould, K.L. (2008). Memory in food caching animals. In: *Learning and Memory: A Comprehensive Reference* (eds. R. Menzel and J.R. Byrne), 419–439. Amsterdam: Academic Press.

Kausler, D.H. (1994). *Learning and Memory in Normal Aging*. London: Academic Press.

Kendrick, K.M., Atkins, K., Hinton, M.R. et al. (1995). Facial and vocal discrimination in sheep. *Animal Behaviour* 49 (6): 1665–1676.

Kendrick, K.M., Haupt, M.A., Hinton, M.R. et al. (2001). Sex differences in the influence of mothers on the sociosexual preferences of their offspring. *Hormones and Behavior* 40 (2): 322–338.

Krebs, J.R. (1973). Social learning and the significance of mixed-species flocks of chickadees (*Parus* spp.). *Canadian Journal of Zoology* 51 (12): 1275–1288.

Krebs, B., Marrin, D., Phelps, A. et al. (2018). Managing aged animals in zoos to promote positive welfare: a review and future directions. *Animals* 8 (7): 116. https://doi.org/10.3390/ani8070116.

Langbein, J., Nürnberg, G., and Manteuffel, G. (2004). Visual discrimination learning in dwarf goats and associated changes in heart rate and heart rate variability. *Physiology and Behavior* 82 (4): 601–609.

Laule, G. (1999). Training laboratory animals. In: *UFAW Handbook on the Care and Management of Laboratory Animals*, 7e, vol. 1 (ed. T. Poole), 21–27. Oxford: Blackwell.

Lohmann, K.J., Lohmann, C.M., Brothers, J.R., and Putman, N.F. (2013). Natal homing and imprinting in sea turtles. In: *The Biology of Sea Turtles*, vol. 3 (eds. J. Wyneken, K.J. Lohmann and J.A. Musick), 59–78. CRC Press.

Mackintosh, N.J. (1994). *Animal Learning and Cognition*. San Diego: Academic Press.

Mallapur, A., Waran, N., and Sinha, A. (2005). Factors influencing the behaviour and welfare of captive lion-tailed macaques in Indian zoos. *Applied Animal Behaviour Science* 91 (3–4): 337–353.

Martin, R. and Handasyde, K.A. (1999). *The Koala: Natural History, Conservation and Management*. UNSW Press.

Mather, J.A. and Anderson, R.C. (1999). Exploration, play and habituation in octopuses (*Octopus dofleini*). *Journal of Comparative Psychology* 113 (3): 333.

Mathis, A., Ferrari, M.C., Windel, N. et al. (2008). Learning by embryos and the ghost of predation future. *Proceedings of the Royal Society of London B: Biological Sciences* 275 (1651): 2603–2607.

Mawdsley, J.R., O'malley, R., and Ojima, D.S. (2009). A review of climate-change adaptation strategies for wildlife management and biodiversity conservation. *Conservation Biology* 23 (5): 1080–1089.

Menzel, E.W. (1973). Chimpanzee spatial memory organization. *Science* 182 (4115): 943–945.

Messer, E.J.E. (2013). Social learning and social behaviour in two mixed-species communities of tufted capuchins (*Sapajus* sp.) and common squirrel monkeys (*Saimiri sciureus*). Doctoral dissertation. University of St Andrews.

Morgan, T., Pluske, J., Miller, D. et al. (2014). Socialising piglets in lactation positively affects their post-weaning behaviour. *Applied Animal Behaviour Science* 158: 23–33.

Nawroth, C. and McElligott, A.G. (2017). Human head orientation and eye visibility as indicators of attention for goats (*Capra hircus*). *PeerJ* 5: e3073.

Oliveira, R.F., McGregor, P.K., and Latruffe, C. (1998). Know thine enemy: fighting fish gather information from observing conspecific interactions. *Proceedings of the Royal Society of London B: Biological Sciences* 265 (1401): 1045–1049.

Orban, D.A., Soltis, J., Perkins, L., and Mellen, J.D. (2017). Sound at the zoo: using animal monitoring, sound measurement, and noise reduction in zoo animal management. *Zoo Biology* 36 (3): 231–236.

Palagi, E., Antonacci, D., and Cordoni, G. (2007). Fine-tuning of social play in juvenile lowland gorillas (*Gorilla gorilla gorilla*). *Developmental Psychobiology: The Journal of the International Society for Developmental Psychobiology* 49 (4): 433–445.

Peake, T.M. (2005). Eavesdropping in communication networks. In: *Animal Communication Networks*. Cambridge University Press.

Pearce, J.M. (2013). *Animal Learning and Cognition: An Introduction*. Psychology Press.

Pellis, S.M. and Pellis, V.C. (2017). What is play fighting and what is it good for? *Learning and Behavior* 45 (4): 355–366.

Pluháček, J., Bartoš, L., Doležalová, M., and Bartošová-Víchová, J. (2007). Sex of the foetus determines the time of weaning of the previous offspring of captive plains zebra (*Equus burchelli*). *Applied Animal Behaviour Science* 105 (1-3): 192–204.

Poole, J.H., Tyack, P.L., Stoeger-Horwath, A.S., and Watwood, S. (2005). Animal behaviour: elephants are capable of vocal learning. *Nature* 434 (7032): 455.

Porton, I. and Niebruegge, K. (2006). The changing role of hand rearing in zoo-based primate breeding programs. In: *Nursery Rearing of Nonhuman Primates in the 21st Century* (eds. G.P. Sackett, G. Ruppenthal and K. Elias), 21–31. Boston: Springer.

Proops, L. and McComb, K. (2010). Attributing attention: the use of human-given cues by domestic horses (*Equus caballus*). *Animal Cognition* 13 (2): 197–205.

Ramirez, K. (2012). Marine mammal training: the history of training animals for medical behaviors and keys to their success. *Veterinary Clinics: Exotic Animal Practice* 15 (3): 413–423.

Riddle, D.R. (2007). *Brain Aging: Models, Methods, and Mechanisms*. CRC Press.

Rogers, L.J. and Kaplan, G.T. (2002). *Songs, Roars, and Rituals: Communication in Birds, Mammals, and Other Animals*. Harvard University Press.

Rosenfeld, S.A. and Van Hoesen, G.W. (1979). Face recognition in the rhesus monkey. *Neuropsychologia* 17 (5): 503–509.

Rowe, C. and Healy, S.D. (2014). Measuring variation in cognition. *Behavioral Ecology* 25 (6): 1287–1292.

Ruehlmann, T.E., Bernstein, I.S., Gordon, T.P., and Balcaen, P. (1988). Wounding patterns in three species of captive macaques. *American Journal of Primatology* 14 (2): 125–134.

Russell, P.A. (1983). Psychological studies of exploration in animals: a reappraisal. In: *Exploration in Animals and Humans* (eds. J. Archer and L.I.A. Birke), 2–54. Van Nostrand Reinhold.

Russon, A.E., Smith, J.J., and Adams, L. (2016). Managing human–orangutan relationships in rehabilitation. In: *Ethnoprimatology. Primate Conservation in the 21st Century*, Developments in Primatology: Progress and Prospects (ed. M. Waller), 233–258. Springer.

Satoh, S., Tanaka, H., and Kohda, M. (2016). Facial recognition in a Discus fish (Cichlidae): experimental approach using digital models. *PLoS One* 11 (5): e0154543.

Shepherd, S.V. (2010). Following gaze: gaze-following behavior as a window into social cognition. *Frontiers in Integrative Neuroscience* 4: 5.

Shettleworth, S.J. (2010). *Cognition, Evolution, and Behavior*. Oxford University Press.

Sibley, M.D. (1994). First hand-rearing of Kakapo *Strigops habroptilus*: at the Auckland Zoological Park. *International Zoo Yearbook* 33 (1): 181–194.

Slagsvold, T., Hansen, B.T., Johannessen, L.E., and Lifjeld, J.T. (2002). Mate choice and imprinting in birds studied by cross-fostering in the wild. *Proceedings of the Royal Society of London B: Biological Sciences* 269 (1499): 1449–1455.

Sluckin, W. (2017). *Imprinting and Early Learning*. Routledge.

Smith, P.K. (1982). Does play matter? Functional and evolutionary aspects of animal and human play. *Behavioral and Brain Sciences* 5 (1): 139–155.

Sneddon, H., Hadden, R., and Hepper, P.G. (1998). Chemosensory learning in the chicken embryo. *Physiology and Behavior* 64 (2): 133–139.

Spinka, M., Newberry, R.C., and Bekoff, M. (2001). Mammalian play: training for the unexpected. *The Quarterly Review of Biology* 76 (2): 141–168.

Stanley, M.E. and Aspey, W.P. (1984). An ethometric analysis in a zoological garden: modification of ungulate behavior by the visual presence of a predator. *Zoo Biology* 3 (2): 89–109.

Stoinski, T.S., Drayton, L.A., and Price, E.E. (2011). Evidence of social learning in black-and-white ruffed lemurs (*Varecia variegata*). *Biology Letters* 7 (3): 376–379.

Tamura, N. (2011). Population differences and learning effects in walnut feeding technique by the Japanese squirrel. *Journal of Ethology* 29 (2): 351–363.

Thornton, A. and McAuliffe, K. (2006). Teaching in wild meerkats. *Science* 313 (5784): 227–229.

Thorpe, W.H. (1963). *Learning and Instinct in Animals*. London: Methuen Press.

Tomasello, M. and Call, J. (1997). *Primate Cognition*. USA: Oxford University Press.

Utt, A.C., Harvey, N.C., Hayes, W.K., and Carter, R.L. (2007). The effects of rearing method on social behaviors of mentored, captive-reared juvenile California condors. *Zoo Biology* 27 (1): 1–18.

Van Leeuwen, E.J., Zimmermann, E., and Ross, M.D. (2011). Responding to inequities: Gorillas try to maintain their competitive advantage during play fights. *Biology Letters* 7 (1): 39–42.

Vorhees, C.V. and Williams, M.T. (2014). Assessing spatial learning and memory in rodents. *Ilar Journal* 55 (2): 310–332.

Watters, J.V. (2014). Searching for behavioral indicators of welfare in zoos: uncovering anticipatory behavior. *Zoo Biology* 33 (4): 251–256.

Weingrill, T., Stanisiere, C., and Noë, R. (2005). Training vervet monkeys to avoid electric wires: is there evidence for social learning? *Zoo Biology* 24 (2): 145–151.

Wells, D.L. and Hepper, P.G. (2006). Prenatal olfactory learning in the domestic dog. *Animal Behaviour* 72 (3): 681–686.

Westlund, K. (2014). Training is enrichment – and beyond. *Applied Animal Behaviour Science* 152: 1–6.

Whiten, A. and Byrne, R.W. (1997). *Machiavellian Intelligence II: Extensions and Evaluations*, vol. 2. Cambridge University Press.

Wielebnowski, N.C., Fletchall, N., Carlstead, K. et al. (2002). Noninvasive assessment of adrenal activity associated with husbandry and behavioral factors in the North American clouded leopard population. *Zoo Biology* 21 (1): 77–98.

Young, R.J. (2013). *Environmental Enrichment for Captive Animals*. Wiley.

Young, R.J. and Cipreste, C.F. (2004). Applying animal learning theory: training captive animals to comply with veterinary and husbandry procedures. *Animal Welfare* 13 (2): 225–232.

Zhang, S.W., Bartsch, K., and Srinivasan, M.V. (1996). Maze learning by honeybees. *Neurobiology of Learning and Memory* 66 (3): 267–282.

6

Environmental Enrichment

The Creation of Opportunities for Informal Learning

Robert John Young, Cristiano Schetini de Azevedo, and Cynthia Fernandes Cipreste

6.1 Introduction

The world of all captive animals should be filled with formal and informal opportunities to learn new associations and contingencies; this is how their wild environment functions and allows animals to operate on it through behavioural expression. It is this very behavioural expression that can be an important factor affecting animal wellbeing. In this chapter we will focus on the informal learning opportunities that arise from environmental enrichment and what their consequences are for the animal.

An animal in a new enclosure has much to learn. What does the keeper's uniform look like? How does my keeper smell and what does their voice sound like? What are the signs that I am about to be fed? Are there any sounds that predict the arrival of food? Which is the quickest way to the outdoor section of the enclosure? Where is the best place to hide if I do not want to see the public? However, after a few weeks the animal has learned all the daily contingencies and learning opportunities may rapidly disappear. The animal may no longer be required to exercise its mind and memory.

In barren environments with unvarying husbandry routines captive animals have little or no opportunities to learn new things whether through formal or informal means. One well-known consequence of this situation is the performance of abnormal behaviour such as stereotypic animal behaviour (e.g. stereotypic route pacing by captive carnivores) (Swaisgood and Shepherdson 2005; Kagan and Veasey 2010). The expression of such abnormal behaviour is widely regarded as an indicator of suboptimal animal welfare (Mason and Latham 2004; Sarrafchi and Blokhuis 2013; Schork and Young 2014).

Studies have shown that even in such barren and unvarying environments operant conditioning, for example, to perform 'tricks' for a show can alleviate the expression of abnormal behaviour (Bloomsmith et al. 2007; Coleman and Maier 2010). It has been argued by many animal trainers that this is because training is in itself an enriching activity (Melfi 2013; Westlund 2014), and that animals may prefer to engage in trained behaviours than to use environmental enrichment (Dorey et al. 2015); this is a debatable point and one that we will not address in this chapter. Instead, our interest will focus on informal learning opportunities, which arise from the most commonly applied method to improve animal welfare in captivity: environmental enrichment.

Environmental enrichment typically involves the addition of novel stimuli to a captive animal's environment in an attempt to improve animal welfare (Shepherdson 2003; Young 2003; Azevedo et al. 2007); for example, the provision of toys to an enclosure. To be effective in the long-term the provision of environmental enrichment needs to be a dynamic and goal-

Zoo Animal Learning and Training, First Edition. Edited by Vicky A. Melfi, Nicole R. Dorey, and Samantha J. Ward.
© 2020 John Wiley & Sons Ltd. Published 2020 by John Wiley & Sons Ltd.

orientated process (Cipreste et al. 2010) in which the stimuli provided are regularly changed and novel stimuli are periodically introduced to the animals. If not, the animals quickly lose interest in the environmental enrichment as there is nothing 'interesting' about it for them, unless the environmental enrichment is associated with food (Vasconcellos et al. 2012; Hosey et al. 2013). A hungry animal will always be interested in food and most animals will always show interest in prized food items (i.e. treats) (Bays 2014). This type of enrichment is really playing on the animal's need to maintain its body in homeostasis; that is, meeting its physiological requirements. It is for this reason that food based environmental enrichment is the most overused category of enrichment (Young 2003).

Research shows that the animal welfare benefits from environmental enrichment come mainly from two sources: (i) the novelty of the stimuli provided; and (ii) control over the environment, which environmental enrichment facilitates (Young 2003). Thus by definition if we are providing animals with novel stimuli then we are creating opportunities for animals to learn about these stimuli, which are being presented in their enclosures. It is worth highlighting that control over the environment involves informal animal learning as well. The public-shy animal, learns that they may go inside to avoid humans staring at them; thus, the animal has learnt how to operate their environment using its behaviour to avoid an aversive situation.

The benefits of environmental enrichment to animal welfare are supported from a wide body of evidence reviewed by Young (2003), which included behavioural to neurological evidence. The behavioural evidence shows improved learning capacity in animals receiving environmental enrichment (Strand et al. 2010; Sorensen et al. 2011); the neurological evidence shows increased dendrite density, dendrite complexity, and increases in size areas of the brain such as the amygdala, which is associated with learning and memory (Rampon et al. 2000; Van Praag et al. 2000; Jung and Herms 2014). It is therefore with good reason to believe that during

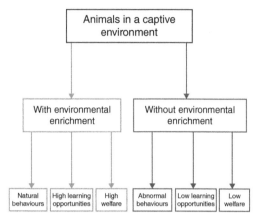

Figure 6.1 Comparison of the consequences of enriched or not enriched environments for captive animals.

the process of receiving environmental enrichment that animals are constantly being presented with learning opportunities (Figure 6.1).

It is these informal opportunities, which we now wish to turn our attention too; it is our aim to explain how best these opportunities can be exploited from an animal learning and animal welfare perspective.

6.2 Environmental Enrichment Categories

Typically environmental enrichment is divided up into five non-mutually exclusive categories (Shepherdson et al. 1999; Young 2003):

1) Social (i.e. social grouping)
2) Occupation or cognitive (e.g. opportunities for mental or physical exercise)
3) Physical (e.g. the use of species appropriate furniture in enclosures)
4) Sensory (i.e. stimulation of the five senses)
5) Nutritional (i.e. the use of food, associated or not to devices that enable animals to use their anatomical and behavioural adaptive features in food handling).

All of these categories of environmental enrichment if managed properly can provide informal learning opportunities for animals

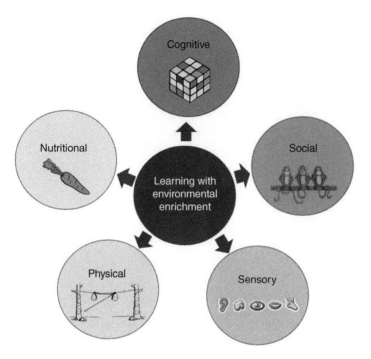

Figure 6.2 How the different types of environmental enrichment act in providing learning opportunities for captive animals. Dark grey means higher learning opportunities and light grey means lower learning opportunities.

(Figure 6.2). However, these learning opportunities must be tailored to a species specific need (Swaisgood 2007; Griffin 2012); for example, creating lots of opportunities for social interactions for an asocial species would clearly be inappropriate from an animal welfare point-of-view even though it may create many learning opportunities.

6.3 Informal Learning Opportunities During Social Enrichment

Social species normally live in stable groups, which may appear to provide the opportunities for captive animals to learn about group members. Obviously, in zoos, captive breeding programmes do necessitate the exchange of animals and breeding creates new individuals in the group (Rees 2011). However, the frequency of animal exchanges is generally low and often is limited to the sex, which would migrate

between groups in the wild for breeding opportunities.

In all species, the arrival of newborn individuals creates opportunities for learning: at its simplest, learning is required to recognise the new individual or at a more complex level, learning how to care for the new individual (e.g. allomothering) (Breed 2014; Vidya 2014) (Figure 6.3). The opportunities of parenthood for learning may be limited by the nature of the species developmental trajectory; for example, hoofstock are generally precocious and develop rapidly, whereas primates are altricial and develop much more slowly. Thus, births in altricial species will have longer term impacts on informal learning opportunities than those in precocious species (Ueno and Suzuki 2014).

Group size and complexity can provide many opportunities for social enrichment. However, often in captivity animals are kept in group sizes, which are significantly smaller than their wild counterparts (e.g. elephants) and this greatly reduces the social complexity

Figure 6.3 Allomothering in collared-peccaries (*Pecari tajacu*). *Source:* Carlos Magno de Faria.

of the situation (Rees 2011). Many species in the wild live in complex social networks, where they need to be constantly monitoring and learning about not only their own social interactions but those of relatives and competitors alike. These social networks are usually dynamic in nature and so individuals must be constantly updating themselves on what is happening within their social group. In certain species, such as chimpanzees this situation is rightly equated with politics (Brosnan 2007), but unfortunately most captive chimpanzee groups are small in terms of the number of individuals. It is important to remember here that as group size increases social complexity increases exponentially (Kempe and Mesoudi 2014). It is for exactly this reason that Humphrey (1976) suggested that the large apes are often very occupied in captivity by the political machinations of their own making.

In the wild, large groups of social animals do not always stay together as a single group, but may form subgroups, which split off for foraging purposes and then return to the main group (Reboreda and Fernandez 1997; Michelena et al. 2009). This adds to social complexity in terms of social monitoring. In captivity, this situation from an enrichment point-of-view has been recreated by providing animals with multiple rooms or even enclosures to travel between (Coe and Dykstra 2010; Coe 2012). For example, the orangutans at the Smithsonian National Zoo in Washington DC may travel by overhead cables to different enclosures to avoid certain individuals, to be with certain individuals or just to be on their own.

The arrival of internet video calling has created a number of extremely interesting social enrichment opportunities; for example, the ability of animals of the same species (or even different species) to interact visually and auditory in a remote manner. This could prove, for some species, to be extremely stimulating and have practical applications. For example, if you were wishing to transfer a female chimpanzee to a new group it would be possible for her to see and interact with individuals in her future group. Thus, facilitating her learning about her new group members and, perhaps, even the layout of her future enclosure (Tibbetts and Dale 2007). Video presentations were also used as environmental enrichment for European starlings, but they were not effective in the prevention of emergence or development of abnormal behaviours (Coulon et al. 2014).

Of course, the use of electronic or any other enrichment devices should be planned, taking into account the animal, not the human perspective; especially those aiming at visual or auditory interactions. Animals, depending on the species, see and hear differently than humans (Jacobs 2009). Canids, as an example, do not have the capacity to see all colours, but old world primates can (Jacobs et al. 1993; Vorobyev 2004; Jacobs 2008). Many birds and reptiles are able to see in the UV spectrum, thus, food enrichment with items reflecting UV could be arranged (Honkavaara et al. 2002). In the same way, auditory environmental enrichment could focus on the whole audible spectrum of the animal, and not only be based on the human hearing spectrum (Wells 2009). More on sensory enrichment can be read below, in the following sections of this chapter.

6.4 Informal Learning Opportunities During Occupational or Cognitive Enrichment

The very idea of cognitive enrichment is to challenge the brains of animals to solve a puzzle and this clearly will involve learning opportunities. Typically, animals are motivated to solve puzzles in order to obtain food rewards (Cheyne 2010). One of the earliest examples of this was the use of artificial termite mounds with captive chimpanzees, which had to learn to dip for honey or yoghurt rather than termites (Nash 1982). Zoos typically provided chimpanzees with branches that needed to be trimmed and shaped for purpose and the chimpanzees learned to do this through trial-and-error or by watching other more experienced individuals dip for food rewards (although the imitation of others in tool use is controversial for chimpanzees, Buttelmann et al. 2013). The first time animals see such a puzzle there

is a real cognitive challenge, but clearly this diminishes over time as the animals learn to perform the task with increasing efficiency. However, it should be noted that some of these tasks may require that the animal practices and improves over long periods of time. For example, it may take wild capuchin monkeys months or even years to fully develop the skill of opening a nut using a rock as a hammer (Ottoni and Izar 2008). For a review on the use and benefits of cognitive environmental enrichment, see Clark (2017).

Where cognitive activities occupy the minds of captive animals, thereby creating many informal learning opportunities, the learning opportunities available from occupational activities such as exercise seem to be much more limited.

Occupational activities such as animals performing exercise can be strongly motivated through the use of food rewards. Fonseca et al. (2014) showed that all tested rats would run several kilometres per day on a running wheel if the distance ran was contingent on the receipt of food rewards. Unfortunately, the learning component in these processes is limited to a short time frame and the animals behaved as if they are inside a Skinner box. The responses appear to become automatic unless variable ratios or variable interval schedules are used for delivering the food rewards (Fonseca et al. 2014). In general, the problem with occupational exercise is that animals will not perform it without being frequently rewarded with food and thus whilst even wild animals have been shown to use a running wheel, they will not use it for prolonged periods of time (Meijer and Robbers 2014).

One other problem of occupational exercise in terms of enrichment is that it tends to be unvarying; the animal learns to express locomotory behaviour at a constant speed. In the wild, animals vary their locomotion according to terrain and the reason they are moving from point A to point B. Studies on physical conditioning in humans have shown that exercise, which is varying in its intensity, is more mentally stimulating and produces higher levels of fitness (Taylor et al. 1985; Swan and Hyland 2012). These concepts have yet to be applied to occupational exercise as environmental enrichment, but it seems possible that providing flexibility would make such an activity much more appealing to animals due to their need to pay greater attention to what is happening (e.g. a sudden increase or decrease in the speed of a running wheel).

6.5 Informal Learning Opportunities During Physical Enrichment

The use of physical environmental enrichment such as the use of furniture and enclosure design to meet the needs of animals in captivity is the one most often favoured by animal keeping institutions such as zoos (Young 2003). The reason for this is that they, incorrectly, believe that this is a one-off cost; for example, you build your penguin enclosure with a large pool and then you do not need to worry anymore about their enrichment. Alternatively, you create an extensive three dimensional climbing structure for your leopards and again the problem of dealing with the animal's welfare is solved. There are two mistakes here: (i) no single piece of enrichment will ever be sufficient to guarantee the welfare of a captive animal; and (ii) environmental enrichment needs to be dynamic.

In the wild, an animal's physical environment is varied and is subject to frequent changes; for example, the margays living in the Atlantic forest of Brazil are highly arboreal cats who must learn how to navigate around their three dimensional home range. In the wet season many branches are blown off trees by storms and some trees within their territory will fall down. Thus, margays need to update their spatial map of their home range at least several times a year. In the zoo environment if the climbing structure is never modified then this updating never occurs. Thus, simple rebranching of an arboreal species' enclosure a few times a year creates learning opportunities as the individuals must learn new routes to be able to move from A to B.

Animals that have been previously housed in unvarying enclosures for long periods of time will require that changes made to the layout of their enclosure are initially made at a slow pace and infrequently, to avoid stressing the animals (Young 2003). Once animals have learnt that change in their enclosure can occur, then modifications to their enclosure can start to be made much more frequently. It is important that animals learn how to cope with change, especially since change will be part of their lives in a zoo (e.g. being moved as part of a breeding exchange).

A new concept in zoo design developed by Jon Coe is the use of tunnels connected to different enclosures in a train track like system at the Philadelphia Zoo. This has taken the concept of changing environment and learning to move around to a new level (Coe 2012). In this concept, all zoo animals have the opportunity to move around a zoo, which could be to a different enclosure or to a viewing point. If all the hundred or so enclosures were connected together through this system (which of course would not always be appropriate as a small primate enclosure would probably be unsuitable for a large carnivore) then animals would be constantly learning new routes around their zoo and the layout, not to mention the smells, of their new enclosure. Thus, one of the most static parts of a zoo environment could, through this system, become one of the most dynamic parts of the zoo environment and create constant learning opportunities for the animals.

6.6 Informal Learning Opportunities During Sensory Enrichment

The captive environment like its wild equivalent can be full of novel sensory stimuli. In captivity, new sensory experiences, in terms of environmental enrichment, are most often provided to stimulate the following modalities: visual, auditory, olfactory, and gustatory (see Modality Boxes); the latter will be considered later in the nutritional enrichment. Each novel sensory stimulus will be registered (cue) in an animal's brain and its characteristics learnt plus any associated contingency (e.g. the sound of a keeper shaking keys and the arrival of food).

Captivity in itself may represent an uncontrolled source of visual, auditory and olfactory stimuli, which may or may not be sources of environmental enrichment. The most obvious sources of visual, auditory and olfactory stimuli in zoos are their visitors that stand in front of an enclosure where they can be seen, heard, and smelt by the animals. Studies have shown that in general, the auditory stimuli (referred to as noise) produced by zoo visitors often impacts negatively on animal welfare (Quadros et al. 2014). Unfortunately, we have no quantitative data on the effects of visual stimuli, other than large groups of visitors appear to be more disturbing than smaller groups (Hosey 2000; Kuhar 2008); but they are confounded with greater noise production. As in the case of visual stimuli from the public, it is highly likely that olfactory stimuli from visitors will affect the behaviour of species with a sharp sense of smell but there are no quantified data available on this. Zoos even without the visitors can produce a range of auditory stimuli (e.g. from activities such as gardening and construction) and olfactory stimuli (e.g. from cafeterias), which may be sensed by the animals at a considerable distance from their source. However, if the sources of these stimuli are never apparent to the animals then the learning opportunities can become diminished. If these stimuli are associated together; for example, the sound a food delivery vehicle and the smell of fish, then this learnt contingency may be a source of stress for animals such as bears (Cremers and Geutjes 2012). They could even induce abnormal behaviour due to food anticipation. An extensive review about the effects of predictability and animal welfare can be found in Bassett and Buchanan-Smith (2007).

The application of environmental enrichment, as it often involves the introduction of a novel stimulus, is a source of visual stimulation and this can enlarge the visual cortex in animals' brains (Sale et al. 2004; Baroncelli et al. 2010). Thus visual stimulation is often a side effect of other types of environmental enrichment and, perhaps, it is for this reason that few people deliberately provide it. From gap analyses of environmental enrichment research it would seem this type of enrichment has been greatly overlooked (Azevedo et al. 2007). In the past, the most common type of deliberately used visual enrichment was televisions for primates (Platt and Novak 1996; Lutz and Novak 2005), but this fell out of favour as zoos felt that it anthropomorphised animals. The use of data projectors permit us to create many different kinds of visual stimuli for animals. A number of species such as birds respond to video images of conspecifics as if they are real and present in their enclosure (Clarke and Jones 2000). For species, which live in large groups in the wild but are housed in small captive groups, such as flamingos, this may be a solution to the stress caused by living in small groups. Flamingos gain protection against predators by living in large groups and normally show synchronism in their reproductive behaviours; thus, in small numbers, such behaviours may never be expressed (Pickering et al. 1992). Asian elephants have complex social structures and maintain small herds, which disrupt these structures causing stress (Rees 2009). Alternatively, low technology solutions to such problems are the use of model birds and mirrors in enclosures (Pickering and Duverge 1992; Azevedo and Faggioli 2004; Sherwin 2004). However, it is more likely that birds will habituate to these stimuli compared to video images or even a live feed of conspecifics in the wild, simply because models and mirrors will never completely imitate live animals, especially when they are added to enclosures of high-cognitive species (Bensom-Amram et al. 2016).

Auditory stimuli often in the form of conspecific vocalisations have been used in environmental enrichment studies (Rukstalis and French 2005; Simonet et al. 2005; Kelling et al. 2012). However, if such calls are used frequently with no contingency to the behavioural response, then animals will learn to habituate to such calls, change its behaviour or even be stressed by the calls (Harris and Haskell 2013; Massen et al. 2014). For example, predator warning calls, which are emitted without showing an appropriate predator model, may soon lose their ability to modify animal behaviour (Griffin et al. 2000), but some species can retain the capacity of predator call recognition over evolutionary time (Hettena et al. 2014). The use of such calls as environmental enrichment is considered by some organisations as questionable in terms of their impact on animal welfare, since they provoke an antipredatory response and predators are a source of stress.

In many species, such as birds, juveniles learn their species-specific vocalisations from listening to other members of their social group and need to practice these vocalisations for them to be perfected (Payne et al. 2000; Catchpole and Slater 2008). Furthermore, many vocalisations will only be expressed once the appropriate stimulus has been presented such as the sight of a predator or the territorial vocalisation of a conspecific (Hollén and Radford 2009). Territorial vocalisations were once popularly used as auditory environmental enrichment for species such as gibbons who would respond to them with their own territorial call (Shepherdson et al. 1999). Beyond the vocal response, animals subjected to such territorial challenges often show enhanced social behaviour and potentially have stronger social bonds.

Finally in terms of auditory enrichment the use of music as a stimulus has been popular ever since studies showed that cows listening to classical music or calf vocalisations produced more milk (McCowan et al. 2002; O'Brien 2014). The structure of music for humans is related to the structure of human speech. It has been suggested that music for animals should be based on the species' vocalisation structures. A study where this was carried out on primates, showed they paid more attention to the 'monkey music' than to human music (Snowdon and Teie 2013). The same was found by Snowdon et al. (2015) for domestic cats.

The use of olfactory enrichment has increased in recent times, despite the fact that humans have relatively poor olfactory abilities (Clark and King 2008; Laska 2017). But as with other sensory stimuli it is important that either there is some natural contingency between the stimulus presented and consequences for the animals (e.g. arrival of food or a model predator) or that each stimulus is used infrequently. Predator scents have been shown to be effective in eliciting antipredatory responses in many prey species without the need for the appearance of a predator model (Apfelbach et al. 2005; Rosell et al. 2013). For example, the smell of a jaguar on llama wool is sufficient for giant anteaters to start rushing around their enclosure in an escape response (Orlando and Fernández 2014). But of course such stimuli must be applied sparingly, for a short period of time (a few minutes only) and in a manner that the stimulus can be fully removed from the animal's enclosure. If this stimulus is prolonged, it could distress the animals since they would not be able to flee from the supposed predator, decreasing their welfare.

Other scents that can be used to enrich the lives of captive animals are scents used to mark the boundaries of territories, such as those used by big cats or primates such as lemurs. The careful placing of such scents at the edges of an enclosure will teach an animal about the limits of its territory (Campbell-Palmer and Rosell 2010; Jackson et al. 2012). In captivity, it is important that scent marking is not constantly removed during the cleaning process because in species, which countermark (mark over other individual's scents) this could be perceived as another individual invading their territory causing significant stress to the enclosure's occupant.

The sense of touch is the most under-used modality in environmental enrichment despite the fact that digits of many animal species are full of nerves able to sense subtle changes in substrates (Lederman 1991). Different substrates provide information to animals about their uses (Marshall et al. 2008); for example, arboreal species learn quickly that very smooth substrates are not easily climbed. The creation of enrichment items with tactile properties for an Asian elephant was attempted by French et al. (2016) with good results. The authors invented a shower for an elephant with simple on/off rope and hessian buttons, and these different materials were attractive to the elephant, which spent a lot of time manipulating them with its trunk. Tactile enrichments should be carefully planned, since each species has its own sensory capabilities. Primates have a great sense of touch in the tip of their fingers, and the use of different materials in the enrichment manufacture will enhance tactile experiences (Dominy et al. 2004). Environmental enrichment items with different temperatures, different pressures, and different hardness of materials will certainly create the opportunities for different sensations and learning experiences.

Sensory stimuli provide a plethora of learning opportunities for captive animals, but currently they are not applied in a systematic manner to enrich the lives of animals.

6.7 Informal Learning Opportunities During Nutritional Enrichment

Food is one of the most powerful reinforcements in the life of captive animals and, therefore, animals quickly learn any contingency related to the delivery of food. Zoo animals, for example, can learn even the sound of the wheelbarrow when the keeper arrives with food and ignore other sounds (Cremers and Geutjes 2012). Thus, the delivery of food to animals in captivity must be done with great care as it can create all kinds of positive and negative contingencies. For example, if starve days are used with large carnivores, signals that food is going to arrive can be a trigger, which initiates the expression of abnormal behaviour such as stereotypic pacing (Bassett and Buchanan-Smith 2007).

Anecdotally, a number of strange behaviours have been observed in zoo animals, which appear to be the result of accidental and unwanted contingencies. At the Belo Horizonte Zoo in Brazil there was a male giraffe, which would touch its neck against the 'hotwire fence' and get an electric shock (Figure 6.4). It would seem that this behaviour had its origin in the animal reaching over the electric fence to reach the succulent leaves outside of its enclosure. Each time the giraffe managed to grasp a mouth full of leaves with its tongue it was shocked by the hotwire; this process was repeated many times and eventually the animal formed an association between being shocked and obtaining food.

The use of food to lure animals to interact with items provided as environmental enrichment is widespread, but is a two edged sword. Once animals associate an enrichment item with food they will not use it once no more food is available. For example, a pig might spend much time rolling a ball around its enclosure, but if food is placed inside that ball, the same individual will only roll the ball around whilst food is available (Young et al. 1994; Young and Lawrence 1996). Both the pig and the giraffe examples can be referred to the Premack principle, which states that if a less desirable activity reinforces a more desirable activity, then an animal will perform the less desirable activity (Bond 2008). Thus, the ball has lost its 'power' as a toy due to the learned contingency with food. It is worth remembering that we cannot use food all the time for environmental enrichment for two reasons. Firstly, the animals will eventually be nutritionally satiated and therefore no longer use any of the environmental enrichment being offered. Secondly, if we over use food as a lure to promote interaction with enrichment we risk captive animals becoming obese

Figure 6.4 Giraffes at Belo Horizonte Zoo, Brazil. One male associated an electric shock with receipt of succulent leaves. Note the hotwire fence just above the wooden fence. *Source:* Fundação de Parques Municipais e Zoobotânica.

and a number of studies have already shown that obesity is on the rise in captive animals (Morfeld et al. 2014).

Food beyond being a primary reinforcement is a source of many sensory stimuli such as visual, olfactory, and tactile which we have already considered. It is also a source of gustatory enrichment. From the colour, smell, feel, and taste of food animals can learn about the relative nutritional value of different food items to them (Rowe and Skelhorn 2005; Werner et al. 2008; Passos et al. 2014). A redder fruit might indicate a sweeter and more energy rich food item to a primate species, for example. However, if zoos or laboratories preprocess food items by peeling fruit and chopping it up, many of these contingencies may be difficult for animals to learn (Sandri et al. 2017).

In general, we have only discussed associative and non-associative learning and not mentioned processes such as imprinting and sensitive periods. Many species learn what to eat during a sensitive period (a learning window) and imprint on that type of food (Burghardt and Hess 1966; Burghardt 1969). Failure to provide appropriate food items during these learning periods can result in captive animals not recognising their prey species if they were released back to the wild. For example, captive cheetah that had been fed cow meat upon reintroduction in Africa tried to hunt inappropriate species such as giraffe and rhinoceroses (Young 1997).

6.8 Discussion

During the process of environmental enrichment animals are learning all the time about the new stimuli in their environment and any contingencies created by these stimuli. If these learning opportunities are carefully managed they add value to the use of environmental enrichment and in the zoo environment this can be through facilitating the four objectives of the modern zoo in terms of conservation, research, education, and leisure.

Much of the benefits of environmental enrichment come from animals learning about their environment, which makes their environment more interesting. In some cases such as the use of nutritional environmental enrichment, food is the primary reinforcement which increases the likelihood of the behaviour (using the enrichment) being performed in the future. In the case of cognitive enrichment, food is often used to lure the animal into using the enrichment; it is then less clear whether the primary reinforcement is the food or the learning opportunity (Sambrook and Buchanan-Smith 1997). The learning which takes place with the cognitive task is however, more spontaneous than seen with nutritional enrichment (Melfi 2013; Clark 2017). It should be remembered that in good environmental enrichment programmes the animals do not receive all their food through using enrichment. But rather through how much time they choose to use enrichment will determine their food income from this source.

Keeping animals' brains active through informal learning opportunities created by environmental enrichment has been shown to increase their learning capacity and ability to deal with changes in their environment (Young 2003; Salvanes et al. 2014; Grimberg-Henrici et al. 2016). In humans, studies have shown that people who are more mentally active usually suffer much less from age-related diseases and illnesses such as dementia (Nithianantharajah and Hannan 2006; Woo and Leon 2013). Due to advances in veterinary care and nutrition provision, animals in captivity are starting to live much longer than their wild counterparts; for example, chimpanzees in the wild normally live to 40+ years old, whereas as in captivity they can reach 60+ years old. Thus, if captive chimpanzees are not mentally stimulated then we can expect to start seeing diseases like dementia; these diseases are already being

seen in pet dogs (Sanabria et al. 2013; Schütt et al. 2015).

It should be remembered that animal training whilst it does promote learning, can only be carried out for short periods every day (Young and Cipreste 2004; Domjan 2005); whereas informal learning can, in correctly enriched and managed environments, be a constant process. It should also be remembered that this type of informal learning is considerably more diverse than the learning opportunities created by operant conditioning. Of course there is a place for both, but it should be clear that their functions are quite different. Therefore, in terms of the aforementioned animal welfare benefits, informal learning opportunities are likely to be much more important.

In conclusion, environmental enrichment items should be provided for captive zoo animals since they can enhance learning opportunities, can increase the expression of natural behaviours, can decrease the expression of abnormal behaviours, and can increase the level of welfare. Animals using environmental enrichment have better performance in cognitive tasks, facilitating their adaptation to new environments. Thus, for animals kept in barren environments, enrichment can be the only cognitive distraction and the main source of learning opportunities.

Acknowledgements

RJY was financially supported by FAPEMIG and CNPq; he is currently supported by SwB (CAPES). CSA is financially supported by FAPEMIG/CEMIG and CAPES.

References

Apfelbach, R., Blanchard, C.D., Blanchard, R.J. et al. (2005). The effects of predator odors in mammalian prey species: a review of field and laboratory studies. *Neuroscience and Biobehavioral Reviews* 29: 1123–1144. https://doi.org/10.1016/j.neubiorev. 2005.05.005.

Azevedo, C.S. and Faggioli, A.B. (2004). Effects of the introduction of mirrors and flamingo statues on the reproductive behaviour of a Chilean flamingo flock. *International Zoo News* 51: 478–483.

Azevedo, C.S., Cipreste, C.F., and Young, R.J. (2007). Environmental enrichment: a gap analysis. *Applied Animal Behaviour Science* 102: 329–343. https://doi.org/10.1016/j. applanim.2006.05.034.

Baroncelli, L., Braschi, C., Spolidoro, M. et al. (2010). Nurturing brain plasticity: impact of environmental enrichment. *Cell Death and Differentiation* 17: 1092–1103. https://doi. org/10.1038/cdd.2009.193.

Bassett, L. and Buchanan-Smith, H.M. (2007). Effects of predictability on the welfare of captive animals. *Applied Animal Behaviour Science* 102: 223–245. https://doi.org/ 10.1016/j.applanim.2006.05.029.

Bays, T.B. (2014). Environmental enrichment for small mammals. *Clinician's Brief* (March), p. 85–89.

Bensom-Amram, S., Dantzer, B., Stricker, G. et al. (2016). Brains size predicts problem-solving ability in mammalian carnivores. *Proceedings of the National Academy of Sciences of the United States of America* 113: 2532–2537. https://doi.org/10.1073/ pnas.1505913113.

Bloomsmith, M.A., Marr, M.J., and Maple, T.L. (2007). Addressing nonhuman primate behavioural problems through the application of operant conditioning: is the human treatment approach a useful model? *Applied Animal Behaviour Science* 102: 205–222. https://doi.org/10.1016/j. applanim.2006.05.028.

Bond, C. (2008). Neurology of learning: an understanding of neurology as the basis of learning and behaviour in the domestic dog. *Journal of Applied Companion Animal Behavior* 2: 50–95.

Breed, M.D. (2014). Kin and nestmate recognition: the influence of W. D. Hamilton on 50 years of research. *Animal Behaviour* 92: 271–279. https://doi.org/10.1016/j. anbehav.2014.02.030.

Brosnan, S.F. (2007). Our social roots. *Nature* 450: 1160–1161. https://doi.org/10.1038/ 4501160a.

Burghardt, G.M. (1969). Effects of early experience on food preference in chicks. *Psychological Science* 14: 7–8.

Burghardt, G.M. and Hess, E.H. (1966). Food imprinting in the snapping turtle, *Chelydra serpentina*. *Science* 151: 108–109. https:// doi.org/10.1126/science.151.3706.108.

Buttelmann, D., Carpenter, M., Call, J., and Tomasello, M. (2013). Chimpanzees, *Pan troglodytes*, recognize successful actions, but fail to imitate them. *Animal Behaviour* 86: 755–761. https://doi.org/10.1016/j. anbehav.2013.07.015.

Campbell-Palmer, R. and Rosell, F. (2010). Conservation of the Eurasian beaver *Castor fiber*: an olfactory perspective. *Mammal Review* 40: 293–312. https://doi. org/10.1111/j.1365-2907.2010.00165.x.

Catchpole, C.K. and Slater, P.J.B. (2008). *Bird Songs: Biological Themes and Variations*, 2e. Cambridge: Cambridge University Press.

Cheyne, S.M. (2010). Studying social development and cognitive abilities in gibbons (*Hylobates* spp.): methods and applications. In: *Primatology: Theories, Methods and Research* (eds. E. Potocki and J. Kraisinski), 1–24. New York: Nova Science Publishers Inc.

Cipreste, C.F., Azevedo, C.S., and Young, R.J. (2010). How to develop a zoo-based environmental enrichment program: incorporating environmental enrichment into exhibits. In: *Wild Mammals in Captivity: Principles and Techniques for Zoo Management*, 2e (eds. D.G. Kleiman, K.V. Thompson and C.K. Baer), 171–180. Chicago: The University of Chicago Press.

Clark, F.E. (2017). Cognitive enrichment and welfare: current approaches and future directions. *Animal Behavior and Cognition* 4: 52–71. https://doi.org/10.12966/ abc.05.02.2017.

Clark, F. and King, A.J. (2008). A critical review of zoo-based olfactory enrichment. In: *Chemical Signals in Vertebrates 11* (eds. J.L. Hurst, R.J. Beynon, S.C. Roberts and T.D. Wyatt), 391–398. New York: Springer.

Clarke, C.H. and Jones, R.B. (2000). Effects of prior video stimulation on open-field behaviour in domestic chicks. *Applied Animal Behaviour Science* 66: 107–117. https://doi.org/10.1016/S0168-1591(99)00071-4.

Coe, J. (2012). *Design and Architecture: Third Generation Conservation, Post-immersion and Beyond*. New York: Buffalo.

Coe, J. and Dykstra, G. (2010). New and sustainable directions in zoo exhibit design. In: *Wild Mammals in Captivity: Principles and Techniques for Zoo Management*, 2e (eds. D.G. Kleiman, K.V. Thompson and C.K. Baer), 202–215. Chicago: The University of Chicago Press.

Coleman, K. and Maier, A. (2010). The use of positive reinforcement training to reduce stereotypic behaviour in rhesus macaques. *Applied Animal Behaviour Science* 124: 142–148. https://doi.org/10.1016/j.applanim.2010.02.008.

Coulon, M., Henry, L., Perret, A. et al. (2014). Assessing video presentations as environmental enrichment for laboratory birds. *PLoS One* 9: e96949. https://doi.org/10.1371/journal.pone.0096949.

Cremers, P.W.F.H. and Geutjes, S.L. (2012). The cause of stereotypic behaviour in a male polar bear (*Ursus maritimus*). In: *Proceedings of Measuring Behavior 2012* (eds. A.J. Spink, F. Grieco, O.E. Krips, et al.). Utrecht, The Netherlands: Measuring Behaviour https://www.measuringbehavior.org/mb2012/files/2012 (accessed 28–31 August 2012).

Dominy, N.J., Ross, C.F., and Smith, T.D. (2004). Evolution of the special senses in primates: past, present, and future. *The Anatomical Record Part A, Discoveries in Molecular, Cellular, and Evolutionary Biology* 281A: 1078–1082. https://doi.org/10.1002/ar.a.20112.

Domjan, M. (2005). *The Essentials of Conditioning and Learning*, 3e. Belmont: Thomson Wadsworth.

Dorey, N.R., Mehrkam, L.R., and Tacey, J. (2015). A method to assess relative preference for training and environmental enrichment in captive wolves (*Canis lupus* and *Canis lupus arctos*). *Zoo Biology* 34: 513–517. https://doi.org/10.1002/zoo.21239.

Fonseca, C.G., Pires, W., Lima, M.R.M. et al. (2014). Hypothalamic temperature of rats subjected to treadmill running in a cold environment. *PLoS One* 9 (11): e111501.

French, F., Mancini, C., and Sharp, H. (2016). Exploring methods for interaction design with animals: a case-study in Valli. *ACI '16 Proceedings of the Third International Conference on Animal-Computer Interaction*, Milton Keynes, United Kingdon (15–17 November 2016). New York, USA: ACM.

Griffin, G. (2012). Evaluating environmental enrichment is essential. *The Enrichment Record* 12: 29–33.

Griffin, A.S., Blumstein, D.T., and Evans, C.S. (2000). Training captive-bred or translocated animals to avoid predators. *Conservation Biology* 14: 1317–1326. https://doi.org/10.1046/j.1523-1739.2000.99326.x.

Grimberg-Henrici, C.G.E., Vermaak, P., Bolhuis, J.E. et al. (2016). Effects of environmental enrichment on cognitive performance of pigs in a spatial holeboard discrimination task. *Animal Cognition* 19: 271–283. https://doi.org/10.1007/s10071-015-0932-7.

Harris, J.B.C. and Haskell, D.G. (2013). Simulated birdwatchers' playback affects the behaviour of two tropical birds. *PLoS One* 8: e77902. https://doi.org/10.1371/journal.pone.0077902.

Hettena, A.M., Munoz, N., and Blumstein, D.T. (2014). Prey response to predator's sound: a review and empirical study. *Ethology* 120: 427–452. https://doi.org/10.1111/eth.12219.

Hollén, L.I. and Radford, A.N. (2009). The development of alarm call behaviour in mammals and birds. *Animal Behaviour* 78: 791–800. https://doi.org/10.1016/j.anbehav.2009.07.021.

Honkavaara, J., Koivula, M., Korpimäki, E. et al. (2002). Ultraviolet vision and foraging in terrestrial vertebrates. *Oikos* 98: 505–511. https://doi.org/10.1034/j.1600-0706.2002.980315.x.

Hosey, G.R. (2000). Zoo animals and their human audiences: what is the visitor effect? *Animal Welfare* 9: 343–357.

Hosey, G., Melfi, V., and Pankhurst, S. (2013). *Zoo Animals: Behaviour, Management, and Welfare*, 2e. Oxford: Oxford University Press.

Humphrey, N. (1976). The social function of intellect. In: *Growing Points in Ethology* (eds. P.P.G. Bateson and R.A. Hinde), 303–317. Cambridge: Cambridge University Press.

Jackson, C.R., McNutt, J.W., and Apps, P.J. (2012). Managing the ranging behaviour of African wild dogs (*Lycaon pictus*) using translocated scent marks. *Wildlife Research* 39: 31–34. https://doi.org/10.1071/WR11070.

Jacobs, G.H. (2008). Primate color vision: a comparative perspective. *Visual Neuroscience* 25: 619–633. https://doi.org/10.1017/S0952523808080760.

Jacobs, G.H. (2009). Evolution of colour vision in mammals. *Philosophical Transactions of the Royal Society B: Biological Sciences* 364: 2957–2967. https://doi.org/10.1098/rstb.2009.0039.

Jacobs, G.H., Deegan, J.F., Crognale, M.A., and Fenwick, J.A. (1993). Photopigments of dogs and foxes and their implications for canid vision. *Visual Neuroscience* 10: 173–180. https://doi.org/10.1017/S0952523800003291.

Jung, C.K.E. and Herms, J. (2014). Structural dynamics of dendritic spines are influenced by environmental enrichment: an in vivo imaging study. *Cerebral Cortex* 24: 377–384. https://doi.org/10.1093/cercor/bhs317.

Kagan, R. and Veasey, J. (2010). Challenges of zoo animal welfare. In: *Wild Mammals in Captivity: Principles and Techniques for Zoo Management*, 2e (eds. D.G. Kleiman, K.V. Thompson and C.K. Baer), 11–21. Chicago: The University of Chicago Press.

Kelling, A.S., Allard, S.M., Kelling, N.J. et al. (2012). Lion, ungulate, and visitor reactions to playbacks of lion roars at Zoo Atlanta. *Journal of Applied Animal Welfare Science* 15: 313–328. https://doi.org/10.1080/10888705.2012.709116.

Kempe, M. and Mesoudi, A. (2014). An experimental demonstration of the effect of group size on cultural accumulation. *Evolution and Human Behavior* 35: 285–290. https://doi.org/10.1016/j.evolhumbehav.2014.02.009.

Kuhar, C.W. (2008). Group differences in captive gorillas' reaction to large crowds. *Applied Animal Behaviour Science* 110: 377–385. https://doi.org/10.1016/j.applanim.2007.04.011.

Laska, M. (2017). Human and animal olfactory capabilities compared. In: *Handbook of Odor* (ed. A. Buettner), 81–82. Berlin: Springer-Verlag.

Lederman, S.J. (1991). Skin and touch. In: *Encyclopaedia of Human Biology*, vol. 7 (ed. R. Dulbecco). Massachusetts: Academic Press Inc.

Lutz, C.K. and Novak, M.A. (2005). Environmental enrichment for nonhuman primates: theory and application. *ILAR Journal* 46: 178–191.

Marshall, J., Pridmore, T., Pound, M., et al. (2008). Pressing the flesh: sensing multiple touch and finger pressure on arbitrary surfaces. *6th International Conference on Pervasive Computing*, Sydney, Australia (19–22 May 2008). Berlin, GER: Springer-Verlag.

Mason, G.J. and Latham, N.R. (2004). Can't stop, won't stop: is stereotypy a reliable animal welfare indicator? *Animal Welfare* 13: 57–69.

Massen, J.J.M., Pasukonis, A., Schmidt, J., and Bugnyar, T. (2014). Ravens notice dominance reversals among conspecifics within and outside their social group. *Nature Communications* 5: 3679. https://doi.org/10.1038/ncomms4679.

McCowan, B., DiLorenzo, A.M., Abichandani, S. et al. (2002). Bioacoustic tools for enhancing animal management and productivity: effects of recorded calf vocalizations on milk production in dairy cows. *Applied Animal Behaviour Science* 77: 13–20. https://doi.org/10.1016/S0168-1591(02)00022-9.

Meijer, J.H. and Robbers, Y. (2014). Wheel running in the wild. *Proceedings of the Royal Society B: Biological Sciences* 281: 20140210. https://doi.org/10.1098/rspb.2014.0210.

Melfi, V. (2013). Is training zoo animals enriching? *Applied Animal Welfare Science* 147: 299–305. https://doi.org/10.1016/j.applanim.2013.04.011.

Michelena, P., Sibbald, A.M., Erhard, H.W., and McLeod, J.E. (2009). Effects of group size and personality on social foraging: the distribution of sheep across patches. *Behavioral Ecology* 20: 145–152. https://doi.org/10.1093/beheco/arn126.

Morfeld, K.A., Lehnhardt, J., Alligood, C. et al. (2014). Development of a body condition scoring index for female African elephants validated by ultrasound measurements of subcutaneous fat. *PLoS One* 9: e93802. https://doi.org/10.1371/journal.pone.0093802.

Nash, V.J. (1982). Tool use of captive chimpanzees at an artificial termite mound. *Zoo Biology* 1: 211–221. https://doi.org/10.1002/zoo.1430010305.

Nithianantharajah, J. and Hannan, A.J. (2006). Enriched environments, experience-dependent plasticity and disorders of the nervous system. *Nature Reviews Neuroscience* 7: 697–709. https://doi.org/10.1038/nrn1970.

O'Brien, A. (2014). Milking to music (10 February). http://modernfarmer.com/2014/02/milking-music (accessed 17 July 2015).

Orlando, C.G. and Fernández, G.J. (2014). Respuesta antipredatoria de osos hormigueros (*Myrmecophaga trydactila*) mantenidos em cautividad. *Edentata* 15: 52–59. https://doi.org/10.5537/020.015.0108.

Ottoni, E.B. and Izar, P. (2008). Capuchin monkey tool use: overview and implications. *Evolutionary Anthropology* 17: 171–178. https://doi.org/10.1002/evan.20185.

Passos, L.F., Santo, H.M.E., and Young, R.J. (2014). Enriching tortoises: assessing color preference. *Journal of Applied Animal Welfare Sciences* 17: 274–281. https://doi.org/10.1080/10888705.2014.917556.

Payne, R.B., Payne, L.L., Woods, J.L., and Sorenson, M.D. (2000). Imprinting and the origin of parasite-host species associations in brood-parasitic indigobirds, *Vidua chalybeate*. *Animal Behaviour* 59: 69–81. https://doi.org/10.1006/anbe.1999.1283.

Pickering, S.P.C. and Duverge, L. (1992). The influence of visual stimuli provided by mirrors on the marching display of lesser flamingos, *Phoeniconais minor*. *Animal Behaviour* 43: 1048–1050. https://doi.org/10.1016/S0003-3472(06)80018-7.

Pickering, S.P.C., Creighton, E., and Stevens-Wood, B. (1992). Flock size and breeding success in flamingos. *Zoo Biology* 11: 229–234. https://doi.org/10.1002/zoo.1430110402.

Platt, D.M. and Novak, M.A. (1996). Videostimulation as enrichment for captive rhesus monkeys (*Macaca mulatta*). *Applied Animal Behaviour Science* 52: 139–155. https://doi.org/10.1016/S0168-1591(96)01093-3.

Quadros, S., Goulart, V.D.L., Passos, L. et al. (2014). Zoo visitor effect on mammal behaviour: does noise matter? *Applied Animal Behaviour Science* 156: 78–84. https://doi.org/10.1016/j.applanim.2014.04.002.

Rampon, C., Jiang, C.H., Dong, H. et al. (2000). Effects of environmental enrichment on gene expression in the brain. *Proceedings of the National Academy of Sciences of the United States of America* 97: 12880–12884. https://doi.org/10.1073/pnas.97.23.12880.

Reboreda, J.C. and Fernandez, G.J. (1997). Sexual, seasonal and group size differences in the allocation of time between vigilance and feeding in greater rhea, *Rhea americana*. *Ethology* 103: 198–207. https://doi.org/10.1111/j.1439-0310.1997.tb00116.x.

Rees, P.A. (2009). The sizes of elephant groups in zoos: implications for elephant welfare. *Journal of Applied Animal Welfare Science* 12: 44–60. https://doi.org/10.1080/10888700802536699.

Rees, P.A. (2011). *An introduction to Zoo Biology and Management*. New York: Wiley.

Rosell, F., Holtan, L.B., Thorsen, J.G., and Heggenes, J. (2013). Predator-naive brown trout (*Salmo trutta*) show antipredator behaviours to scent from an introduced piscivorous mammalian predator fed conspecifics. *Ethology* 119: 303–308. https://doi.org/10.1111/eth.12065.

Rowe, C. and Skelhorn, J. (2005). Colour biases are a question of taste. *Animal Behaviour*

69: 587–594. https://doi.org/10.1016/j. anbehav.2004.06.010.

Rukstalis, M. and French, J.A. (2005). Vocal buffering of the stress response: exposure to conspecific vocalizations moderates urinary cortisol excretion in isolated marmosets. *Hormones and Behavior* 47: 1–7. https://doi. org/10.1016/j.yhbeh.2004.09.004.

Sale, A., Putignano, E., Cancedda, L. et al. (2004). Enriched environment and acceleration of visual system development. *Neuropharmacology* 47: 649–660. https:// doi.org/10.1016/j.neuropharm.2004.07.008.

Salvanes, A.G.V., Moberg, O., Ebbesson, L.O.E. et al. (2014). Environmental enrichment promotes neural plasticity and cognitive ability in fish. *Proceedings of the Royal Society B: Biological Sciences* 280: 20131331. https://doi.org/10.1098/rspb.2013.1331.

Sambrook, T.D. and Buchanan-Smith, H. (1997). Control and complexity in novel object enrichment. *Animal Welfare* 6: 207–216.

Sanabria, C.O., Olea, F., and Rojas, M. (2013). Cognitive dysfunction syndrome in senior dogs. In: *Neurodegenerative Diseases* (ed. U. Kishore). InTech. Available from: https:// www.intechopen.com/books/ neurodegenerative-diseases/cognitive- dysfunction-syndrome-in-senior-dogs (accessed 1 February 2018).

Sandri, C., Regaiolli, B., Vespiniani, A., and Spiezio, C. (2017). New food provision strategy for a colony of Barbary macaques (*Macaca sylvanus*): effects of social hierarchy? *Integrative Food Nutrition and Metabolism* 4: 1–8. https://doi. org/10.15761/IFNM.1000181.

Sarrafchi, A. and Blokhuis, H.J. (2013). Equine stereotypic behaviors: causation, occurrence, and prevention. *Journal of Veterinary Behavior* 8: 386–394. https://doi. org/10.1016/j.jveb.2013.04.068.

Schork, I.G. and Young, R.J. (2014). Rapid animal welfare assessment: an archaeological approach. *Biology Letters* 10: 20140390. https://doi.org/10.1098/rsbl. 2014.0390.

Schütt, T., Toft, N., and Berendt, M. (2015). Cognitive function, progression of

age-related behavioural changes, biomarkers, and survival in dogs more than 8 years old. *Journal of Veterinary Internal Medicine* 29: 1569–1577. https://doi.org/10.1111/ jvim.13633.

Shepherdson, D.J. (2003). Environmental enrichment: past, present and future. *International Zoo Yearbook* 38: 118–124. https://doi.org/10.1111/j.1748-1090.2003. tb02071.x.

Shepherdson, D.J., Mellen, J.D., and Hutchins, M. (1999). *Second Nature: Environmental Enrichment for Captive Animals*. Washington DC: Smithsonian Institution Press.

Sherwin, C.M. (2004). Mirrors as potential environmental enrichment for individually housed laboratory mice. *Applied Animal Behaviour Science* 87: 95–103. https://doi. org/10.1016/j.applanim.2003.12.014.

Simonet, P., Versteeg, D., and Storie, D. (2005). Dog-laughter: recorded playback reduces stress related behaviour in shelter dogs. *Proceedings of the 7th International Conference on Environmental Enrichment*, New York, USA (31 July–5 August 2005). New York, USA: Wildlife Conservation Society.

Snowdon, C.T. and Teie, D. (2013). Emotional communication in monkeys: music to their ears? In: *Evolution of Emotional Communication: From Sounds in Nonhuman Mammals to Speech and Music in Man* (eds. E. Altenmüller, S. Schmidt and E. Zimmermann), 133–151. Oxford: Oxford University Press.

Snowdon, C.T., Teie, D., and Savage, M. (2015). Cats prefer species-appropriate music. *Applied Animal Behaviour Science* 166: 106–111. https://doi.org/10.1016/j. applanim.2015.02.012.

Sorensen, D.B., Mikkelsen, L.F., Nielsen, S.G. et al. (2011). The influence of enriched environments on learning and memory abilities in group-housed SD rats. *Scandinavian Journal of Laboratory Animal Science* 38: 5–17.

Strand, D.A., Utne-Palm, A.C., Jakobsen, P.J. et al. (2010). Enrichment promotes learning in fish. *Marine Ecology Progress Series* 412: 273–282. http://dx.doi.org/10.3354/meps08682.

Swaisgood, R.R. (2007). Current status and future directions of applied behavioral research for animal welfare and conservation. *Applied Animal Behaviour Science* 102: 139–162. https://doi.org/10.1016/j.applanim.2006.05.027.

Swaisgood, R.R. and Shepherdson, D.J. (2005). Scientific approaches to enrichment and stereotypies in zoo animals: what's been done and where should we go next? *Zoo Biology* 24: 499–518. https://doi.org/10.1002/zoo.20066.

Swan, J. and Hyland, P. (2012). A review of the beneficial mental health effects of exercise and recommendations for future studies. *Psychology and Society* 5: 1–15.

Taylor, C.B., Sallis, J.F., and Needle, R. (1985). The relation of physical activity and exercise to mental health. *Public Health* 100: 195–202.

Tibbetts, E.A. and Dale, J. (2007). Individual recognition: it is good to be different. *Trends in Ecology and Evolution* 22: 529–537. https://doi.org/10.1016/j.tree.2007.09.001.

Ueno, A. and Suzuki, K. (2014). Comparison of learning ability and memory retention in altricial (Bengalese finch, *Lonchura striata var. domestica*) and precocial (blue-breasted quail, *Coturnix chinensis*) birds using a color discrimination task. *Animal Science Journal* 85: 186–192. https://doi.org/10.1111/asj.12092.

Van Praag, H., Kempermann, G., and Gage, F.H. (2000). Neural consequences of environmental enrichment. *Nature Reviews Neuroscience* 1: 191–198. https://doi.org/10.1038/35044558.

Vasconcellos, A.S., Adania, C.H., and Ades, C. (2012). Contrafreeloading in maned wolves: implications for their management and welfare. *Applied Animal Behaviour Science* 140: 85–91. https://doi.org/10.1016/j.applanim.2012.04.012.

Vidya, T.N.C. (2014). Novel behavior shown by an Asian elephant in the context of allomothering. *Acta Ethologica* 17: 123–127. https://doi.org/10.1007/s10211-013-0168-y.

Vorobyev, M. (2004). Ecology and evolution of primate colour vision. *Clinical and Experimental Optometry* 87: 230–238. https://doi.org/10.1111/j.1444-0938.2004.tb05053.x.

Wells, D.L. (2009). Sensory stimulation as environmental enrichment for captive animals: a review. *Applied Animal Behaviour Science* 118: 1–11. https://doi.org/10.1016/j.applanim.2009.01.002.

Werner, S.J., Kimball, B.A., and Provenza, F.D. (2008). Food color, flavor, and conditioned avoidance among red-winged blackbirds. *Physiology and Behavior* 93: 110–117. https://doi.org/10.1016/j.physbeh.2007.08.002.

Westlund, K. (2014). Training is enrichment – and beyond. *Applied Animal Behaviour Science* 152: 1–6. https://doi.org/10.1016/j.applanim.2013.12.009.

Woo, C.C. and Leon, M. (2013). Environmental enrichment as an effective treatment for autism: a randomized controlled trial. *Behavioral Neuroscience* 127: 487–497. https://doi.org/10.1037/a0033010.

Young, R.J. (1997). The importance of food presentation for animal welfare and conservation. *Proceedings of the Nutrition Society* 56: 1095–1104. https://doi.org/10.1079/PNS19970113.

Young, R.J. (2003). *Environmental Enrichment for Captive Animals*. Oxford: Blackwell.

Young, R.J. and Cipreste, C.F. (2004). Applying animal learning theory: training captive animals to comply with veterinary and husbandry procedures. *Animal Welfare* 13: 225–232.

Young, R.J. and Lawrence, A.B. (1996). The effects of high and low rates of food reinforced on the behaviour of pigs. *Applied Animal Behaviour Science* 49: 365–374. https://doi.org/10.1016/0168-1591(96)01052-0.

Young, R.J., Carruthers, J., and Lawrence, A.B. (1994). The effect of a foraging device (the 'Edinburgh Foodball') on the behaviour of pigs. *Applied Animal Behaviour Science* 39: 237–247. https://doi.org/10.1016/0168-1591(94)90159-7.

Questions

1 Why should environmental enrichment be considered as a source of learning for captive animals?

2 For physical enrichment to be efficient in providing learning opportunities to animals, it needs to be dynamic. How does dynamism work and why is it important?

3 The use of food as environmental enrichment is common across the world. Why is this type of enrichment preferred over other types and how can food be used to facilitate informal learning?

4 During social encounters, animals have the opportunity to learn important lessons. Explain.

5 Explain how to plan an environmental enrichment regime for captive animals which promotes informal learning?

7

The Art of 'Active' Training
Steve Martin

7.1 Introduction

An artistic animal trainer goes beyond the basics of positive reinforcement training to fine tune antecedents and consequences to promote a level of learning that transcends basic animal training practices. The 'art' of animal training can be described as the intuitive non-scientific application of animal training 'rules', which have developed through practical experience working with animals, and the people who care for them. The focus of this chapter is to describe animal training from this perspective, and provide a background of how it can be achieved in zoos and aquaria.

7.2 Motivation

At the heart of all animal training is motivation. Professionals in Applied Behaviour Analysis (a sub-discipline of psychology) refer to a motivating operation as something that changes the effectiveness of a consequence. For instance, warming meat before a training session can sufficiently change a lion's perception of the meat (reinforcer), so that it results in an increase in the lion's motivation to perform a cued behaviour for that particular reinforcer. Motivating operations that increase behaviour are called establishing operations and motivating operations that decrease behaviour are called abolishing operations. For instance, an establishing operation may involve slightly increasing the distance between the trainer and a nervous lion in an enclosure to increase motivation for the animal to participate in a training.

Some zoo professionals have a myriad of tools in their training toolbox to influence motivation. These tools are generally associated with careful attention to antecedent arrangement and establishing operations, such as limiting distractions, careful attention to an animal's body language and responding appropriately, using a high rate of high-value and a wide variety of reinforcers, and much more. Unfortunately, some have yet to learn the vast array of motivating operations available to them and often focus only on food reinforcers in their positive reinforcement training programme.

Zoo professionals have recently begun to understand and quantify an extensive list of motivating operations that influence training sessions with their animals. Of the countless motivating operations that affect the behavioural choices of the animals in training programmes, the following are amongst the most influential.

7.2.1 Relationship

A trusting relationship between animal and trainer is an important influence on motivation. Trust levels are on a continuum specific to each animal and the conditions in which it

Zoo Animal Learning and Training, First Edition. Edited by Vicky A. Melfi, Nicole R. Dorey, and Samantha J. Ward.
© 2020 John Wiley & Sons Ltd. Published 2020 by John Wiley & Sons Ltd.

behaves. Though we often focus on trust between animals and humans, trust also relates to how the animal perceives and responds to various objects and situations; which are equally important to the animals and training programmes. Animals build trust in exhibit features, housing furniture, other animals, and even enrichment items, in a similar way that they build trust in humans. When a gibbon jumps on a 3-inch thick branch in a tree and the branch supports its weight, the behaviour is reinforced and the animal builds trust in similar sized branches. When a keeper first shows a target stick to a zebra, it may take several approximations of approach behaviour without aversive consequences before the animal builds enough trust in the target to touch it with its nose to gain a reinforcer. The higher the level of trust an animal has in a person, the more likely the animal is to participate in interactions with that person.

The relationship between a trainer and the animals she/he trains might be considered as a trust account at their bank of relationships. Each time a trainer does something the animal likes, i.e. provides something desirable that an animal will work to gain, the trainer is making a deposit into a trust account. If the trainer does something the animal dislikes, i.e. will work to avoid, the trainer makes a withdrawal from the trust account. Restricting a tiger's movements in a squeeze chute to give an injection may be a withdrawal from the trust the tiger has in the trainer and the squeeze chute behaviour. After only a few repetitions of the squeeze chute behaviour, the resulting withdrawals may bankrupt the trust account, resulting in decrease or termination of future approach behaviour with both the chute and the trainer. Using positive reinforcement to teach the tiger to accept injections may take longer, but the training will pay dividends as the multiple deposits into the trust account results in reliable behaviour that will endure the occasional withdrawal.

Past experiences become antecedent conditions for future behaviour. Zoo professionals often wonder if they should be in the room with one of their animals during a stressful experience such as a vet immobilising the animal. Usually, the best answer to this question is, 'It depends. How is your trust account with that animal?' If a person has a high trust account with an animal, he or she may provide some level of comfort to the animal once it is darted, caught in a net, or otherwise put in a stressful situation. We often see animals with high trust accounts go directly to a familiar trainer after a stressful experience. A male gorilla *(Gorilla gorilla)* at Cheyenne Mountain Zoo presented his shoulder for an injection by the same keeper who delivered a successful hand-injection anaesthetic the day before. At Columbus Zoo the keepers have hand-injected vaccinations with their kinkajou and two wart hogs, and all three animals returned immediately to the trainer and participated in additional injection training behaviours. Both examples demonstrate that a level of trust remained despite the previous negative interaction, and there are countless other examples of animals returning to a keeper the day after that keeper was involved with an injection and anaesthetic the day before. However, there are also many examples of a stressful experience completely bankrupting a keeper's trust account simply by being in the room when an aversive event occurred.

7.2.2 Ability

Animals build skill and behavioural fluency through reinforced practice. Some behaviours require more effort than others and are therefore more difficult for the animal to perform. As an animal develops its skills with a specific behaviour, the motivation to perform that behaviour increases. For instance, a leopard that had no access to trees whilst growing up at one facility may be transferred to another facility with hopes of it being the star attraction in their new exhibit that has the perfect branch for it to lay on over the heads of the guests. With no previous tree-climbing skill the animal may be poorly motivated to attempt to climb trees. In this case a keeper may need to shape the climbing behaviour by first teaching the leopard to touch a target outside the mesh and move the target to where the cat needs to jump on the log to touch the target. Or, a trainer can teach the animal to touch a laser dot as a target on the log, and then gradually approximate the dot up the tree branch. In some cases it might be best to make the behaviour easier by teaching the leopard to touch the dot on a log laying on the ground or leaning on a rock then generalise the behaviour to increasingly steeper angles, raising the tree a foot or two each day, until the tree is vertical.

7.2.3 Learning History

Past consequences have strong antecedent influence on motivation for future behaviour. Many trainers have experienced the frustration associated with trying to teach a parrot with previously clipped wings the skill of flying from one place to another. If the bird's wings were clipped during its first year of life, its flight attempts will have been punished by repeated crashes to the floor or running into walls or other objects, thus reducing its motivation to attempt to fly later in life. It is certainly possible for a parrot to learn to fly later in life after its wing feathers have grown in. However, it will require much more time and effort for the bird to acquire the skills than it would have if it learned to fly in its first few months of life. It will take an enormous amount of reinforced repetitions to counteract the bird's punishing history of flight attempts. Additionally, teaching an older bird to fly requires a very capable trainer to lead the bird through the repetition of small approximations to acquire the ability and confidence in the action of flying (Figure 7.1).

Learning history associated with the type and amount of reinforcers is an important motivating operation. If the amount and type of food reinforcer is consistent each time, the

Figure 7.1 Bird flight has been used as a central theme in many captive bird displays, like this one at Disney's Animal Kingdom. *Source: Steve Martin.*

motivation to perform a behaviour will likely decrease over time. If you received the same lunch items each day the behaviour of opening the lunch box will likely decrease over time, especially when you have other alternatives. However, if the reinforcer for behaviour varies randomly, the motivation to perform the behaviour will likely increase. If your lunch contains different food items each day, especially if someone else packed your lunch, the behaviour of opening the box to discover what is inside might increase. Varying the type and quantity of reinforcers is often key to motivating animals to participate in training.

The benefit of delivering a variety of reinforcers is evident in the free-flight macaws at Disney's Animal Kingdom. Three groups of 20 macaws make two flights each day across the park. They are released from their holding facility, and then fly half a mile past the Tree of Life to a perch on the other side of the park. Once they are at the designated landing area, the macaws receive a variety of food items, such as pellets, nuts, and fruits. After a 10 minute interpretive programme the macaws are cued to return to their holding facility where they make their way into individual cages and receive a mixture of high value food items that change with every flight. These food items include pellets, nuts, fruits, vegetables, and even a few healthy human food items such as granola bars, crackers, trail mix, etc. After each bird eats their food the birds are let out into the large flight pen where a table is spread with additional fruits, vegetables, and other treats. Additionally, after each flight the trainers add a wide variety of items ranging from browse, to hidden treats, to large bins filled with chewable objects. These birds perform these flights twice a day at ad-lib weight. They choose to come to the holding facility because the reinforcers inside their facility out-compete the myriad of other reinforcers available to them in the park, including unlimited browse, great views, and the acorns that some of them learned to eat in the trees along the way.

7.2.4 Control

Watson found (as cited in Friedman 2005) control of one's outcomes to be a primary reinforcer for behaviour, and loss of control can punish, or reduce, behaviour. Challenges that occur when training animals to enter crates, chutes, and other confined areas are often associated with the animal's perception of a loss of control. By giving animals a higher level of control over their environment trainers can solve many shifting problems in zoological and aquarium settings. For instance, shifting problems are often associated with the consequence of the moving-inside behaviour being the loss of access to outside. Locking an animal inside an enclosure can punish future behaviour of coming inside. If the consequences of coming inside results in both a food reinforcer and opening the door for the animal to go back outside, future behaviour of coming inside is likely to increase. However, at some point a keeper needs to lock the animal inside. In that case, after several repetitions of coming in and going back out, a trainer can then offer increased quantity of high-value reinforcers to offset the possible aversive nature of losing access to outside (Figure 7.2).

At Givskud Zoo in Denmark, the keepers wanted to teach their chimpanzees *(Pan troglodytes)* to participate in husbandry and medical behaviours in a chute leading from the night holding area out to the exhibit. The chimp's previous history of being locked in the chute where it was darted and anesthetised punished the behaviour of coming inside when a keeper was anywhere near the shift door. A new plan was designed for the keepers to give the chimps 'control' of the door. The trainer started the session by reinforcing the dominant female for taking small steps towards the inside of the chute. Each time the animal moved forward the trainer reinforced the behaviour with the animal's favourite food. The next step involved the trainer and a second keeper giving the animal 'the power', through her body language, to control the behaviour of the keeper near the

Figure 7.2 Chimpanzee at Givskud Zoo sits inside the chute after learning she could control the keeper's behavior of closing the shift door by looking at him. *Source:* Steve Martin.

door. Any time the chimp looked at the keeper, he would back away from the door. However, when the chimp 'allowed' the keeper to come forward, the chimp received a high value reinforcer from the trainer. Soon, the chimp walked fully inside the chute, sat in front of the trainer, and allowed the keeper to put his hand on the slide door. As he moved the door, the chimp received a treat from the trainer. If the chimp looked at the keeper, he backed off. Through several repetitions, the training progressed to the point that reinforcement became contingent upon door closing. The chimp's level of control also progressed from looking at the keeper to get him to back up, to moving towards the door as the cue for the keeper to open the door. Through the entire process the chimpanzee maintained control of the door through her body language, which was a huge reinforcer for her behaviour of staying in the chute whilst the door was closed.

The above events may lead some to think this was a training process conducted over several weeks or months by an expert trainer.

Actually, the person doing the training had little or no training experience (however she was coached by an experienced trainer) and the animal had no previous training history. Additionally, the training was conducted over one, 20-minute session! Not only did the chimpanzee learn to sit in the chute with the door closed in the very first session, she also learned three other behaviours on cue. She learned to offer three body parts for the keeper to touch on the verbal cues: 'finger, arm, and head', all in the first 20-minute training session. This is an excellent example of the progress that can be made in training sessions when animals have control of their environment.

7.2.5 Environmental Influences During Training

Training sessions occur in a variety of locations, from the relative quiet of indoor holding facilities to the noisy unpredictability of on-exhibit training areas. No matter where training occurs there will be opportunities for a wide variety of stimuli in the environment to interrupt training sessions and impact an animal's motivation to participate.

A contact call from a conspecific can disrupt a session as an animal stops what it's doing to listen or establish communication with other animals. An alarm call can send an animal bolting from the session just as it would in the wild where the alarm call might signal the presence of a predator. Some animals show more motivation when they can see other animals in the group. For others, it might be distracting for one animal to see another during a training session. Training environment should be adjusted according to each animal's behaviour to maximise motivation.

Some trainers prefer to train in a quiet and controlled environment with few distractions. This enables the animal and trainer to increase focus on the session at hand and reduce distractions to the point that it actually disrupts their motivation and performance in the future. What might be small distractions for most animals can be huge distractions for animals trained in uncharacteristically tranquil settings. Quiet settings are helpful to establishing new behaviour, but once an animal has learned to perform a behaviour without hesitation in response to a cue, the next step should be to generalise that behaviour to novel environments, including new people, locations, and degree of distractions.

7.2.6 Rate of Reinforcement

Maintaining motivation to participate in training is often related to the rate at which an animal earns reinforcers. The rate of reinforcement can be thought of as how many reinforcers per minute the animal receives during a training session. Trainers may thin the reinforcement ratio when using an intermittent schedule of reinforcement to shape various behaviours, such as duration of a target hold or open-mouth behaviour. Stretching the ratio too far or too fast can lead to performance breaking down, a phenomenon called ratio strain (Chance 2014).

The rate of reinforcement can also decline when trainers make too large an approximation when shaping a particular behaviour. For instance, teaching an animal to step inside a travel crate typically involves many successive approximations. There is often a critical point in the process when the animal's whole body is inside the crate, except its back feet. Up to this point, the animal has performed each approximation without hesitation, but is now stalled and reluctant to make that final step. If a trainer holds out too long for just 'one more step' the animal may lose motivation and just walk away, no matter how much cueing and prompting the trainer uses. The best approach in this case is to return to an earlier successful approximation and reinforce smaller approximations at a higher rate of reinforcement. Instead of waiting for both hind feet to be inside, reinforce approximations associated with lifting one foot, moving that foot towards the crate, touching the crate, etc. The size of the approximations is determined by the progress of the animal. If the animal is slow to move forward, make the approximations smaller until each approximation is performed without hesitation. With this approach a trainer builds behavioural momentum (Mace et al. 1988), which often helps the animal move past the point in the approximations where it hesitated before.

7.3 Two-way Communication

Contemporary trainers operating at the highest level are skilled observers of animal body language and give the animal a strong voice in their relationship. They form partnerships with animals that supersede the dominance-based relationship that was once so prevalent in the zoological world.

Through careful observation of an animal's body language a trainer can empower the animal with a level of control in its environment where its 'voice' (through its body language) is as meaningful as the trainer's voice and actions. A trainer gives a cue for the animal to perform a specific behaviour, then the animal's body language provides feedback about its level of motivation to participate or not. If the body language shows the animal is not motivated, the trainer can change the antecedents and/or consequences to encourage the animal, or stop the training session and try again later. When cues and criteria for behaviour are clear, the animal may learn quicker and motivation may increase. When animals don't learn as expected, a trainer should review antecedent conditions including cues, criteria, and reinforcers.

The most successful animal trainers are often the ones who are the most sensitive to animals' body language. As a trainer approaches an animal's enclosure there is an important opportunity for the trainer to observe the animal's body language and determine the most helpful speed of approach, or if approach is even advisable. Too often, keepers just march up to a training area with little or no concern for what the animal's behaviour might be telling them. Like humans, each animal has its own personal space and the data flows the moment the animal can perceive the trainer's presence.

7.3.1 Personal Space

The concept of personal space was introduced by Edward T. Hall in his book *The Hidden Dimension*. He said, 'Most people value their personal space and feel discomfort, anger, or anxiety when their personal space is encroached' (Hall 1966). Personal space is also considered 'flight distance' in animals and can be seen as the distance at which an animal shows comfortable body language as a person or other animal approaches. Judith Bardwick, author of *Danger in the Comfort Zone*, defines the comfort zone as a behaviour state where a person operates in an anxiety-neutral position (Bardwick 1995). Flight distance and comfort zones are directly related to the relationships we have with animals in our care and are strongly influenced by current conditions. A lion *(Panthera leo)* at Assiniboine Zoo was new to training and just beginning to learn target training. As the keeper moved

the target forward the lion snarled and aggressed repeatedly with feet hitting the wire mesh. As soon as the keeper backed up no more than one foot, the lion stopped the aggression and sat calmly in the presence of the keeper. Those 12 inches made a world of difference in the attention span and motivation of the lion.

A trainer can have a high trust account with a particular animal then one day, with no apparent reason from the trainer's point of view, the animal might retreat at the trainer's approach. Though we will never know what an animal is thinking, we can observe what the animal does. When an animal's body language shows distrust or concern, a trainer should stop what he or she is doing, quickly evaluate the conditions, and move back to the animal's comfort zone. No matter how high the trust account or how much history a person has with an animal, trainers should always approach an animal with careful observation of its body language, and only enter an animal's personal space when the animal's body language invites them in. If we give animals control of our behaviour though their body language we will gain trust, improve personal space or flight distance, increase their comfort zone, and have more productive training sessions.

Even when a trainer has established a high trust account with an animal, standing too close can be a problem that disrupts training sessions. Some animals focus intently on a person's hand, target, food container, etc., and stop paying attention to the other important aspects of the training environment. There are times when we want an animal to hold its nose or other body part against a target. However, there are also times when a target held too close to the mesh could cause an animal to try to bite or lick the target instead of moving as indicated to the target. Additionally, when an animal is 'cross-eyed' on a close target it may not see the rest of the environment including prompts and cues. When you look cross-eyed at something close to your face everything else goes out of focus. The same happens when trainers use prompts, cues, targets, bait sticks, etc. too close to an animal. For instance, giving an open mouth cue right at the bars or mesh barrier might cause an animal to focus intently on the hand to try and bite or lick it (Figure 7.3).

Move the hand back 3 or 4 inches and the animal gets a different perspective on the situation and will be more likely to perform a behaviour in response to the cue. Expert trainers understand the critical distances associated with presenting cues and prompts in ways that provide clear communication of contingencies.

7.3.2 Shared Information

As trainers we bring only part of the information to the training session. No matter how much time, effort, and thought we put into the training plan – or how many managers have approved it – the animal's contribution to the information in the session is just as important as the information we bring to the session. The best training occurs when the trainer is flexible to change the plan when an animal's body language suggests an alternative approach will produce better results.

Some trainers say they don't want to change their plan because they want to avoid confusing the animal, or they want to build on what they have started, or they are sure their plan will work if given enough time. If the animal is motivated to participate in training and the plan isn't working then chances are the animal is already confused. To continue on the same path may lead to more confusion, frustration, aggression, or the animal simply walking away from the session. A skilled trainer can see when the animal's body language encourages moving forward according to the plan, or to jettison the plan and start a new one. Training plans should be dynamic not dogmatic.

7.3.3 Event Markers

Whistles and clickers may be effective auditory event markers when an animal is working away from a trainer or in a crowded or noisy environment where a trainer's voice may not be heard. Visual event markers, such as movement of the hand or even turning the body or walking a certain direction can be strong bridging stimuli for animals across an exhibit that are taught to hold on a particular

Figure 7.3 An excellent open-mouth behaviour, but moving the hand away from the mesh a bit may allow this lion to see more of it's environment, including the training and the cue. *Source:* Steve Martin.

station until a visual bridge is given. Tactile markers are good communication tools when animals are in close proximity to the trainer and not looking at the trainer, such as when a chimp performs a back inspection behaviour, or a sea lion has its head underwater but its body is near the trainer. The most appropriate event marker for the animal and current conditions should be used.

Many trainers have said things like 'I tried clicker training and it didn't work', or 'We haven't started our training programme yet because we don't have a clicker'. There is no magic in the clicker; the magic is in the very act of marking the behaviour and closing the gap between criterion performance and the backup reinforcer. Tightening the contiguity between the behaviour and consequence is clear communication that seems to help the animal understand the function of its behaviour. In my experience, event markers like the clicker are great in many conditions. However, in some conditions a whistle, verbal, visual, or even a tactile event marker might be a better tool for the job. For instance, when training great apes or carnivores behind barriers, some trainers find a verbal event marker can be a better tool than a clicker or whistle. The verbal marker frees up the trainer's hands to hold targets, give hand cues, and deliver reinforcers. Plus, without a whistle in the mouth it is easier for a trainer to give verbal cues and prompts along with the event marker.

In some circles it has become routine to use a marker without a well-established backup reinforcer. For instance, the animal may perform three behaviours to criteria and receive an event marker for each behaviour, but only receive a food reinforcer after the third or fourth behaviour. This is a very common reinforcement strategy at zoos, and it is also one of the most confusing reinforcement strategies for animals. I have found that animals trained with this type of inconsistent pairing of bridge and backup reinforcer often lose motivation to participate in the training session, exhibit a high level of incorrect responses to cues, and show frustration-induced aggression. When this happens trainers often blame the animal for the poor performance, labelling the animal as distracted, aloof, messing with their minds, etc. By placing the blame on the animal, trainers relieve themselves of responsibility for the outcomes, but miss valuable information about how to increase motivation through clear communication of cues and high rates of reinforcement.

Some trainers believe since the marker is reinforcement for behaviour they don't have to provide a backup reinforcer. They incorrectly call this a 'variable schedule of reinforcement'. However, if the marker is an effective conditioned reinforcer (as evidenced by its ability to increase or maintain behaviour on its own), the trainer is using a continuous schedule of reinforcement – at least until the reinforcing strength of the bridge extinguishes. Each time the bridge is given without a backup reinforcer can thus be logically viewed as a respondent extinction trial. Just as Pavlov paired the metronome sound with meat powder to elicit the dog's salivation at the sound of the metronome alone, trainers pair the marker stimulus to a backup reinforcer, often food. When Pavlov stopped backing up the sound of the metronome with the meat powder, the metronome extinguished as an elicitor. Similarly, when trainers unpair the bridge and the backup reinforcer, animals eventually stop listening to the marker and begin focusing on more reliable, salient signals of the backup reinforcer to come. This is often the action of the trainer's hand moving to the backup reinforcer, e.g. hand to feedbag. This visual bridging stimulus can serve the function of the intended marker, keeping the animal in the training environment a bit longer. However, eventually the low rate of backup reinforcement reduces motivation and the animal either leaves the session, shows aggression, or the behavioural response deteriorates to the point the trainer ends the session.

A keeper with a long history of training a male gorilla *(G. gorilla)* with this inconsistent pairing of the marker and backup reinforcer agreed to participate in a small experiment. She first demonstrated her usual training session with the animal performing several behaviours in rapid-fire succession and each correct performance receiving a click but only every third or fourth click being backed up with a food reinforcer. The animal performed several behaviours correctly and several behaviours incorrectly, often after multiple cues. About half way into the session the gorilla stood up and hit the top of the door with all his might and ran off to the adjoining den.

A few weeks later the same keeper and the same gorilla did another small experiment. This time the keeper backed up each bridging stimulus with a food reinforcer. The keeper even mentioned that this session might be difficult for her because she has never paired every bridge with a food reinforcer before. The training session went perfectly. Every cue brought a correct response, and every correct response resulted in a marker followed by a food treat. The session went on and on and neither the keeper nor the gorilla looked like they wanted to stop. The high rate of reinforcement provided clear communication of the behaviour–consequence contingencies and increased the gorilla's performance and motivation to stay in the training session.

Some of the most common reasons keepers say they use their bridging stimuli without the backup reinforcer include the following erroneous rationale: not backing up each bridge makes training more interesting for the animal; it builds stronger behaviour because it is a variable schedule of reinforcement; the clicker is a reinforcer so you don't need two reinforcers; it reduces frustration elicited aggression if a trainer runs out of food; it means good job and keep going. Each of these are not scientifically sound. A more detailed discussion of these points is found in the article called, 'Blazing Clickers' (Dorey and Cox 2018; Martin and Friedman 2011).

7.3.4 Prompts

Prompts are antecedent stimuli trainers use to increase the likelihood that an animal will perform a specific behaviour. At the Denver Zoo a trainer taught a jaguar *(Panthera onca)* to roll over by kneeling in front of the cat's cage, leaning to her side, tipping her head over, and motioning with her hand as if to guide the cat's head. The jaguar followed the trainer's body language, and the trainer reinforced approximations of the roll behaviour until the cat finally rolled all the way over. The trainer then began to systematically fade out each of the prompts until the cat rolled over given the verbal cue 'roll'.

A trainer at Cheyenne Mountain Zoo taught a Wolf's Guenon *(Cercopithecus wolfi)* to brachiate around the exhibit by walking along the edge of the exhibit and reinforcing small approximations of the animal following her along the barrier. Many trainers have taught animals to open their mouths, lift their arms, move from one area to another and countless other behaviours using their own body language to prompt the behaviour. Prompts can be very important tools to help shape behaviour. However, the rule of prompts is to fade them out as soon as the behaviour will allow. When prompts are not faded the animal's behaviour will become dependent on the prompt. If the trainer at Denver Zoo had not faded the prompts the jaguar would not perform the behaviour unless the trainer knelt down, leaned to her side, tilted her head and gestured with her hand. As the trainer systematically fades prompts, the animal gains a better understanding of criteria, cue, and consequence. Ultimately, past consequences should drive future behaviour, not antecedent prompts, and the signal for the behaviour is a deliberate one.

Baiting, or luring, is also a prompt. Many keepers show an animal food to get it to come into a holding area, go into a crate, or move from one area to another. Baiting can be a helpful prompt to establish new behaviour, but baiting can be a liability if not faded out of the training programme early. When an animal comes to a shift door and sees the food already placed on the floor inside, it can decide if that type or amount of food is worth going into the holding area. If the animal decides to stay outside, many keepers will up the ante and offer more food. This only compounds the problem as some animals learn waiting will result in higher value reinforcers being offered inside. As with all reinforcers, food, especially favourite food items, should be presented after the animal has come inside the area and the door is closed.

Trainers at free-flight bird programmes often experience problems associated with prolonged use of the baiting strategy. To encourage a hawk to fly to the glove, many trainers will show the bird a particular food item, such as a small piece of lean meat. From the perch where the bird sits it can see the food and decide if that type or quantity or reinforcer is worth the effort of flying to the glove. Often, after a short delay, and whilst the trainer is ad-libbing dialogue, the trainer reaches into the bait bag to add a piece of food to the offering. Too often if the bird delays longer the trainer will offer even more food, maybe even a whole, dead mouse, the bird's favourite treat. I often wonder if these birds are thinking, 'these humans are so easy to train!'

If baiting is used to help encourage behaviour, the trainer should provide an additional backup reinforcer after the bird has landed on the glove. If the trainer needs to prompt with a small piece of food in the early stages of training, additional food items should be hidden in the glove. Occasionally, the bird should discover a whole mouse or other high value reinforcer when it lands on the glove. As the consequence reinforcer hidden in the glove encourages quick flights, the bait should be faded out completely.

7.3.5 Jackpots

Trainers often talk about delivering 'magnitude reinforcement' or 'jackpots' for particularly high levels of performance of behaviour. Their hope is to increase the likelihood the animal will recognise the large quantity of food is delivered in response to a supercriterion performance of behaviour, thus increasing performance in subsequent trials. However, that may not be the case. Two

important aspects of shaping are the fluidity of movement from one approximation to the next and the contiguity in the cue-behaviour-consequence sequence. The longer an animal takes with a reinforcer, the farther away in time the animal is from the next behaviour. For instance, an otter can eat a 3-g piece of fish in a couple seconds, but if the animal is given a 20-g piece of fish, it might take 10 seconds to eat the fish. Those extra eight seconds can upset the flow resulting in the need to move backward in the approximations to regain the momentum and get back on track.

The shaping process is most successful when the animal moves quickly across approximations to the goal behaviour. The size and type of food reinforcer should promote quick consumption to keep the animal progressing smoothly through the approximations and focused on criteria and consequences. The jackpot reinforcer not only disrupts the flow, it distracts the animal as it takes extra time to eat the increased quantity of food, and results in faster satiation. This small distraction may not ruin a training session or even cause problems, but a jackpot in the middle of a shaping session may not help the animal learn quicker or better either. Jackpots and magnitude reinforcers may be best used at the end of the training session to reinforce calm behaviour when the trainer leaves the training session.

7.3.6 Fluency and Speed

During shaping many trainers follow an 80% rule, requiring 8 correct responses out of every 10 trials (or 4 out of 5, etc.) before moving to the next approximation towards the goal behaviour. Other trainers think of the 80% rule in a more subjective manner, such as a behaviour being 80% perfect before moving to the next approximation. Either way, many people follow a shaping strategy that involves the animal performing multiple correct behaviours of a specific approximation before moving to the next approximation. In this manner a trainer puts a previously reinforced behaviour on extinction and selectively reinforces the next correct closer response to the goal behaviour.

A common problem with shaping behaviour in this manner results when a trainer reinforces too many repetitions of one approximation. Performing multiple repetitions of one approximation creates reinforcement history that can slow progress and make it more difficult to move from one approximation to the next. To create the best flow of behaviour, a trainer should move to the next approximation when the current approximation is performed *without hesitation*. With this strategy animals may perform approximations only one time when the trainer and animal are really in sync. If the animal hesitates, the trainer can always repeat that approximation or even move back in the approximations to gain behavioural momentum.

Shaping should involve letting the animal's behaviour determine the size and speed a trainer should move through the approximations. Through careful observation, a skilled trainer can recognise the subtle changes in the natural variation of an animal's behaviour and reinforce small increases in movements towards a goal behaviour. Other times an animal might offer larger approximations that are more obvious to trainers. Either way, the animal's behaviour combined with the trainer's observational and mechanical skills determines the speed of training.

Animals can learn faster than most trainers give them credit for, which is generally faster than most trainers are comfortable training. Where one trainer might believe it will take a year to teach a primate to accept an injection, another trainer might believe it will take two weeks. Some animals learn faster than others and some trainers teach faster than others. However, the trainer who believes it will take a year to teach the behaviour will almost certainly take significantly longer to train the behaviour than the person who believes it can happen in two weeks. Training at the animal's speed is a valuable skill that is developed over time and with much practice.

7.3.7 Adding the Cue

Some trainers prefer to add the cue after the behaviour is performed at a high level of fluency. In this manner the cue is associated with only the mastered behaviour instead of approximations leading up to the goal behaviour. For instance, teaching a giraffe to walk into a chute or giraffe restraint devise (GRD) may involve shaping the behaviour through a baiting strategy involving the trainer holding browse just out of reach of the giraffe and reinforcing each step closer to the GRD. After several repetitions the giraffe may walk straight into the GRD without seeing the browse because it has learned through its reinforcement history to expect the browse once it is fully into the GRD. Now that the goal behaviour is performed with a high level of fluency, it is time to help the animal understand reinforcers will only be available for the GRD behaviour in certain conditions. The trainer will then begin associating the cue, which is generally a visual or audible stimulus. The trainer may say 'Shift' or point her finger straight out to her side as the giraffe enters the GRD to receive a reinforcer. After the giraffe comes out of the GRD, the trainer can reinforce the behaviour of approaching the trainer, then cue the GRD behaviour. With each repetition the trainer starts a bit farther away from the GRD until she can give the cue across the room and the animal will leave her to walk across the room and enter the GRD.

Other trainers prefer to add the cue as they shape the behaviour. For instance, some mouth-open behaviours are taught using a target on the top of an animal's nose and then a second target on the animal's chin. If these two targets are the trainer's finger and thumb, the prompts of spreading the finger and thumb gradually farther apart become the cue for the open-mouth behaviour. Even when shaping a turn behaviour for a macaw, a lay down behaviour for a cheetah, or any other behaviour for that matter, some trainers add a verbal or hand cue during the shaping process.

The merits associated with adding the cue after mastery of a behaviour or during the shaping process are often debated in various circles. However, it is important to remember every animal is an individual, every trainer is an individual and conditions are constantly changing. What works for one trainer and animal may not work for another trainer and animal. As with most questions about the specifics of training approaches, it is often best to use your good judgement and 'test it'. If it works, do it again. Except in welfare issues, it is often best to not worry too much about what others say and discover for yourself which strategy works best.

7.3.8 Consequences Drive Future Behaviour

The skilled and artistic application of scientific principles is seen in trainers who understand consequences, not antecedents, are the most important factors in driving future behaviour. Antecedent cues tell an animal that the opportunity for reinforcement is available contingent upon their performing a specific behaviour. It is the reinforcing consequence that follows the behaviour that increases the likelihood that the behaviour will occur again.

The trainers at Natural Encounters Inc., trained a young Grey Crowned Crane *(Balearica regulorum)* to fly from one trainer to the other about 200 m apart. After the crane ate the food reinforcers from a trainer's hand, the trainer at the other end of the field began calling (prompting) and gesturing with hand cues for the bird to fly back to the first trainer. After a few minutes of picking at the grass, exploring objects on the ground, and looking at the trainer by its side, the crane finally looked towards the trainer at the other end of the field. Only then was the crane ready to receive the cue, and quickly took off in the direction of the trainer who had been giving the cue and prompts for the past two minutes. This repeated cuing and prompting had no effect on behaviour and quite possibly served to decrease the meaning of the cues and prompts since they weren't immediately followed by reinforcement.

To address this latency in behaviour and strengthen the cue-behaviour response the trainers shortened the distance between them to about 20 m. The trainer standing next to the bird held completely still so not to distract the crane. The trainer away from the bird waited for the crane to start moving in her direction before delivering the cue. When the crane arrived at the trainer it received the reinforcers and the sequence was repeated going back to the other trainer. The trainers realised the crane took time to investigate its surroundings before it was ready to receive the cue. They also realised they did not even need to call the bird because its reinforcement history motivated the animal to go from one trainer to the other. Once the behaviour was performed fluently, the trainers inserted the cue into the antecedent position.

Even in the best circumstances, animals will not always respond immediately to the cue and this is where skill and experience are most valuable. When an animal fails to perform after a cue, a skilled trainer will read the animal's body language and determine if the animal saw the cue or not. If the trainer is sure the animal saw the cue but is momentarily distracted, she may wait for a few seconds, then recue the behaviour when the animal is looking at her. If the trainer does not think the animal saw the cue, she will present another cue even if it was only one or two seconds after the first cue. Lack of contiguity between the presentation of the cue and performance of the behaviour may weaken stimulus control, promote offering of the behaviour at random times, and encourage latency of performance in response to the cue.

Once an animal has learned to perform a behaviour at a high level of stimulus control, the behaviour is usually put on a maintenance schedule, i.e. the trainer only cues and reinforces the behaviour occasionally, or the behaviour is performed as part of its daily routine, such as shifting on and off exhibit. Some trainers run through an animal's entire repertoire of behaviours on a daily basis, as if they are worried that the animal might forget the behaviours. Animals remember behaviour in proportion to their practice and reinforcement history. They rarely forget behaviour with strong reinforcement history, but they do lack motivation under certain conditions. When trainers run through an animal's repertoire on a daily basis, often these training sessions involve the animal doing quick, short duration, performances of many behaviours in succession. This rapid-fire performance of behaviours serves little useful purpose for husbandry or medical procedures and may even decrease performance of these important behaviours in the future. Short duration practice of a behaviour that requires long duration performance to be useful for medical procedures can be counterproductive. It is far better to do fewer behaviours at long duration performance than to do several behaviours at short duration, which are below criterion for intended purposes of the behaviour.

During maintenance schedules is when some behaviours regress and performance drops off to a point that a bit of 'tune-up' may be required to get the behaviour back to a high level of stimulus control. For instance, there is a tendency in some trainers to allow the animal additional time to perform a behaviour that has lost some of its fluency. From a primate that used to perform with crisp response now taking 30 seconds to present its shoulder, to a bear taking 30 minutes or longer to shift inside, latency is a common problem in training programmes at zoological facilities.

Poor performance of behaviour is often caused by lack of practice, poor motivation, or trainer errors associated with reinforcing behaviour below criteria. It is often the case that trainers will unwittingly reinforce gradually decreasing criteria of behaviour. For instance, an animal that once shifted within 30 seconds of the door opening, maybe takes 45 seconds to come inside because it was distracted by animal activity in the next exhibit. The following day the animal may be sleeping in the soft grass on a sunny day and take two minutes to come inside. The next week the animal might take 10 minutes to come out of the pool to go inside. Each time the animal takes longer, but the keeper still reinforces the behaviour, it increases the chances the latency will continue or get worse. Similarly, an animal that once held its shoulder against the bars for three minutes pulls away at two minutes but may still receive reinforcement for 'a good effort'. Soon the animal is pulling away at one minute, then 30 seconds, and soon the behaviour is no longer useful for injections. It is important for trainers to hold to criteria and only reinforce full-criteria behaviour, even if they have to devote an entire training session to rebuilding duration behaviour through successive approximations of increasingly long duration of a behaviour.

7.3.9 Short Window of Opportunity

Motivating operations for animals in human care are different from their wild counterparts that respond quickly to environmental stimuli, often to obtain food or to keep from becoming someone else's food. Consider a brown bear *(Ursus arctos)* hunting salmon in a stream. If the bear moved like many of its counterparts in zoos it surely would not catch many fish. The only thing keeping a brown bear in a zoo from moving as fast as its wild counterparts is motivation, which is created through reinforcement history. The bear at the zoo has learned the keeper will leave the door open for at least 30 minutes, and the same type and quantity of food will probably be waiting inside, so what's the hurry? However, if the keeper opens the door of the shift cage and then closes the door after one minute, the bear will lose its opportunity to come inside for the food reinforcer. Through this 'limited hold' contingency (Pierce and Cheney 2013), the bear learns the consequence of staying outside when the door opens is the lost opportunity to eat the food. This short window of opportunity to gain the food resource will give the bear a reason to perform more quickly in the future.

Plus, the trainer in this case has experienced a 1-minute training session instead of a 30-minute session.

Some keepers report they do not have time to train with all of the other work required of them. In some cases, trainers can shorten training sessions simply by giving animals a shorter window of time to perform behaviour (limited hold). A hawk sitting in a tree watching a mouse scurry about in a meadow has a short window of opportunity to perform the mouse-catching behaviour before the mouse disappears down a hole. However, a hawk in a zoo, like most other animals, has little reason to perform behaviour with any sense of urgency because it knows the food will always be waiting when it decides to shift inside. Shortening the window of opportunity by reducing access to reinforcers can help speed up training sessions and give keepers valuable time to work on other projects.

7.3.10   Ending on a Good Note

Some training sessions last longer than an animal's attention span as trainers try to squeeze out one more good repetition because they believe they should 'always end on a good note'. Ending a session with quick behavioural response and fluent behaviour is what many keepers shoot for. However, it can be an unrealistic goal for every session, condition, or animal. When an animal walks away from a training session it is often a good sign that the animal has ended the session. In this case, a trainer is best advised to let his or her partner in the training session – the animal – determine the end of the session. Attempts to bring the animal back to station for one more successful repetition may actually hurt training progress as cues and prompts are ignored and criteria for behaviour is compromised. On the other hand, ending a session when an animal is highly engaged and motivated to work for reinforcers may also be ill advised. In that case it might be best to end the session on a good performance of behaviour and an extra-large amount of food reinforcers, desirable enrichment item, or

another conditioned reinforcer that will keep the animal engaged as you walk away.

7.3.11   End of Session Cue

In addition to walking away from the training area, some trainers use an additional 'end of session' cue to convey to an animal that the session is over. Other trainers wonder if they should use an end of session cue or not. The best answer is 'it depends'. If the animal responds to the end of session cue with calm behaviour, especially turning and leaving the training area, the end of session cue can be a useful tool. However, if the animal shows aggression in response to the cue it may be best to withhold the cue and develop a plan to replace the aggression with a more desirable behaviour.

For some animals, an end of session cue can lead to frustration and ultimately aggression. This aggression can range from subtle body language to more dramatic behaviour such as spitting, attacking the barrier or loud, aggressive vocalisations. An animal that practices aggression will often get better at it. Therefore it is generally better to replace unwanted behaviour, such as this aggression, with a more desirable behaviour, such as sitting calmly at the barrier.

Several years ago, the male sea lion (*Zalophus californianus*) at Singapore Zoo would block the trainer's exit path and approach the trainer aggressively when the trainer gave the end of session cue. To keep from being bitten the trainer had to toss several fish into the pool to encourage the sea lion to leave the path and go into the pool for the fish. The trainer also had to have a stick in his boot to defend himself if the sea lion attacked him. After some discussion, the trainers developed a plan involving differential reinforcement of incompatible behaviour (DRI) to use half of the animal's food for a stationing behaviour that was incompatible with attacking the trainer. They also changed the meaning of the end of session cue to signal the animal to go to a particular station for reinforcement. They first taught the sea lion to swim to a rock on the island across the pool from the trainer's exit. Using a variable duration schedule the trainer increased the amount of time between reinforcers as the sea lion sat on the rock. The trainer then took a couple of steps towards the exit and tossed the sea lion a fish to reinforce the behaviour of sitting on the rock as the trainer moved towards the exit. The trainer made a few steps forward, tossed the sea lion another fish and then went back on the spot on the path where he started. He repeated these steps until he could leave through the gate, and then come back to reinforce the stationing behaviour of the sea lion. Soon, the end of session cue meant go to the rock across the pool and wait for the trainer to go to the gate. As the sea lion sat on the rock, the trainer came back and reinforced the behaviour with varying type and quantity of primary and secondary reinforcers.

Two male lions at another zoo worked perfectly in their training sessions until the keeper gave the end of session cue, which resulted in the lions' loud vocalisations and feet clawing at the bars. This behaviour continued for years because the trainers were instructed to always give an end of session cue to their animals when they ended a session. On a suggestion from another trainer, they developed a plan to replace aggressive behaviour with calm behaviour in response to the end of session cue. The trainer gave the end of session cue and immediately gave a piece of food to the lions. After a few repetitions of associating the cue with the reinforcers, the trainer had shaped calm behaviour that replaced the previous aggressive behaviour when the cue was presented. The trainer then used a variable duration schedule of reinforcement to extend the amount of time the lions performed the calm behaviour before reinforcement. After calm behaviour of about five seconds was on a high level of stimulus control, the trainer began moving away from the barrier after giving the cue and reinforced approximations of calm behaviour as he walked away from the lions. In this one training session the keeper was able to give the end of session cue, leave the area of the lions and go fully inside the office, then come back to reinforce the lion's calm behaviour.

The trainer then found other ways to reinforce calm behaviour. Before each indoor training session, he scattered enrichment items and hid food treats in the exhibit. At the end of the training session, the trainer gave an end of session cue and when the lions showed calm behaviour he opened the shift door to the outdoor exhibit. Access to the variety and value of treats and enrichment in the exhibit became reinforcers for calm behaviour after the end of session cue.

7.4 Training Specific Behaviours

7.4.1 Target Training

One of the first behaviours many trainers teach animals in zoos it to touch its nose or other body part on a target (Figure 7.4). This target can be the end of a stick, a ball on the end of a stick, a balled-up fist, or any other object that the trainer can move around from place to place. Animals can learn to touch a target with their nose, hand, foot, side, or almost any body part. Targets can be used in protected contact by placing them just outside the barrier and the animal will touch the inside of the barrier at the location of the target. Targets increase motivation for some animals' approach behaviour, give clear directional cues, provide safer training environments, and generally give trainers a tool for moving animals with more precision, comfort, and fluidity. Laser dots are also used as targets when animals are taught to touch the dot. Teaching an animal to touch a laser dot with its nose can be dangerous if the laser hits the animal's eye for a prolonged period of time, which is why some trainers prefer to teach animals to touch laser dots with their hand, foot, or finger.

A long duration hold is an important part of target training. Using a variable duration

Figure 7.4 This hippo is being trained to touch its nose to a target; this is often one of the first behaviours included in a training programme. *Source:* Steve Martin.

schedule of reinforcement a trainer systematically teaches the animal to hold progressively longer duration of hold behaviour. The keepers at Denver Zoo taught a Spotted hyena *(Crocuta crocuta)* to target its nose on a hand target (closed fist) though the mesh barrier. They extended the target hold behaviour to several minutes. They used the target behaviour to raise the hyena's head straight up whilst placing its shoulders against the mesh. In this position the hyena holds perfectly still with its nose targeting the keeper's hand through the mesh as the vet tech draws blood from the hyena's jugular vein. Many primates are taught to hold clips or carabineers attached to the mesh and positioned so that the animal's arms are fully outstretched when performing the behaviour. The location of these targets can be adjusted to position the animal's arms for body examinations or even as a safety measure to position the animal's hands away from the keeper. The trainers at Maryland Zoo taught several of their chimpanzees to station their arm away from their body so the keepers can draw blood from the animal's arm without an arm sleeve (Figure 7.5).

During shaping it is important to carefully observe the topography of the target behaviour that is being shaped. Some animals will try to bite a target during the shaping process, others will try to nudge or push the target with their nose or mouth. One poorly timed reinforcer can result in persistent biting at a target. Precise timing of the bridging stimulus will help trainers shape for gentle contact and constant pressure of the body part on the target.

7.4.2 Station Training

Some trainers make the distinction between target training and station training in that targets are moveable and stations are stationary objects, as their name suggests. A station can be a bench, rock, tree, stool, patch of grass, or any area that an animal can learn to go to and hold a position in relation to that station. At Northwest Trek a keeper taught the North American porcupine *(Erethizon dorsatum)* and North American beaver *(Castor canadensis)* to station on tree stumps instead of standing at the door when she

Figure 7.5 The chimpanzees at Maryland Zoo were trained to station with an arm extended away from the body, which facilitated further husbandry behaviours like taking blood samples. *Source:* Steve Martin.

approached. Keepers at Columbus Zoo taught a silvery-cheeked hornbill *(Bycanistes brevis)* to fly from her cage across the African exhibit and land on a tree station in the middle of the exhibit. The keepers at Cheyenne Mountain Zoo used a laser target to teach an orangutan *(Pongo abelii)* to climb to a specific high point in the exhibit. When the laser prompt was faded that location in the exhibit became the station for the termination of the animal's climbing behaviour and a place where the animal waited for the bridging stimulus, then returned to the trainer for the food reinforcer.

7.4.3 Medical and Husbandry Behaviours

There is no limit to the range and number of behaviours animals can learn to perform in zoological facilities. What people could barely imagine only a few years ago is now old news in contemporary zoological settings. There was a time not long ago that people were amazed a primate was trained to put its arm in a sleeve and let a veterinarian draw blood. Today, the number and types of

animals taught to participate in voluntary injections, blood draws, ultrasounds, and more, exceed the expectations of most zoological professionals.

The strategies used to teach these medical and husbandry behaviours vary between each institution, and even each trainer. However, there is a growing list of best practices disseminated in publications, conferences, and records exchanged between institutions with animal transfers, some of which are described below.

7.4.4 Injection Training

As with training any new behaviour, teaching an animal to participate in injection or blood draws involves a partnership where the animal has choice and control. Forcing an animal to accept an injection by squeezing it in a chute can destroy trust that took weeks to build. By shutting an animal in a restraint devise, a keeper takes away an animal's choice and control in the situation, which often reduces an animal's motivation to participate in training. Unless the animal's behaviour of being closed in the restraint devise is exceptionally well generalised to novel conditions, it is best to train the animal whilst the doors are open giving it the ability to leave. Positive consequences should drive future behaviour of staying in the restraint devise, not closed doors and reduced access. Making it the animal's choice to stay in the chute for training should be the goal, which is best accomplished by giving it the power to leave. Many skilled trainers have learned that when an animal can leave it is more likely to stay, when the positive consequences are worth it.

The first step to teaching an animal to accept an injection is to desensitise the animal to the training area and to the syringe, and possibly to an alcohol swab. If the animal shows fearful body language a systematic desensitisation process involving counterconditioning will often be the best approach to building trust in the syringe. The veterinarian or technician may also be part of this counterconditioning programme to promote trusting relationships that will be important for medical behaviours in the future.

The successive approximations for training the injection behaviour will vary from one animal to the next as each animal brings its own learning history to the session. Most injection training is conducted in a protected contact environment. Even for animals that are tame and tractable, it is often best to teach the injection behaviour through wire mesh, bars, or some other barrier to give the animal more control and ability to move away. Additionally, wire mesh or other barriers provide a station for animals to press into, which will steady the animal's body and make the injection behaviour easier to accomplish.

Teach the lean-in behaviour only after the animal shows comfortable body language in the training area. The lean-in behaviour often starts with some form of targeting behaviour to guide the animal's head into a position where its side is near the training barrier. A trainer can then lightly touch the side of the animal with a hand, finger, dowel, stick, or any other object that will fit through the barrier. We will use a stick for the examples below. Some animals are comfortable with tactile contact and some have to learn the

association of the contact with positive reinforcement. Either way, shaping the behaviour of touching the stick usually begins with letting the animal see and smell the stick as the trainer reinforces closer approximations of the stick towards the animal. Teaching the animal to move towards the stick, instead of the trainer moving the stick towards the animal, is the best approach because the animal has more control. The trainers at Cheyenne Mountain Zoo taught their reticulated giraffe *(Giraffa camelopardalis reticulata)* to touch its shoulder on an 8-inch piece of 1-inch thick plastic garden hose fixed on the inside of the barrier at shoulder height for the giraffe. They used a target pole to direct the giraffe into a position where it touched the plastic hose with its shoulder. After a few repetitions, the giraffe learned that touching the plastic hose was associated with food reinforcers and the behaviour of leaning into the hose increased. Within a few more repetitions the trainers were able to generalise the shoulder lean-in behaviour to a trainer's hand and ultimately to other objects.

Often it is best to start with a touch of the stick on the neck or shoulder of an animal and approximate the touch farther down the animal's body until reaching the hip. Other animals are comfortable being touched on the hip and a trainer can start there. The animal's behaviour will determine the placement of the touch and rate of progress. A trainer can hurt or even destroy trust by trying to sneak the stick in to touch the animal when it is not looking. Always let the animal have the opportunity to see what is happening and pair that experience with positive reinforcement as you establish the conditioned reinforcer.

Once an animal is comfortable with a light touch of the blunt stick (or other object) on its shoulder or hip, the trainer can shape an increase in pressure through successive approximations. Consider a range of pressure on a scale of 1 to 10, with 10 being almost as hard as a person can push. The pressure of the stick against the animal's muscle should increase to about level four, as the trainer reinforces each approximation of increased pressure. After a few repetitions (approximately four or five) at an intensity of level four the animal should begin to understand reinforcement is contingent on pressure at level four. This is when the trainer should reduce the pressure to about level one and wait for the animal to lean-in to apply pressure on the stick. This allows the animal to gain control of the pressure and the trainer can selectively reinforce successive approximations of increased pressure that the animal puts on the stick.

Introduce the syringe only when the animal is putting pressure on the stick at a level six or seven, and the animal's body is up against the barrier. It's important to show the syringe to the animal in the same way the stick behaviour was shaped earlier. The animal will quickly learn to generalise the lean-in behaviour to the needless syringe and the trainer can shape increased pressure to seven or eight. Some keepers add a step at this point and use a paper clip, ballpoint pen, or other semi sharp object to help generalise the behaviour to sensations other than the force of the needless syringe. Blunted needles have also been used but can be dangerous as they are more likely to penetrate the skin causing a higher level of pain and significant setback in the training. It is important to keep in mind that the animal creates the pressure, not the trainer. Finally, the goal is for the amount of discomfort associated with the pressure the animal puts on the needleless syringe or other objects should be greater than the discomfort of the needle during the actual injection.

With clear contingencies and reinforcing consequences there are few limits to what an animal can accomplish with injection training. Trainers at Denver Zoo taught their Titi monkey *(Paralouatta Aureipalatii)* to accept insulin injections, and he has performed the behaviour fluently twice a day for the past four years. An Andean bear *(Tremarctos ornatus)* at Cheyenne Mountain Zoo took his painful rabies vaccination and walked away only to come right back for another vaccination injection. He has also been hand-injected and immobilised, but he always returns to his trainers for more injection training after these stressful experiences. There are many more examples of animals trained to voluntarily receive injections and return immediately for more training. Most often these success stories are associated with high trust accounts, empowered animals, and strong reinforcement history.

7.4.5 Blood Draw Behaviour

Training an animal to participate in blood draw behaviour is similar to teaching an animal to accept an injection. However, with blood draw training there is often a need to shave the hair from the injection site. Desensitising an animal to electric clippers can be more challenging than teaching an injection, however, the shaping process is the same. Counterconditioning the fear response with incrementally closer approach of the clippers is a good approach. Each approximation should be associated with a high value reinforcer. The keepers at Cheyenne Mountain zoo taught their Grizzly bear *(Ursus arctos horribilis)* to voluntarily put its foot through a door in the training panel and accept its foot being shaved. They then taught it to accept scrubbing with an alcohol swab, then to voluntarily participate in blood draws from the top of its foot whilst remaining in position calmly throughout the procedure. These behaviours are performed routinely at the on-exhibit training wall in view of the guests.

Many animals, especially primates, are taught to put their arm through a sleeve and hold onto an object such as a peg or bolt at the far end of the sleeve. Holding the peg gives the animal a target to hold and helps stabilise the arm. A veterinarian can then access the animal's arm through a small opening at the top of the sleeve. A critical behaviour in this sequence is the long duration hold of the peg. Some trainers reinforce the peg-holding behaviour then provide multiple treats with no particular behavioural criteria. This type of 'distraction training' is meant to simply encourage the animal to keep its arm in the sleeve whilst being distracted with the food. An approach to training the behaviour with more clear contingencies involves shaping longer duration of peg-holding behaviour. In this procedure the animal learns reinforcers are contingent on longer duration of the hold behaviour.

7.4.6 Foot-work

Participating in hoof trims is an especially important behaviour for many exhibit animals. Where animals had to be darted and anesthetised to have their hooves trimmed in the past, animals are now taught to present their hooves and allow keepers and farriers to perform this necessary work. Elephants and even rhinos are also taught to present their feet and hold for long duration as a keeper provides valuable routine inspections and footwork. There are few limits to what types of animals can be taught to perform voluntary foot inspections, plus husbandry and medical behaviours.

As with blood draws, there is a tendency with some trainers to simply allow an animal to feed from a bucket whilst a person raises its leg and trims its hoof or works on a foot. If the animal's motivation to eat the food is high enough, the act of eating may distract it from the work being done on its foot. This approach can work in some situations; however the behaviour will be less reliable than the same behaviour shaped through approximations of longer duration holds using positive reinforcement. The shaping process builds trust and behavioural fluency, unlike distraction training that is not built on approximations to fall back on if the behaviour breaks down.

Sixteen reticulated giraffe (*G. camelopardalis reticulata*) at the Cheyenne Mountain Zoo are trained to lift each foot onto a station, curl their hoof, and hold still as a keeper works on their foot. Using a variable duration schedule of reinforcement, the trainers taught the animals to place and

hold their foot on a station for several minutes as the trainer, or a farrier, trims its hooves. The animals learned reinforcement is contingent on placing and holding their foot in a specific spot for a variable amount of time. If the animals had been trained by distracting them with their head in a bucket of food they would not understand the criteria for reinforcement and the work on the foot may seem like an annoyance to them, like a fly biting their leg, which could make things dangerous for the person working on the foot.

7.5 The Right Tool for the Job

Training animals in a zoological setting is far more than clicking a clicker and giving an animal some food. Most contemporary animal trainers work with multiple species, in a wide variety of training environments and are responsible for meeting demanding programme goals. To accomplish the training tasks at the highest level a keeper must have a good working knowledge of the science of behaviour change principles combined with outstanding mechanical and observational skills.

Unfortunately, the majority of animal caregivers learn training skills on the job, often just by doing it on their own with little or no guidance. The number of zoological institutions with formal behaviour management programmes is small compared to the number of facilities where keepers are involved with training animals each day.

Directors, veterinarians, curators, and a wide variety of zoological managers boast of having expert animal trainers at their facilities. However, what education, experience, or knowledge prepared them with the ability to tell an expert from an average animal trainer? Without standards with which to judge a trainer's performance, the title of 'expert' can be used to describe a wide range of performance. Operationalising what an expert trainer *does* should be helpful to the zoological community in general.

The following is a list of a few of the most important behaviours commonly observed in 'expert' animal trainers:

Uses positive reinforcement
Teaching animals to perform some behaviours can be quickly accomplished through negative reinforcement, such as shifting a rhino out onto exhibit by squirting it with cold water, or shifting gazelle inside by chasing them with push boards. Expert trainers understand the value of using positive reinforcement whenever possible to teach behaviours even though it may take a bit more time and effort.

Avoids punishment
Many people fail to realise that anything they do that results in a decrease in behaviour is punishment, including the timeout procedure. Expert trainers avoid punishment whenever possible, partly because they understand the potential for detrimental side effects such as aggression, apathy, escape/avoidance, and generalised aversion to the environment. Expert trainers replace unwanted behaviour with desirable behaviour through differential reinforcement of alternative or incompatible behaviour.

Takes responsibility for behaviour
Expert trainers avoid blaming their animal and understand poor performance is most often associated with antecedent and consequence conditions that they have failed to effectively establish.

Demonstrates flexibility
Expert trainers understand no matter how hard they worked on the training plan, it is only about half the information; the animal brings the other half of the information. Adjusting the training plan in response to the animal's behaviour is a trait of expert trainers.

Carefully arranges antecedent and consequence conditions
The condition in which learning occurs varies from moment to moment and antecedent stimuli can make or break a training session.

Expert trainers carefully evaluate and adjust current antecedent conditions to encourage desirable behaviour, and provide reinforcing consequences that motivate animals to participate in the learning experience.

7.5.1 A Reinforcing Future

Contemporary animal training starts with the skilful and ethical application of the scientific principles of behaviour change. A commitment to the most positive least intrusive training strategies, improves behavioural outcomes and welfare for animals in human care. The fields of animal welfare and animal training should advance together as practitioners improve their knowledge and skills to teach animals to willingly participate in husbandry and medical procedures thereby reducing stress and danger that resulted from immobilisations and restraint in the past.

Where trainers once used force, coercion, and hunger to motivate animals, trainers now create motivation by first forming trusting relationships and empowering animals with choice and control. By allowing communication to flow in both directions, trainers give their animals a 'voice', expressed through the animal's body language resulting in reciprocal influence.

Trainers who once understandably worried about clocking out on time because their animals would not shift inside can now teach quick response to shift cues at any time of day. Animals once labelled 'slow, obstinate, aggressive, untrainable' and more, now respond quickly to cues presented by trainers who have changed their behaviour and maybe their lives through clear communication, antecedent arrangement, and reinforcing consequences. Problem behaviours are opportunities for skilful trainers to understand that all behaviour has function and by changing conditions, we can replace problem behaviours with more appropriate behaviours.

Visitors to zoological facilities also benefit when they see programmes displaying animals empowered to use their senses and adaptations to earn a living in similar ways as their wild counterparts. By demonstrating these species-appropriate behaviours in exhibits, trainers help people learn more about the species and its relationship with the natural world. Through demonstrating behaviours (instead of lecturing about behaviours), keepers improve their ability to inspire caring and conservation action in guests viewing public programmes.

Even though the field of animal training has greatly improved, there is still much room for growth. It is helpful for zoological supervisors, managers, veterinarians, and directors to improve their knowledge of the science of behaviour change principles and gain insights into how to tell an average trainer from an expert, or artistic trainer. A deeper understanding of the skills seen in high-performing trainers will help managers empower their staff and provide better welfare for the animals in their care. Animal welfare in zoological facilities is directly related to the training competencies of animal caregivers.

References

Bardwick, J.M. (1995). *Danger in the Comfort Zone.* New York: AMACOM (American Management Association).

Chance, P. (2014). *Learning and Behavior.* Belmont, CA: Wadsworth.

Dorey, N.R. and Cox, D.J. (2018). Function matters: a review of terminological differences in applied and basic clicker training research. *PeerJ* 6: e5621. https://doi.org/10.7717/peerj.5621.

Friedman, S.G. (2005). He said, she said, science says. *The APDT Chronicle of the Dog* (Nov/Dec, Vol XIII, No. 6), pp. 19–26.

Hall, E.T. (1966). *The Hidden Dimension.* Garden City, N.Y.: Doubleday.

Mace, F.C., Hock, M.L., Lalli, J.S. et al. (1988). Behavioral momentum in the treatment of noncompliance. *Journal of Applied Behavior Analysis* 21 (2): 123–141. https://doi.org/10.1901/jaba.1988.21-123.

Martin, S. and Friedman, S.G. (2011). Blazing clickers. Denver, CO: Animal Behavior Management Alliance Conference.

Pierce, D. and Cheney, C.P. (2013). *Behavior Analysis and Learning.* New York, NY: Psychology Press.

8

Integrating Training into Animal Husbandry

Marty Sevenich-MacPhee

This chapter will discuss how to integrate training into your animal husbandry programme and will include some practical considerations for this process. I say integrate, but perhaps a better term is recognition, as training is happening all the time. Animal learning, in fact, is impossible to prevent. A responsible zoo professional not only recognises this feature of the animals in the collection but plans and properly directs this learning according to each animal's need.

8.1 Husbandry Buy-in

When interacting with a living being, whether it is an animal or another person, you have to make yourself open to, flexible with, and able to react to the current circumstance and environment in which you are placed. You need the ability to take in information and respond to questions posed to you. With that in mind, I ask you this question:

Would you rather eat oatmeal with toe nails in it, or eat a hair sandwich?

This was the crazy question my kids posed to me when I picked them up from school one day. They burst into the car, breathless with excitement, and awaited my answer.

I am not sure how you would respond to this question, but my answer was 'neither'.

When told I had to choose, I felt backed into a corner and started to negotiate my options. *How many toe nails? How much hair?* It turns out that they were playing a game, which poses choices, called 'Would You Rather'. You may be asking yourself what this story has to do with integrating training into animal husbandry. Well, it seems that many times the opportunity to train option is put to a team like an unsavoury choice in the 'Would You Rather' game. 'We are going to train our animals' is so often sprung onto a team with breathless excitement and with the expectation that everyone will magically want to come along with the vision or risk not being part of the team. Instead, you will probably be met with a variety of reactions when your new programme concept is launched with your team because people inevitably have varying levels of experience with training, comfort with programme change, and even levels of trust with their peers or management. When hearing about new programme expectations, some people may feel backed into a corner, unsure, uninformed, or fearful and, in turn, may try to shut down the change or negotiate its terms. On the other hand, some people may be excited to participate, but unsure of what it means for them, and some may be gung-ho. Some may want to participate but feel that training is an individual venture and will want to participate on their own terms. Regardless, lack of clarity within the team from the inception of the

Zoo Animal Learning and Training, First Edition. Edited by Vicky A. Melfi, Nicole R. Dorey, and Samantha J. Ward.
© 2020 John Wiley & Sons Ltd. Published 2020 by John Wiley & Sons Ltd.

training programme and with regard to the direction of the proposal can lead to an unsuccessful programme for both the people involved and/or the targeted animals.

When approaching the training of a behaviour, we meet animals where they are and approximate them forward to reach the target behaviour. This same approach can be effective with people, as well: meet people where they are and approximate the team forward to the goal. In many cases, the success of a training programme hinges more on the buy-in of the people than on the participation of animals. Demanding a willingness to participate wholeheartedly as a team can sometimes feel like the equivalent of asking someone to take a bite out of that hair sandwich. One solution to remedy such a feeling for the entire team is addressed in Simon Sinek's book *Start with Why* (2009). Here, Sinek embraces the philosophy that humans are inspired by why they do things, not how or what. This applies to animal husbandry professionals as their buy-in and inspiration is directly connected to the care and welfare of their animals. When introducing new training goals, by employing the 'why' they are being instituted as opposed to just the 'how' they will be done, or even the 'what' keepers will have to do, a leader can powerfully imbue each keeper with the purpose of understanding. The power of starting with your 'why' can be the guiding principle to the training practices you utilise, the goals you adopt, hiring and training the team to implement your programme, and as a bar for ongoing assessment.

8.2 Starting Assessment

Developing an action plan and approach requires an honest look at your team, your animals, your facility, your resources, and the vision for your programme. Throughout this book you will find guidance on assessing various aspects of animal learning and training, but this chapter will focus on training the

team who will implement the activities required to make the programme successful.

A gap assessment, in its most simple form, is the process of identifying your current situation and comparing it to your desired situation (your vision). The gap that exists between the two is what you need to fill with new staff, new skill development, focused staff time, facility modifications, and whatever else is necessary to meet the vision. Because training programmes involve living beings, they must be dynamic and flexible.

8.2.1 Assessing Your Staff

When assessing your staff on their current skills, you should take into account their individual skills about animal training. You must take into account a keeper's ability to communicate and share ideas, as well as a keeper's ability to approximate a behaviour. You may also want to consider performing a team assessment, wherein you will review the team dynamics, formal/informal leaders in the group, the team's openness to change, consistency in training methods both individually and in groups of staff members, and success over their career with assessing and addressing animal welfare using innovative techniques.

The two most challenging individuals to lead in a team, in my opinion, are those who do not share the same programme vision and those who are not as skilled as they believe themselves to be. If you are dealing with the former, the answer is simple. The offending keeper will be an ongoing drain to the effectiveness of your team and should be guided to an opportunity elsewhere, more fitting of their interests. In the case of the latter, it can be very difficult for these individuals, who deem themselves highly skilled, to see areas of opportunity to improve themselves or to hear suggestions from someone else. Creating opportunities for self-assessment, reviewing training resources as a team, and videotaping sessions can be useful to guiding these individuals to a truer sense of their ability. Some individuals, however, can be so committed to the story of their own greatness that even with overwhelming evidence of their lack of progress or videos of their lacklustre performance in training sessions, they will find excuses for their poor performance. The hardest thing is being honest and upfront at all times with 'difficult' individuals. When leading a team, it is important that your expectations, in regard to developing skills, are clearly communicated, even if some individuals may continually argue that they already possess these skills. It is important when leading a team that ample opportunities to practice and focus on skill sets that have clear, measurable outcomes are provided – especially where skills are lacking. In some cases, having 'difficult' individuals attend external training courses that have a practical hands-on component taught by someone they respect may help them improve.

In addition to an assessment of team members (Figure 8.1) and how the team operates as a whole, it is also important to look ahead to the future potential of the team. If you have a history of leading a team, you may be able to accurately predict how well they will pick up new skill sets or work together as a team on the desired task. With this assessment, you can decide if adjustments should be made to your vision or timeline. For example, let's say you want to train a family group of western lowland gorillas (*Gorilla gorilla*) to accept a cardiac ultrasound and they are currently unwilling to separate from one another in order to be trained. According to the vet, ultrasounds with the two adults in the family group are needed within the next four months to monitor the new medication they have been given.

To start, you can use information gained during staff gap assessments, both individual and team, to help determine how long a training programme might take that would result in successful cardiac ultrasounds of each gorilla. If you discover that the staff have a non-receptive attitude towards the proposed training programme and that individuals are currently unskilled, you may predict it could take six to eight months to realistically achieve the training goal. If you also look at the team's future potential and discover that they could be open to this training vision with sufficient training, and/or

Figure 8.1 Keeper observing another team member training an Asian elephant *Elephas maximus*. Source: Denver Zoo.

that individuals in the team have a tremendous natural talent for this type of training, you may be able to reduce the previous six to eight months' timeframe to three to six months, or less. This type of gap assessment can help you effectively manage the expectations of all involved in a project.

Undertaking a gap assessment to understand skills and attitudes is a first step in 'meeting your team where they are' and is necessary in moving them to where you'd like them to be. Your team may not agree on what specific training to integrate into your animal husbandry, but with both individual and group assessments you can help achieve the vision with a strategic initial approach to the team. The outcomes from these types of gap assessments can help drive key decisions, such as what teams or team members are approached first, or used at all, and the training methods which should be adopted.

8.2.2 Creating Successful Teams

Creating an environment that is conducive to change, and in which people feel confident that the training vision can be achieved, is fundamental when integrating training into animal husbandry. Covey (2004) discusses ways to sharpen the leadership skills needed to create a dynamic environment, separated into seven habits. Habit five is 'seek first to understand, then to be understood'. This is a valuable 'habit' to keep in mind when initially approaching your team regarding any significant changes, whether training or related to other activities. It is tempting to share all of your big ideas in full detail; however, heed the advice to suspend your own perspectives and opinions when initially talking with others and don't feel compelled to have all the answers right away. Although you will want to clearly articulate the training vision of the programme, presenting fully-formed ideas or opinions may actually hinder your ability to be open to the thoughts and ideas from others and/or an open exchange of ideas between others if they feel that a singular direction has already been decided. Instead, during this initial phase, engage your team and get them excited about the *idea*. You need them to feel that they are a part of the decision-making process and can shape the idea, as they are the ones who will be implementing it. Make sure your training vision is compelling and share it with the team, so they can envision what the future could look like, without the detail that you may have in your mind. The formula here is simple: lead with Sinek's 'why', and then listen with Covey's fifth habit.

Once you've clearly articulated the vision of the programme, let your team enlighten you. Whether your ideas become the construct you build your work around is irrelevant, as either way, buy-in goes up.

Soliciting honest thoughts and opinions may not be as easy as just simply asking for them. The level of trust in a team and the personality types of the team members will alter their willingness to share what they are thinking. For teams that may be experiencing low trust within the group, you may need to spend time with individual members, as opposed to organising group forums, to get the information needed. You should also be prepared for the implementation of your project to take longer than it probably would with a team experiencing high trust.

A high trust team may embrace a new training initiative quickly and move forward with it more rapidly than those in a low trust team. I have observed teams on both ends of this trust spectrum who had drastically different training experiences and results. One team, who was working with plains zebras *(Equus quagga)*, had low trust amongst them, which caused deliberation over every decision. They would second-guess each other and had no consistency in their training methods because they had difficulty sticking to the decisions that they *did* make. The zebras' training went very slowly and was frustrating for the team because they were putting a lot of time into their sessions. By contrast, another team working with the red river hog *(Potmochoerus porcus)* were a high trust team that could make decisions and stick to them. They trusted each other so much that they supported decisions that were made, even when the whole team was not present. They valued consistency in the actions delivered to the animals above their own individual desires. This team and the red river hog were able to make quick progress with their training programme and were always setting new goals. Not only did the high trust enable this team to meet their goals, they were able to tackle any challenge that came up that would have crippled a low trust team.

You may also find that some members of your team are introverts and may have different wants/needs for sharing information than other team members, as teams are often dominated by extroverts. It may level the playing field if you send out questions for the team to ponder ahead of time, allow members to share their ideas before a meeting on note cards or discuss ideas in small groups before a larger team meeting. These techniques may provide a platform for all team members to feel they have a voice. Cain (2012) discusses introversion personality preferences. It is an excellent book that provides insight that may allow you to tap into your team's problem solving and creative potential in new ways.

The point is, when seeking everyone's opinion and buy-in, you may need to adjust your leadership style to meet the needs of your individual team members and listen to their ideas. If you take a 'I'm the leader and we're doing it my way' approach with your team, they may make an effort to appear engaged with the training programme in your presence, but it will not likely 'stick' when you are not there. You want everyone on the team to feel that they have a vested interest in the success of the training programme and the only way to do that is to have them feel the sense of ownership that comes with helping to build the programme from the ground up.

8.2.3 Moving Forward

Based on the team's initial feedback regarding your training vision for the training programme, you will need to develop a measurable plan for moving forward. These plans should include very clearly identified expectations of who will do what by when. Your initial plan may be focused on laying the groundwork for your programme, and may include elements such as reviewing records, observing current behaviour, reviewing relevant literature on such things as the species that will be trained, facility design, and/or training techniques.

8.2.4 Facing Resistance

Resistance to embracing a training programme can stem from a variety of sources, including considering the act of training as being manipulative and changing an animal's wild state, thinking that training creates a situation of poor animal welfare due to food deprivation or use of aversive techniques, or that training takes up too much time in the day. When encountered with this resistance, do not react to these statements of concern in haste or with closed mindedness, even if you strongly believe a counter-perspective. You must remember that everyone has different life experiences that they bring to the table. One person's definition and history of training may be very different from your own. Engaging in discussion about these fears and what is intended in developing a training programme and how success will be measured can be helpful, as it can alleviate some fears and will allow people to talk openly about them.

Discussing the potential positive animal welfare implications of training can be an asset to the development of a training programme. Understanding how animals' behaviour and the team's behaviour can be shaped can open people's minds to understand that animals in our care are always learning and responding to their environment, regardless of whether you have a formal training programme in place or not. As a leader, it is your job to create environments for both the animals and the team to be successful. For some people who are resistant to formal training, recognising all of the training they are currently doing, but not labelled as training, may help them embrace a more structured approach. They may then see that a more structured programme is not such a big change to current practices.

8.2.5 Awareness Is Key

If successful in overcoming resistance, team members might begin to see how the training can be a tool that could help to achieve desired behaviours and address undesirable animal behaviour they may be experiencing within their collection. Animals are always learning at times their learning opportunities are directed by those that care for them however, animals are also able to learn through other interfaces in their environment, such as food supply, social groupings, guests, etc. For example, a flock of wild turkeys (*Meleagris gallopavo*) had become 'attack' turkeys when keepers entered the exhibit yard to fill enrichment devices with food for other animals sharing the exhibit. After a problem-solving session and a complete review of animal training records, the keepers were able to decipher that they had inadvertently trained this 'attack' behaviour. Their problem-solving revealed that the food canisters the keepers initially used for enrichment had leaked small amounts of pellets when they walked across the yard housing the turkeys. The turkeys ate the pellets, associated the pellets with keepers, and then started to follow the keepers. The canisters were changed to ones that did not leak pellets, but the turkeys continued to follow the keepers, still expecting a snack. When no snack appeared, slower keepers were pecked on the legs, causing them to jump, and pellets were eventually released from jostled canisters, feeding (and, therefore, rewarding) the turkeys for pecking the keepers' legs in the first place. Eventually, the turkeys would hunt down keepers who entered their yard and viciously peck their legs without mercy, resulting in the keepers perceiving them to be 'attack' turkeys. Understanding this sit-

uation took some time, given the number of different keepers who walked through the yard and how consistently the turkeys were able to get their desired outcome (food for pecking). In this case, the keepers trained the turkeys to peck and the turkeys trained the keepers to dance. After the problem-solving exercise, the keepers were able to remedy the situation by training the turkeys an incompatible behaviour (stationing at a feeder device that the turkeys could peck for pellets) when they entered the yard and they solved the problem. Eventually, the keepers were able to enter the yard without incident. This turkey example appears obvious to those of us with the advantage of hindsight provided after a problem-solving exercise and when it is presented clearly in written form. There are, however, numerous examples of undesirable behaviours that are inadvertently shaped. Typically, when the behaviours are undesirable, the animals are given some sort of label and consequently dealt with as animals with a 'behaviour problem'. Animals are always learning and both desirable and undesirable behaviours are constantly shaped by the husbandry practices we currently use. When you have the responsibility to care for an animal, you need to take personal responsibility for the animal's behaviour and know that you are training behaviours, whether you call yourself a trainer or not.

On the subject of problem-solving, it is important to mention that in a programme where training is integrated into husbandry practices, that training is one tool of many. When problem-solving sessions occur, training may or may not be a useful part of a solution. Teams should always approach their problem solving with a very wide scope and be open to multifaceted solutions for a best result.

8.2.6 Welfare Questions

When discussions regarding the animals' welfare arise due to issues with food quantities offered or training methods employed, realise that these discussions can become highly emotional. Don't shy away from these discussions just because they may be a little uncomfortable. You need to give people the space to express their feelings and concerns, even if they are not in alignment with your knowledge and/or experience. It is important to keep these discussions out in the open so that you know how a team is feeling about the training. These conversations keep the focus on the animal's welfare, not to mention if you don't allow these discussions to occur, they will simply be driven 'underground' and could undermine your programme later.

8.2.7 Time Commitment

Undoubtedly, you have noticed many of the aforementioned suggestions for effective programmes are time-consuming endeavours and, of course, the use of time should be guarded because it is one of the most precious commodities that we have. That said, time spent in strategic development up front can avoid troubling time commitments on the back end. Time is one of the top reasons why there may be push back against embracing a training programme from either senior management or the people working directly with the animals. Training does take time – there is no way around that. You need to consider time realistically and know that moving people effectively through a change process is going to take a lot of time. In most cases, the initial time invested will ultimately *save* time and other resources in the long run.

Training animals can take a variable amount of time, depending on both the individual animal and staff involved. With this initial investment of time can come a large payoff in the future daily husbandry, non-routine husbandry, medical procedures, your ability to train individual animals for additional behaviours, and emergency response. For example, a large group of cownose rays (*Rhinoptera bonasus*) living in a very large multi-species aquarium were trained to swim up to and across a platform of lattice work that was just below the surface of the water for their regular feed. This one trained behaviour enabled aquarists to get a close look at the animals on a regular basis. The platform could be easily raised up to the surface to gain full access to the animals for physical examinations and other procedures, when necessary. After a procedure, the rays were still very willing to swim back onto the platform for their food. Training this behaviour opened up the opportunity for a level of husbandry that was not possible prior to investing in training.

In a different example, initial training progress was slow, but after a simple change, rapid progress was made. In this situation, the training goal was to shift a very large American crocodile (*Crocodylus acutus*) from a pool, across a dry corridor exhibit space, and into a back area. The constraint for this team was that they had to do their training sessions in the morning before guests arrived at the park. After a month, with twice-weekly sessions, the crocodile was making little progress and the length of training sessions were long, waiting for the animal to respond. When the time of day restriction was lifted, and sessions were moved to afternoons when the sun had warmed up the animal, very rapid progress was made at every session. Without the time of day constraint, the team was able to take advantage of the natural history of this species and training sessions became more efficient.

In summary, your training programme and vision will need to be customised to your situation and prioritised by the animal's needs. You may not be able to do everything you and your team would like to at once. Always keep in mind how much capacity people have in their daily workload for any additional training requirements, as well as how much change the individual team members can handle at any given time. Slow, consistent progress is better than pushing for large, initial progress only to watch it degrade over time because it never took hold in the first place.

8.3 Identifying the Philosophy and Expectations for Your Team

8.3.1 Philosophy

It was previously mentioned that we may all have different definitions of training and ideas of what methods are appropriate to use when training animals. It is best practice to discuss and write down specifically what is appropriate for your team and the animals in your care. This document will become your training philosophy and should be considered a 'living' document that can be updated and changed as needed, continuously compared to actual practice, used for discussion and selection of training methods and act as a reference for staff training. This also means that all expectations should be included in this document. For example, if your aim is to integrate training into daily husbandry for the entire collection and not just some selected species, this should also be included into the philosophy. If the foremost focus of your training method is to be positive, then this should also be clearly spelled out in the document.

Discussing the importance of the document and its contents as a key element of the success of the training programme and the welfare of the collection is imperative. In daily practice, the members of the team will many times be making choices alone about how they treat the animals in their care. Your philosophy document will only provide guidance as to how to best make and communicate these choices. Everyone may not agree with the choices required to adhere to the philosophy document and training programme, and this is to be expected. Your challenge as their leader is to lead in a way that the team feels comfortable challenging the training programme with their input at meetings, but always providing consistency of care to the animals that adheres to the group's commitments to the training programme, even when commitments may veer from personally-held views.

The success of your programme is ultimately a result of how individuals behave when no one is watching. The commitment that your team members have to the team's philosophy needs to be deeper than a generic 'head nod' at a meeting: it must be something that each person truly feels a part of and believes in to maintain commitment and consistency. Some members of your team may slip into using methods from their past, as they are more comfortable and/or confident using them. For example, if the philosophy aims to adopt positive training methods, but an individual trainer reverts to using an aversive stimulus to motivate a behaviour instead, the individual's choice may seem only mildly out of line (change in their body position) but could erode trust with the animal and the team and could completely alter the course of the training programme.

Your philosophy document should also cover who is responsible for training the animals and a timeline. The document should answer questions, such as:

- What roles will the various staff members play in the training programme (Figure 8.2)?
- What level of mastery in knowledge of animal natural history/behaviour, knowledge/skills associated with training theory, and understanding of safety expectations are needed in order to participate?
- Who will train staff *how* to train?

When introducing the need for this sort of documentation, focus on how it is a tool for communication for all members of the team, including yourself. Having a written philosophy document that outlines training expectations provides the team with a tool to aid in their understanding of what they should expect from themselves and each other.

8.3.2 Expectations

Every person in a team should understand what their role is and what they are expected to do with regards to the new training programme. Discussing or providing a written explanation of how training fits into individual job descriptions or daily schedules can provide a clear understanding of expectations. In

Figure 8.2 Team of zoo professionals working together with the Asian elephant *Elephas maximus* trained to voluntarily accept foot laser therapy. Source: Denver Zoo.

addition to ongoing coaching and feedback, providing specific and measurable written goals for the coming year and feedback on past performance at employee reviews should be common practice. Expectations can include specifics, such as: 'Bill will: shift all hoofstock over the scale platform three times a week during the normal shifting routine; weigh and record all animal weights monthly; review progress at monthly team meetings' or 'Kelli will train hippo "Buttons" for a tusk trim to be completed by January 15, which requires that a training plan is submitted to the manager by October 1 for review, all session documentation and milestone updates will be entered into the daily report system. For the training goal to be considered complete, the hippo should be comfortable with Jennifer also able to successfully request/cue this behaviour and perform the trim'.

If a team or leader has never gone through the process of creating written expectations, they may be initially hesitant, feeling that the process represents micromanaging or an interfering boss. This process may be more difficult still if that person has been put in charge of running a training programme but

is not in a management position. Having clear expectations can ultimately be a relief for all involved. Providing expectations can help align a team, when, for example, some members want to spend all of their time training and interacting with the animals and other members do not want to spend *any* time training or interacting with the animals. Written expectations for realising your philosophy can also provide a big sense of accomplishment when goals are ultimately achieved.

Training philosophy documents and programmes will look different at different facilities, with different teams, and with different animal species. There is no single answer for how long a training session should be, or how many training sessions per day should take place. For example, a team working with a collection of primates, carnivores, and birds who have the facilities and access to the animals throughout the day may have extensive training goals outlined with multiple training opportunities provided throughout the week with individual animals. By contrast, another team may have multiple groups of hoofstock, kept on a savanna 'outdoor' exhibit all day, and are only available to be trained before being released from their barn in the morning or in the evening when they are brought back inside. Their training programme may have few training goals and longer timelines to achieve specific behaviours. A collection of crocodilians that live outdoors may be trained frequently during their feeding times when the weather is warmer, and then less frequently when the temperatures are cooler, and the animals' feeds are decreased. This could result in a cyclical training programme that changes according to season.

8.3.3 Written Goals

The desired goal for an integrated training programme is to view training as a part of husbandry for all animals in a collection. It must be clear to all team members what behaviours are going to be trained with regards to each animal. Having a written list of goal behaviours will provide a clear direction and expectation and serve as a communication tool to facilitate your training programme. When developing a list of behavioural goals for an individual animal, you should take into account behaviours needed for the animal's daily husbandry, non-routine husbandry, and medical care. These lists will need to be balanced with a realistic assessment of resources, human health and safety, animal safety, and a keen knowledge of the needs of the collection and the individual animals.

It may be helpful for you to use a list of questions to guide your team members' arrival at a set of goals (Mellen and MacPhee 2012). Some sample questions are as follows:

- What is the animal's normal daily routine?
- What are the routine husbandry procedures that are desirable for this animal to take part in?
- What procedures are necessary for an annual medical examination?
- What are some of the medical conditions common to this species that need to be monitored?

Typically, when initially generated by the team, these lists can be long and daunting. Prioritising the list can help reduce the number of behaviours that need to be trained to make it manageable for the team.

Warning: it can be tempting to move forward and start training without solidifying the desired goals with the team and agree on training methods, particularly for those who are excited about training. When this happens, behaviours may be trained that are not necessarily useful to animal husbandry and are, therefore, a waste of precious resources, including time, and does not contribute to the overall welfare of the animal. In a worst-case scenario, well-intentioned team members who train outside of a 'list' may be training behaviours that could actually end up being unsafe and/or undesirable to the overall training programme. For example, a keeper once trained a young male zebra to approach her whilst she was unprotected in a stall with the purpose of getting the zebra to tolerate tactile manipulation of its head, mane, and neck for husbandry purposes. Although this behaviour seemed beneficial in the short term, as the male zebra aged there were multiple safety issues with this animal approaching keepers whilst they were working in with this animal. As a consequence, this zebra was labelled as a nuisance or aggressive animal by other keepers – which might not be a reflection of his actual temperament – all because he was trained to approach people. With a list of behaviours which trainers can work from, investment can be planned to train appropriate behaviours identified for the individual animal that should result in both safe and beneficial long-term husbandry of the animal.

When identifying goal behaviours, it is important to think about not only the immediate impact these behaviours will have on the animal, but also how they will impact the animal in accordance with your long-term vision and the future of the training programme. In some cases, you may not have specific needs that must be met now, but you can train behaviours that provide a foundation for future needs that may arise. For example, training behaviours that require animals to move from one point to another, stationing, or the concept of targeting, can be useful first steps to many other behaviours you may want later.

When developing behaviour training goals at an institution-wide level for husbandry, particularly those that require veterinary involvement, it may be most efficient to start with a prioritisation process, asking questions like:

- What are the animals that we do not want to immobilise or restrain?
- What behaviours would be most valuable to train our collection animals?

Questions like these can provide some much-needed focus to a new training programme.

Remember that even on a large-scale institution level, integrating training into every animal's husbandry plan is of top consequence and a prioritised plan should be established for all animals in the collection.

8.4 Staff Training

Years ago, I attended an animal training presentation by Bob Bailey, CEO of Eclectic Science Productions. One of the things he said was a very profound and truthful statement: 'Training is simple, but not easy.' I have pondered that statement many times over, from a multitude of perspectives, and have found that it can especially ring true when contemplating designing a plan for training your staff. If you think about learning to drive a car, you generally learn the fundamentals and gather confidence by starting on the familiar, less-travelled neighbourhood roads. This experience feels nothing like being on an unfamiliar, busy, city expressway during a blizzard, but both activities are called driving. Driving can be simple, but it is not easy. How do you prepare zoo professionals with skills that cover the broad range of situations that they may encounter?

When I first started training, one of the first assignments I received was to train a particular behaviour to an Atlantic walrus (*Odobenus rosmarus*). I was able to work with my attention focused on one walrus and this particular animal picked up behaviours very quickly. Now compare this experience to another, during which I worked with a group of five dolphins at once and managed all of the complex social nuances that occurred simultaneously. Training the walrus was like learning to drive on the neighbourhood road. Training the more complex social group of dolphins was akin to eventually becoming skilled enough to drive in the blizzard.

Like any skill set one might master, animal trainers must practice, then practice more, and then practice even more. Your job as team leader is to find or create multiple opportunities for the initial phase of learning about animal training; as in the analogy above, create those 'neighbourhood streets' and then find or create additional opportunities within more challenging scenarios until the team feels confident navigating the 'expressway during the blizzard.'

8.4.1 Knowledge/Language

Having an extensive knowledge of the natural history of a species and the individual history of the animal that you'll be training is a critical first step to a zoo professional's training journey. Providing staff with access to all of the resources that they need to gain this knowledge, so that they become comfortable with their role and expectations, will be pivotal in their future success. In addition to reading information, it is important to take the time to observe the animal's behaviours in a variety of different situations and settings. These observations will help provide a broader view of the animal's behaviour and physical abilities.

A grounding in the fundamental principles of animal learning theory will help your team grasp the concepts of what they are or will be doing with their animals. You may select readings for them to do on their own or present the materials to them as a group. The discussion around the practical application of the theory, i.e. 'how to actually do it,' is important. Translating theory to real life situations will provide a deeper understanding of animal learning theory and increase the teams' retention of the information.

Developing a common language around the theory will ensure that everyone is on the same page during formal and informal discussions about animal training. You will want to create a dictionary or formal list of the terms, and their meaning(s), used by your team (see Glossary). Whilst there is wide variation in the preferred length of the terminology lists for animal training, there is no ideal length of list. The right list length is whatever will work for your culture, in your institution, coupled with the professional norms for the species that you work with. Established terminology enables accurate communication within the team, and with peers, about the training. If all of this can be accomplished, then the terminology list is appropriate.

Common terms for quick reference are best practice, but all leaders must be cautious of the inefficiency pitfalls that accompany copious or hazy team terms. Terms can become cumbersome and get in the way of a team's goals, when unspecified or unnecessary terms highjack a discussion of what's occurring in a programme and lead the team into a linguistic quagmire.

Methods that can be adopted to help train you and your team include observations, professional development, and practical application. If it can be arranged, observing someone else train and the discussions you have thereafter about timing, techniques, approximations to the target behaviour, etc. are quite valuable. Top athletes spend copious amounts of time reviewing and studying game films – both of their own teams and others – observing and learning. Borrowing a page out of the professional sports playbook may be in order to help your team reach their highest potential. Make sure your team is knowledgeable on the latest research and publications on animal training and encourage

them to keep in contact with peers at multiple institutions with whom they can discuss animal training. Give your team the opportunity to train animals or species they don't normally have the opportunity to work with. Of course, all of these activities should be done only after consideration of human and animal safety and the trained animal's welfare. All activities will need forethought and planning in order to be successful.

8.4.2 Observation

Training is a skill of observation and timing. To train well, you need to know what behaviour you want the animal to perform and respond quickly to the animal when it performs this behaviour appropriately. Before training an animal, new zoo professionals should work on their ability to observe and identify behaviour of animals and the timing of their responses to these animals by observing staff who are more seasoned and skilled in animal training. In addition to observing staff, team members should also observe the specific animals they will be training to see how they move and what motivates their behaviour.

Observing others train is important in the learning process. Everything about observing a training session can be useful, including where and how to stand and position your body, how to interpret the animals' behaviour, what safety precautions to use, timing, reinforcement delivery, what behaviour is being reinforced, how to begin and end a training session, how to use the various training tools available, and anything else you see or ask about after the session. If the person training is experienced and would not be distracted, and if it is safe to do so, the trainer may be able to talk and explain what they are doing during the training session.

When a new trainer begins training an animal for the first time, it is helpful to first cue behaviours that have been previously trained by others. This provides an opportunity to work on safety awareness, observation skills, timing, food delivery, and tool coordination.

When confidence and training skills have developed with existing trained behaviours, you can move on to training new behaviours.

It is best to select behaviours for a new person to train that will set them up for success. Quick wins tend to build confidence for more difficult situations that might arise later. Ideally, you would have a new team member write a training plan prior to their first training session that would identify what behaviour they were aiming to reinforce during the first and subsequent sessions. This would allow the new team member to visualise the desired behaviour and help their timing during the session. The training plan can also help identify areas for coaching staff throughout the training process.

8.4.3 Coaching Trainers

Many people do not like to be watched and coached when training animals, especially when they are new, and it may take some time for everyone to get comfortable. This adjustment period is fine and is a part of the process. A suggestion that may help people get comfortable quickly is to talk about the coaching process ahead of time. A discussion about learning styles and expectations can help put people at ease. Setting up a consistent coaching format is helpful, as well. For example, it may be decided that before each session you will have a brief chat about what behaviour will be visualised, which is then reinforced. You may also discuss what will happen if the coach sees something unsafe or feels they have a suggestion that is critical to the session. After the session, the new team member can begin with a debrief, talking about what they thought went well and what they would like to have done differently. The coach can then follow up with what they thought went well, questions, and what different choices could have been made. Allowing the new team member to share first allows them to self-report what they would do differently, which can alleviate some of the potential embarrassment of having someone else list known challenges.

At first, this conversation may feel stiff, but after some time it can become part of the culture. These pre- and post-training session chats should be kept brief, but, if time allows, additional team members can be included for input. This level of communication may be difficult to persuade some teams to commit to, but if the group can stick with it, it shortly becomes 'just how we do things around here' and people will seek out multiple perspectives and input, if it is not first volunteered.

8.4.4 Consistency for People and Animals

Assigning a single trainer to shape a behaviour provides the best consistency; however, animals can learn more than one behaviour at a time from different trainers. Close communication between team members needs to occur to assure confusion is not occurring.

At times it may, be necessary for a behaviour to be shaped by more than one trainer. At other times, multiple team members may be working with a group of animals. Although multiple people and multiple animals may be involved, everyone should be following one action plan. In these cases, again, communication is paramount to provide consistency to animals in the learning process.

Consistency is a cornerstone in a training programme. Consistent communication and feedback does take time and can feel daunting at first, but is well worth the effort and will make a difference in the training programme. In addition to pre- and post-training session meetings, having a written list of cues, behaviour criteria, and protocols for all to follow will also help with consistency. Furthermore, photos or video which support the written documents can improve the team's visualisation of your training philosophy expectations. This sort of documentation is invaluable for maintaining behaviour integrity, as well as existing as a tool for training new team members.

There may be some training programmes where everyone is starting fresh: the animals are all untrained, the individuals are all new to the training process or methods chosen and those in the lead may feel they have minimal experience. Depending on the goals of the training programme, the team may have to look to an outside consultant for assistance in getting started down the right path. In other situations, the team may feel that the goals of their training programme are within their reach and their need will be to maintain a focus on skill building and making progress. In yet other situations, a leader may be in a position where they have less training experience than someone on their team. This should not minimise the leader's ability to set goals, observe sessions, seek consistency within the team, and maintain programme focus and progress. No one should be expected to have all of the answers. It is better to have a leader who is secure in what they do know and who seeks proper input to make decisions than someone who makes decisions without the information that is needed out of fear of being 'exposed'.

8.4.5 Sharpening Skills

Gladwell (2008) studied the concept of being an expert in a field of discipline. There is a repeated theme about building expertise suggesting that to master any skill, you need to practice – and not just a little bit. To truly reach a level of mastery in any skill, it takes approximately 10,000 hours of focused, purposeful practice (Gladwell 2008). Likewise, building expertise in the field of animal training takes many hours of focused and purposeful practice, both for you as a leader and for your team. Just as athletes perform drills to develop specific muscle groups or broaden specific nuanced abilities within a sport, you will need to focus efforts on specific areas within animal training. For example, if you or your team have experience with an event marker (e.g. whistle), time needs to be dedicated to building skills with other bridging stimuli, such as event markers, verbal markers, tactile markers, and/or others that may be appropriate for the animals you work with – and not just for a few sessions, but for

an extended period of time until use of multiple bridging stimuli can be achieved with comfort and confidence.

Regardless of the training methods used, all skills needed must be practiced out of the presence of the animal: it *will* make a difference. For example, if the skill is the precision of throwing a meatball a distance, your team members should practice the ability to throw it into a small bucket from a number of paces away and in a certain window of time. Such consistent accuracy is important to the overall training of your animals. For example, a team member who erroneously throws a meatball inaccurately will send their animal in pursuit of the errant meatball. In a social animal setting, this error can cause chaos as it may draw the attention of other animals in the area who may go after the meatball, as well. To further exemplify, a specific group of tigers (*Panthera tigris*) needed to station individually in areas of their exhibit in order to safely allow remote control doors to close. The only access to these cats was to throw them meatballs from a tower high above the exhibit. It was very important that the handlers have great throwing accuracy to reinforce the tigers' behaviour for being in their proper position. An errant meatball could cause a fight between tigers or lure them away from their safe position in the enclosure when the remote doors were moved. The zoo professionals in this example took their training seriously and had pride in their good aim, resulting in precision meatball throws.

Training your brain to see the behaviour you or others are training and using a bridging stimulus correctly and at the right time can be practiced, even outside of a training session. You can be creative and discover your own games to work on timing, such as using children's toys that light up in various colours, show displays, or play music at random. You can mark an aspect of a toy's repertoire. You can mark actions or phrases that are repetitive on TV shows. Playing the training game, by communicating only with the event markers (e.g. whistles or clickers), can be great practice using people on your team and great fun, as well. You can also train your pets at home. Every attempt should be made to practice these bridging skills before working with animals. Creating games that are non-intimidating and light-hearted can keep the learning atmosphere welcoming and enjoyable and minimises the fear of appearing incompetent or foolish. Providing ideas for people to do activities at home on their own or to share their own skill-building games can give them more control over their learning and give anyone particularly worried about learning in front of others a risk-free opportunity to practice ahead of time. These games are useful no matter how experienced someone is and can be good to refresh skills as a team.

In addition to practicing the timing of a marker, getting experience with other tools used in training is helpful. If pouches are to be worn or props manipulated, such as a meat stick, targets, food buckets, a chute wall, crate doors and/or any other props, experience should be gained with them all out of the presence of the animal first. The time to figure out the locking mechanism on the door, the sound the wall makes when you slide it and how to hold on to the target is not with the animal at your side.

Another example of building skills is to extinguish an animal's trained behaviour and then retrain it, or to change a cue for a behaviour. Spending time practising any of these skills, and especially nuances that you or your team may not have had exposure to, or have simply never tried, is important.

It is possible to train animals for years and never actually grow in terms of your skillset. This reality is especially true if you do not continually challenge yourself. Sometimes it seems like an animal reaches a plateau during a training programme where it no longer appears to learn. This lack of continual growth may be due to the training programme or to the person training having become 'stale', no longer interesting, or engaging. Animal training programmes filled with such 'stale' team members can arrive at this point for many reasons: a lack of clear, compelling,

or new goals, unrealistic time constraints, fear of change, lack of creativity, and many additional highly understandable reasons. No matter the cause, it will take effort to keep your animal training programme moving in a positive direction. Here are some ideas how you might do this:

- If time is limited, only make small programme changes, such as changing one cue for a behaviour or practise using a new bridging stimulus. Find brief opportunities during the day to review a short training paper as a team or a chapter in a book.
- Get the team engaged in a regular review of all training goals, assessing progress, and new goals or variations of behaviours that could be added or take the place of existing goals.
- Get the team engaged in self assessments, where they are introspective and come up with ways they can improve on their personal performance. Individuals can bring books or DVD's back to the team that they have reviewed and found helpful.
- Sharing progress or roadblocks with training projects at team meetings is a way to get input and inspire each other. This will get team members to use different techniques that have been successful for others on their team. Encouraging team problem-solving can also use the larger brain trust of the team. Also, consider bringing in people from outside the team who can provide a fresh perspective.
- Bring in a guest speaker or training consultant for a presentation or review of your animal training programme. This can create stimulating conversation and discussions of currently held beliefs. It is important that this is done in an animal training programme that is stable, with a leader that is secure enough to handle this level of input and can facilitate positive discussion.

The main goal of 'sharpening skills' is to keep your team in a highly engaged, adaptable state that encourages growth and helps to maintain a desire to learn and take part in the training programme. Under these conditions, your team will not only grow in sensitivity and learn the arts of how to approach an animal training session, how to shape behaviour, and even recognise the animals' smallest efforts to behave in the desired direction, but they will also learn to recognise precursors to undesired behaviours and how to end a training session at the right time before it has gone on for too long.

8.5 Active Training

Frequency of training and animal participation can be present in a programme, and yet due to the lack of rigorous mental stimulation on the part of trainer and/or trained animal, the training performance might not reflect the operational needs of the programme. This type of approach can result in an ineffective programme for both trainer and animal.

This outcome can occur when no new behaviours are being trained and team members are repetitively requesting that animals only perform previously trained behaviours. It can feel rewarding to the team to conduct training sessions where animals know all of the 'right' responses to every request but the feeling can be misleading when an animal's ability to inculcate new trained behaviours deteriorates from lack of practice.

Another situation where a training programme might become complacent and thereby inefficient is when there is a degradation in the team members' expectation of what the animal's behaviour should look like. The result is a behaviour that does not serve the purpose for which it was intended. For example, initially an animal may have been trained to hold a specific body part in a specific position for a specific period of time. Over time, the period of time which the animal is expected to 'hold' the behaviour becomes shorter, simply out of expediency on behalf of the team. The result is that the animal now performs a behaviour which no longer allows for the full examination of that body part, which was the intention of the trained behaviour in the first place.

In order to avoid such unhappy outcomes, a programme leader should ask their team to remain intentional during each training session and reinforce the approach by encouraging team members to have goals for 'routine' daily training sessions. This should be common practice even when new approximations for new behaviours are not necessarily being trained. An example of how your team might do this is to create husbandry cards that are selected a few times a month from a deck. Cards could say things like "the monkey has a small cut between the toes on the left foot", "the animal needs to be held for close inspection", "the animal needs drainage from ear", or "a sample needs to be collected". The cards may represent old situations that have come up in the past, or may be creative, representing ideas the team has come up with of possible future situations. Results from these training sessions can then be shared at team meetings. This system will challenge the team to be able to work very specifically with an animal in order to gain behaviours which may be needed in the future.

If you notice sessions are filled with multiple requests for a previously trained behaviour, this could indicate that new behaviours have not been identified for training, a team member does not have the experience to train new behaviours, is fearful about how to approach training, or may not be aware of how they are spending time during a training session. If any of these are the case, it is your responsibility as the team leader, to build confidence and skills in your team by giving them the opportunity to practice, which will build confidence over time, just as you would with a new team member. It may also be helpful for you to model the best way to use the time set aside for a training session. For example, working on new approximations and more challenging training in the early part of the session, when the animal's attention may be at its highest may produce quicker, more successful results and saving the previously learned or high confidence behaviours for later in the session, when the animal's attention span may be waning.

Every team member, animal, and situation is different, so you, the team leader, should work to fully understand why a programme might be going 'stale'. The important thing is to make sure that the time set aside for training is being used wisely. By making the effort to keep team members focused and intentional about their actions, a culture of ongoing learning with both the animal and the team will be promoted.

8.5.1 When Have You Completed Your Goal Behaviour?

This question seems like it should have an easy answer ... when the animal does the behaviour under stimulus control, when the animal shifts off exhibit when they hear a bell, when the blood sample can be voluntarily drawn or when the hoof can be voluntarily trimmed. The answer actually can be more complex. For the trained behaviour to have the maximum positive effect on the animal's welfare it needs to meet the operational or husbandry needs of that animal. This need should be fully defined at the onset. Teams must be able to anticipate changes in the ani-

mal or operational changes to the best of their ability so that approximations can be planned for and made.

For shifting, does the animal need to shift on/off exhibit any time during the day, for any of their keepers? Is there a social order that animals have to maintain during the shifting process? Do they need to come off exhibit immediately in the case of an emergency when there are distractions?

Does the husbandry or medical behaviour (Figure 8.3) need to be done any day of the week, from various locations, with a diverse set of veterinarians/technicians participating and requested by various keepers? Is there specialised equipment involved? Does the team and animals have easy access to the equipment for training? The expectations need to be defined and constraints identified.

You may find points in the approximation process where the animal's behaviours may plateau. This lack of progression can be at times when the trainer perceives that there may be some risk of regression in the behaviour being trained. In these case, considering taking steps such as introducing new equipment or including another keeper,

Figure 8.3 Medical training – African lion *Panthera leo* trained to voluntarily accept injections. Source: Denver Zoo.

veterinarian, or technician into the training process. All new training approximations can have some initial regression involved and should be considered part of the process. A leader needs to help their teams recognise, accept and push through this aspect of training to achieve their goals and complete the behaviour. Support can be given by looking ahead to future approximations, discussions in team meetings, and continual follow-up.

A phenomenon that may occur during the training process is some team members may feel a strong sense of ownership, of both the animal they are working with and their trained behaviours. This type of feeling can create significant dysfunction in a team if this team member is more focused on their own needs than that of the animal. One way to curb this behaviour is to ensure that the team continues to refocus the training programme goals on the needs of the animals, on how best they can work to serve the animal, and that what is best will promote good animal welfare. A team leader will be balancing the promotion of positive human/animal relationships without allowing unproductive boundaries to be crossed.

This continual refocus on the training programme goals can help ensure that, at the end of the day, animals needing medical treatment are trained the behaviours they need to be treated effectively and efficiently. If animals are in need of urgent medical care, there may not be time to 'complete behaviour' and behaviours that are in some state of regression might be called upon to ensure the animal gets the medical care it needs. If an animal is cued for a behaviour before training has been properly completed, or cued for a regressed behaviour, additional regression can occur, and it may be difficult to retrain the desired behaviour according to the training programme criterion for completion. Whether further regression occurs depends on the behaviour that is required to ensure medical care can be given and the level of trust between the team member and the animal, working to ensure the behaviour is performed. When immediate medical attention is required, the team must take a close look at an animal's history and assess the risk and benefits of breaking that trust by leveraging an uncompleted or regressed trained behaviour to achieve a medical need over restraint or immobilisation.

Transitioning animals to new facilities or to new keepers/aquarists can also be a challenging and stressful time for both the animals and the people involved. When moving or taking receipt of an animal, it is sensible to make sure you either provide or ask for records and details of the specificities around how an animal was trained. This type of information can assist the new team with retraining any behaviours that may have regressed as a result of the move. Any video of prior training sessions with that animal can also be extremely helpful to ensure that there is as minimal a disruption in animal (and staff) learning as possible. Providing as much information as possible about the animal and its historical experiences relating to training (familiar props, etc.) can be very helpful and serve to lessen the anxiety for all involved.

8.6 Facilities

Facility design should fit hand in hand with the goals for the husbandry programme, which can make a big difference in the ability to meet your training goals safely, efficiently, or even at all. For example, having large crocodilians swim through a waterway between enclosures may be a behaviour that takes far less time to train than trying to train them to walk across land between enclosures. A giraffe chute (restraint) that is level with the enclosure floor, rather than one which requires the reticulated giraffe (*Giraffa reticulata*) to step up into it, may be the difference for some animals moving into the chute calmly on a regular basis verses baulking daily at that step. It is becoming more common that when facilities are designed, animal training requirements are considered during the design process. When facility design is done well, animals move through the facility, through doors, into chutes and approach training panels with much less effort and stress and more safely for all involved. For many, animal facilities may need to be retrofitted to provide proper comfort and access for the animals. Depending on the species, individual temperament and a host of other factors, the design of your facility will dictate if your animal will be in a space that will facilitate training or hinder it. In addition to the facility layout, extra attention should be given to where teams place themselves, how they move, the noise they make, where they place their equipment, and other environmental stimuli they may use in the animal's world (facility). Observing the animal's behaviour throughout the day and in all different situations can allow a team to recognise behavioural signs of comfort; however, this does not suggest that animals should be trained in sterile environments, or that animals should not be desensitised and habituated to environmental stimuli. There are many animals that may easily tolerate a lot of motion, sound, smells, and novelty and remain calm and focused on training. It is necessary to be aware of individual animal's needs and the effect the surroundings can have on an animal's ability to learn.

Care should be taken when introducing anything new to an animal and within its environment (facility). Certain husbandry behaviours may require equipment for training and final implementation of the behaviour. Individual animals may have very different reactions to the same item. When introducing ultrasound gel to a group of gorillas, for example, one gorilla accepted it readily, one backed off, another inspected it closely for a long period before acceptance and yet another continually ate the gel when it was placed on the body. It is important to have a lengthy discussion with the veterinarian prior to training and get as much information as possible from them about what is needed and a description of the goal behaviours so that an appropriate training programme can be created for an animal. If it can be done, put yourself in the animal's position so you can experience the equipment and requirements of the goal behaviour, as this can be very useful. Crawling into chutes, under an X-ray, having an ultrasound, stepping onto the scale platform, etc., will give you more insight into the necessary goal behaviour and what will be expected of the training programme. Consider the following: do lights and shadows come into the chute? Is there a flashing light or buzz when the X-ray machine is turned on? How much pressure will the animal feel on its skin when the ultrasound is given (Figure 8.4)? Is the scale slippery? Does anything make a sound or wiggle? Can the end be pulled off of the target? Is it safe if the animal is able to touch the target? The animal should not be the first to learn the answers to these questions. Smells that are medicinal or from other animals on equipment can also be a distraction, so consider olfactory cues, as well. It is the responsibility of the team to know as much about the training requirements as possible before the animal is involved. The animal will be far more sensitive than you and your team, but the process is still helpful.

Researching facility designs that others have used is good practice. Having the designs from other institutions and their training experiences may save you time and money. Most individuals are very willing to share this information, and there is a lot of information available on facility design on the ZooLex website (www.zoolex.org). For chutes and crates, some animals may step more readily into an open mesh crate, whilst, for other animals, the opposite is true. Facility designs do not have to be very expensive to work well for animals and teams. The lattice work ramp for the cownose rays mentioned earlier was picked up at a local hardware store and cheap to purchase and install. In another example, a wooden training panel for hoofstock was made, which was very functional for the animals and the team referred to it as 'the lemonade stand'. That particular team liked the wooden design as much as their more expensively built-in chute. They needed both the 'lemonade stand' and the expensive chute for

Figure 8.4 Greater one-horned rhinoceros *Rhinoceros unicornis* trained to accept transrectal ultrasound for reproductive monitoring. *Source:* Denver Zoo.

Figure 8.5 Training of smaller species, keeper target training two Asian small-clawed otters *Aonyx cinereus* simultaneously. *Source:* Denver Zoo.

their programme but did not feel one was of lesser importance. There might also be creative ways to use the space that exists in your facility by opening up areas of a chain link fence and working through door panels, as long as it doesn't change the fence's integrity. There are also many other animal facility 'hacks' to facilitate animal training.

The smaller animals in our collections should not be forgotten (Figure 8.5). These animals can also benefit from designing a facility with consideration of the training that might be necessary, as well as their behavioural needs. It can be tempting to just hand capture, manipulate with body positioning, or simply ignore behavioural signs of distress performed by smaller animals in your collection, as it is often easier to do nothing rather than take the time to train behaviours. By adding more complexity to the animals' environment with visual barriers, or by training behaviours such as stationing, shifting, or crating we can improve the management and welfare of animals. An example of providing good conditions for a social group is illustrated in the management of a group of snowflake eels *(Echidna nebulosi)* in a shared small exhibit. One animal was growing a lot bigger than the rest of the group because it dominated the other eels during feeding times and ate larger amounts of food. The creative aquarium team made stations out of clear tubing set side by side for the eels to swim into and to feed, unrestrained. This provided the eels with equal opportunities to feed. The result was a balance in food intake and eventually equal sized eels sharing the enclosure.

No matter how beautiful or functional a new facility may appear, animals need to have ample time and appropriate introduction to their new surroundings to be successful. The goal is for the animals to be comfortable and open to new learning opportunities and training in these areas. Great care should be taken to assure that this is the outcome. Grandin et al. (1995) discusses the use of animal facilities and how impactful the animals' first introduction to it is on future learning/training. If you have a new animal arrive at your facility, restraining the animal in a chute in the barn as their first experience of their new facility may lead to long term and future issues with its comfort in the barn. Strategically planning how your animals will have positive interactions with all areas of your facility, especially when they first encounter them, will be beneficial for the future of your husbandry programme.

Ongoing assessment of how you are utilising your facilities is important. An area, such as a chute or crate, that may have been identified initially as a training space for accessing animals more safely may, by inadvertent changes in training techniques, begin to be mildly or fully restraining or used to 'trick' animals to achieve compliance with training goals. Although the team may feel the session is positive because they are providing food, they may unintentionally also have other motivators at play. An animal's participation in the training session is not entirely voluntary if they cannot move away or choose to not take part. It is wise for the team to fully recognise what is occurring during the training session and how they use the facility.

Overall, the facility design and use of the facility needs to create a comfortable, safe, and confident setting for both the team and the animals. No facility should be considered completely safe and teams always need to be well trained in all safety precautions (see Chapter 13). Team members should be trained to work within the facility and have reasonable abilities that they are confident in fulfilling. This will provide the ideal setting for making the best training decisions and for animal learning.

8.7 Relationships and Goals

The relationship and history that a team member has with an animal can be one of the biggest factors that will affect the speed of the shaping process or the outcome. Maintaining a trusting relationship should be a top priority when approaching every interaction. In some cases, there is a need to invest very little actual training time to achieve some husbandry goals. Due to such a positive relationship between trainer and animal, it might be possible for a keeper to touch and palpate a fresh wound or even give an injection to an animal they have worked with for a long time. For example, in one situation a Malayan tapir *(Tapirus indicus)*, after being scratched for a brief period, would lay down and remain immobile – whilst still being scratched – at the same time another keeper was able to draw a blood sample from its leg. These sorts of accomplishments are great examples of how keeper/animal relationships (see more in Chapter 9) definitely are a huge help to assisting the husbandry of an animal and should not be undervalued; however, because the performance of behaviour in this situation, and other similar situations, relies so heavily on the specific keeper/animal relationship and are not under stimulus control, they cannot be considered reliable or trained to the point where they are necessary for a husbandry programme. The keeper–animal relationship and behaviours which can be elicited as a consequence of it, can be used as a foundation for the husbandry programme. These behaviours can be retrained introducing formal cues associated with the behaviour, and other steps taken, to enable other team members the ability to cue behaviours successfully. A transition can be made to take a series of behaviours beneficial to animal husbandry, which could only be achieved by certain team members or only under certain narrow conditions, to behaviours that can be successfully and reliably cued by all necessary team members, and thus meet more of the animal's husbandry needs.

8.7.1 It Is Personal

Leading or being a participant in these training programmes usually means that some challenging discussions may be a part of a team's future. Husbandry training is a slice of animal care that some people tend to take very personally. It involves many elements that are equated with the welfare of our animals: feeding, close interaction, meeting medical needs, and other decisions about animals' daily management and comfort. For the training programme to be successful, a level of agreement about the purpose and methods used are necessary to achieve consistency. This places people in situations where discussions inevitably arise about what they think is best for the animal. Having meeting agendas, problem-solving formats, and communication norms for teams can help channel dialogue into appropriate discussions and facilitate open lines of communication. Heath and Heath (2011) discuss the difficulty of changing people's opinions. A point they make is that people are driven more strongly by their emotions in decisions that they care about than about the knowledge content. It is important to recognise that just laying out a list of bulleted facts about why something should or should not be done may not sway your team when you need everyone on board. You may need to dig to understand what is at the heart of their concerns, and the emotions involved, if you are interested in influencing their decision-making.

8.8 Conclusion

A commitment to a husbandry training programme takes an investment in programme planning, staff training, facility design, and ongoing maintenance and leadership. We have yet to determine or measure the many ways these training programmes benefit our animals. The goal of this chapter is to be a resource to initiate a training programme or reignite a training programme in which you take part or lead.

References

Cain, S. (2012). *Quiet: The Power of Introverts in a World That Can't Stop Talking*. New York: Crown Publishers.

Covey, S.R. (2004). *The 7 Habits of Highly Effective People: Restoring The Character Ethic*. New York, NY: Free Press.

Gladwell, M. (2008). *Outliers: The Story of Success*. USA: Little Brown and Company.

Grandin, T., Rooney, M.B., Phillips, M. et al. (1995). Conditioning of nyala (*Tragelaphus angasi*) to blood sampling in a crate with positive reinforcement. *Zoo Biology* 14: 261–273.

Heath, C. and Heath, D. (2011). *Switch: How to Change When Change Is Hard*. Waterville, ME: Thorndike Press.

Mellen, J. and MacPhee, M. (2012). Animal learning and husbandry training for management. In: *Wild Mammals in Captivity: Principles and Techniques* (eds. D. Kleinman, K. Thompson and C. Kirk Baer), 314–328. Chicago, IL: University of Chicago Press.

Sinek, S. (2009). *Start with Why: How Great Leaders Inspire Everyone to Take Action*. New York, NY: Portfolio.

9

Us and Them

Human–Animal Interactions as Learning Events

Geoff Hosey and Vicky A. Melfi

9.1   Introduction

Until the mid-1980s there were no systematic studies of how zoo animals responded to people, and anecdotal reports often concentrated on the negative interactions that visitors initiated, such as teasing, poking with sticks, feeding inappropriate objects, and sometimes even worse (Stemmler-Morath 1968; Hediger 1970); or the inventive ways that animals used visitors as a source of stimulation (Morris 1964). It was assumed that animals in zoos risked becoming 'mal-imprinted' through processes such as hand-rearing, leading to animals who considered themselves to be human (Morris 1964); and that provided visitors kept to areas behind barriers within normal zoo opening hours, the animals would disregard or ignore them (Snyder 1975). These views should be seen within the context of the times: enclosures were small and barren by modern standards, and in many cases the public were allowed closer contact with the animals than is permissible now (see Section 9.4); many more animals were hand-reared then (Morris 1964); and the prevailing Tinbergian–Lorenzian theoretical framework for explaining animal behaviour was more dominated by drive-instinct concepts than is modern behavioural biology. Nevertheless, underlying these observations

was the assumption that the animals were learning novel responses to people within the zoo environment.

Forty years on and zoo housing and husbandry have changed dramatically; so, we hope, has visitor behaviour. Few empirical studies survey general zoo visitor behaviour directed towards animals systematically. Instead, studies are generally biassed towards quantifying and ameliorating 'bad visitor behaviour' (e.g. Kemp et al. submitted, Parker et al. 2018). As such we know some zoo visitors behave negatively towards animals, but what the nature, frequency, duration, and valence of these behaviours are, viewed within the zoo visitor population, and whether they have changed is also unknown. The underlying belief that contact with people in a human-dominated environment provides a context for captive animals to learn new responses is still a valid one, and the growth of systematic studies on zoo animals allows us to examine this in greater detail than was previously possible. A good starting point is to recognise two dichotomies within the arena of human–animal interactions (HAI) in zoos: firstly, that the responses animals learn to familiar people (particularly keepers) are likely to be different from those they learn to unfamiliar people (such as visitors), especially since they have

Zoo Animal Learning and Training, First Edition. Edited by Vicky A. Melfi, Nicole R. Dorey, and Samantha J. Ward.
© 2020 John Wiley & Sons Ltd. Published 2020 by John Wiley & Sons Ltd.

greater contact and more repeated interactions with familiar people, which allows a relationship (human–animal relationship) to build up. Secondly, the interactions might be direct interactions between person and animal, where an initiating behaviour and a response to it can be detected by an observer, or may be indirect, in the sense that the animal's response is to a situation that involves the presence of people, but where no observable interaction takes place. Although we have referred to these here as dichotomies, it is important to understand that this is used here as an aid to understanding, and that in reality both are the end points of a continuum. Thus, some people, such as vets, may be a bit more familiar than visitors, but not as familiar as keepers. Similarly, a keeper undertaking routine husbandry procedures might inadvertently send direct signals to the animal, to which the animal will then respond. With these caveats, we can construct a table (Table 9.1) to guide our discussion.

9.2 Learning to Discriminate Different Kinds of People

Nobody who has ever shared their home with a companion animal would be surprised at the assertion that many animals can learn to distinguish different people. When we consider whether animals learn to discriminate between people in the zoo, we can ask whether they can learn to distinguish between familiar and unfamiliar people, as well as within these different categories or groups of people. Answering this is of theoretical interest in terms of what it can tell us about the discrimination and categorisation abilities of different animals. It is also of more applied interest, in that it may help us to understand some of the variability we see in zoo animal responses to visitors.

There is good empirical evidence that animals learn to distinguish familiar from unfamiliar people on the farm (Boivin et al. 1998; Rousing et al. 2005), in the laboratory (Davis 2002), and in fact, in the zoo (Mitchell et al. 1991; Martin and Melfi 2016). One study of zoo animals observed that several species were more likely to approach familiar people compared to unfamiliar people, even when both categories of people were similarly dressed and in the same context, i.e. cleaning an enclosure (Martin and Melfi 2016). The ability of agricultural animals to discriminate between different categories of people has been attributed to them learning to recognise the differences in peoples' behaviour and clothing (Munksgaard et al. 1997). Tanida and Nagano (1998), also considered that agricultural animals (young pigs) could discriminate between people using visual, auditory, and/or olfactory cues. It seems reasonable to consider that zoo animals,

Table 9.1 Situations involving people which might provide learning opportunities for zoo animals. Human action towards the animal is considered to be part of the learning process, and the types of people are seen as a variable affecting the learning process.

	Types of people	
Interaction	Familiar	Unfamiliar

Indirect	Keeper observations of animals
	Husbandry provision including
	  enrichment, cleaning and feeding
		Visitors watching at exhibit
		  'stand and stare'

Direct	Handling and catch-ups Health checks and veterinary treatment Provision of training activities Education activities	Interactive educational activities Keeper for a Day

whether as a population or some species/ individuals within this group, also learn to discriminate between different categories of people using these same cues. Learning from previous experience of being handled has also been found to aid agricultural animals in discriminating between different people (de Passillé et al. 1996; Munksgaard et al. 1997; Csatádi et al. 2007). Unlike agricultural settings, where direct handling or least physical contact occurs frequently between animals and familiar people, zoo housing and husbandry is often set up to avoid and/or limit direct handling and physical contact between zoo keepers and animals; though there are exceptions, e.g. when hand-rearing. This means that for zoo animals, handling doesn't represent a good source for learning about humans as it is infrequent. The same reasoning could be used to argue that due to the infrequent nature of handling and physical contact in zoos, that if and when it does occur, it might represent a significant learning opportunity. It is likely zoo animals also learn to discriminate between different categories of people by other types of interactions which they share with them. For example learning may occur from indirect human interactions towards the animals via participation and the outcome of husbandry activities that different people provide. For example, Melfi and Thomas (2005) observed that zoo-housed colobus monkeys (*Colobus guereza*) were able to discriminate, and behaved differently towards, three categories of people they observed in front of their enclosures; keepers who looked after their daily care, keepers/zoo staff in the same uniform who didn't look after their daily care, and zoo visitors (those people not in uniform!). Interestingly this study observed that after the initiation of a training programme to facilitate health checks, the rate of behaviours directed towards people of all categories declined significantly. The authors suggested that the colobus monkeys had learnt that directing behaviour towards people was more productive during the training session, rather than outside of this time and

as a consequence, the rate of behaviours directed to people of all categories dropped once training had been initiated.

Wild-living animals have also been seen to discriminate between different people, for example magpies (*Pica pica*), recognise people who have accessed their nests on previous occasions, and direct aggressive responses at them, which they do not do to people who have not accessed their nest (Lee et al. 2011). Male golden-bellied mangabeys (*Cercocebus chrysogaster*) at Sacramento Zoo threatened adult male visitors particularly, but rarely threatened infants, senior men, and women, whereas female mangabeys threatened women twice as often as they threatened men (Mitchell et al. 1992). These could, of course, be species-specific responses triggered by recognition by the animals that their human targets were an equivalent age and sex group to themselves; but could also be based on previous experience with people in those categories, since the study also found that men and boys harassed the male mangabeys more than the females. The mangabeys appeared to be responding to particular categories of people, where category discrimination could be based on a number of visual and behavioural cues, but the magpies in the earlier example were discriminating between individuals who apparently differed only in facial features. Agricultural animals are also able to distinguish people according to their facial features (pigs: Koba and Tanida 2001; cows: Rybarczyk et al. 2001; horses: Stone 2010). Intriguingly, wild American crows (*Corvus brachyrhynchos*) scold and mob people wearing a mask portraying a 'dangerous' face regardless of age, sex, size, or appearance of those people, but not people wearing a neutral mask (Marzluff et al. 2010). Scary masks (a vampire face) were shown by Sinnot et al. (2012) to a variety of zoo animals (primates, carnivores, hoofstock, and birds) and compared to a non-scary mask (a Bill Clinton mask), an aversive response to just the scary mask was found only in the primates. So discrimination of people based on the scariness of their face could be a conditioned response, but could also reflect underlying cognitive or taxonomic differences. In any case there is scope for more research here.

9.2.1 Familiar People

Interestingly most research which has been undertaken to study HAI in zoos has thus far focussed on interactions with unfamiliar people; the zoo visitor (see Section 9.4). Largely the impact of familiar people, comprising zoo keepers and other zoo professionals which spend so much of their time interacting with zoo animals, has been largely unstudied. These familiar people are ubiquitous in the lives of zoo animals, being present in all zoos and necessary in the creation and maintenance of good zoo animal welfare. Nevertheless, the impact they have as people on the learning landscape of zoo animals, rather than as providers of good care, is rarely considered.

Familiar people in the lives of zoo animals are those that interact and play a part in the animals' lives on a frequent basis. The frequency with which familiar people have these interactions, direct, or indirect, with animals can vary; but what is important in the context of this book is that all these interactions provide learning opportunities for the animals. Probably the most familiar people in the lives of zoo animals are their keepers. Often referred to as stockpeople within agriculture, the study of stockmanship has found that interactions between familiar people and the animals in their care can have far reaching ramifications. Positive HAI in agricultural settings have been associated with improved production and welfare, measured as improved growth, fecundity, and reduced morbidity (reviewed by Waiblinger 2019). It is unsurprising therefore that zoo professionals too can have profound impacts on the animals in their care (reviewed Ward and Sherwen 2019).

There is a high level of discrimination reported in the ability of zoo animals to differentiate between familiar people. Recognising specific individuals is evidenced by specific (positive or negative) behavioural reactions towards those people, and some report that this individual recognition and specific reaction to keepers can still be seen in animal-keepers separated from the animals for some years. Unfortunately, the degree of sophistication displayed by animals in their ability to discriminate between familiar people, the examples included above, are for the most part anecdotal reports. There are however a small number of empirical studies which have observed how zoo animals can discriminate between familiar people. For example, as already described, zoo animals have been observed to distinguish between keepers that maintain their daily needs versus those who work elsewhere in the zoos (Melfi and Thomas 2005). These data suggest that zoo animals are able to distinguish between those zoo professionals who provide daily care versus those that are less frequent carers or that provide different services, for example vet staff, researchers, and other uniformed staff. What is clear is that zoo animals learn to discriminate between these people, indicating that there are likely differences in the interactions occurring between the animals and these people, which impact whether a positive, neutral, or negative relationship is formed (Hosey 2008, 2013). Nature, frequency and type of these human–zoo-animal interactions all provide fruitful opportunities for learning.

Familiar people in the lives of zoo animals set up their animals' environments to facilitate learning opportunities. Zoo keepers provide learning opportunities through permanent or temporary enclosure design and changes (see Chapter 5), regular husbandry tasks, the provision of environmental enrichment (see Chapter 6), and of course formal training programmes (the topic of this book). For example, Ward and Melfi (2013) investigated the impacts of positive reinforcement training on the response rates for other non-trained behaviours. They found that animals which underwent a formal training regime, responded more quickly to non-trained cues than their non-trained counterparts. Authors suggested this finding resulted from a positive relationship being developed from regular and consistent interactions formed as part of the positive reinforcement training. Some of these learning opportunities might not be perceived positively. Animals may learn that some interactions with familiar people are worth avoiding because they may be associated with negative ramifications. Visits from vets have traditionally been synonymous with the necessary, but sometimes, unpleasant activities associated with maintaining good health, including vaccinations and restraints for health checks. These negative interactions with familiar people, have been hugely impacted by the addition of formal training programmes which aim to ensure that both, animals and people, can learn what to expect during these procedures and thus change their perception of them (see Chapter 11).

A key process which results from learning is the development of human–animal relationships; which rely on an animal–human dyad learning from multiple interactions what to expect from these interactions. If these interactions are generally positive it is likely that the animal will learn to respond accordingly and a positive human–animal relationship will be formed. Take this a step further, where both parties learn to anticipate a positive emotional experience alongside the positive HAI and you have the development of a human–animal bond. Human–animal bonds have been reported to occur, by a number of zoo keepers (Hosey and Melfi 2012); it is difficult to determine what if any emotional experience animals feel. Most descriptions of an animal's emotional experience during interactions with people are anecdotal, for example animals seeking interactions with familiar people (e.g. Masson and McCarthy 1996; Figure 9.1).

9.3 Unfamiliar People

The majority of people that animals encounter in the zoo are unfamiliar to the animals, in the sense that the animal has either not previously encountered them, or has not had opportunities to develop a relationship with any of them. As unfamiliar people are present both in great numbers (potentially

Figure 9.1 An illustration of keeper–animal interactions, which can impact both parties mediated through a variety of different interactions, and might be initiated by either party. *Source: Katharina Herrmann.*

several thousand in the zoo each day) and over a long period (up to eight or more hours each day), they represent a significant array of stimuli to which the animal can respond, and we would expect that the animals learn various characteristics of these people and change their responses to them accordingly. However, behaviours that we see directed at unfamiliar people, or which appear to be responses to them, can mostly only be inferred to be the result of learning. Firstly, we are by no means sure what the baseline behaviours prior to learning about unfamiliar people should look like. Comparison with free-living animals can be helpful, but these animals too have almost certainly come into contact with unfamiliar people previously, possibly in a way which is quantitatively and qualitatively different from what happens in the zoo. Secondly, as far as we are aware, no learning experiments have taken place in zoos in which unfamiliar people constitute the independent variable, so although we see what appear to be learned responses to the presence of people, we have not seen the process which leads to those responses. Thus it is possible that some of the responses or discriminations that we see are unlearned species-typical behaviours; an example is the way in which mangabeys direct different responses to male and female members of the viewing public, referred to in Section 9.2. To what extent they are the result of learning is potentially important, both for reasons of welfare (e.g. modifying what appear to be adverse responses to people in order to improve welfare) and of conservation (e.g. modifying responses back to a more 'wild-type' condition in animals which are due for release into the wild).

Keeping this in mind, we can list some of the behaviours that we are likely to see animals perform in the presence of unfamiliar people at the zoo, together with our inference about the learning process which has resulted in these (Table 9.2). Few of these have been studied systematically, and none has been studied as a learning process.

Table 9.2 Behaviours we might see zoo animals perform in the presence of unfamiliar people, and our inferences about the learning processes which have resulted in them.

Behaviour observed	Inferred learning process
Getting used to people; ignoring people; taking little or no notice of people.	Habituation
Soliciting food from people; attempting to interact with people; using people as a source of positive stimulation and potentially 'enrichment'.	Classical conditioning
Avoidance of people and exposed public areas; hiding; increasing aggression.	Operant conditioning
Behaving differently to different categories of people: keepers vs visitors, men vs women, children vs adults, etc.	Discrimination learning

9.3.1 Learning to Disregard Visitors

It is widely assumed that animals disregard zoo visitors and anecdotally most of us have noticed that many animals in zoos appear to take no notice of us as we walk past, or stand watching them. So we should ask whether animals in zoos really do disregard visitors, and if so, whether this is due to learning. A number of empirical studies have investigated zoo animal responses to the mere presence of people, as well as their responses to visitors who are noisy, active, or who attempt to interact with the animals. Several of these have recorded little or no responsiveness to people. For example, Choo et al. (2011) found that orangutans (*Pongo pygmaeus*) at Singapore Zoo did show some changes in behaviour in response to the presence and behaviour of the public, but that these changes were much less severe than expected, which they suggested could be due to habituation to people. Similarly, Sherwen et al. (2014) found no change in their measures of meerkat (*Suricata suricatta*) responses to visitors after the visitors had been asked to be less noisy, and again suggested habituation as a possible reason. Burrell and Altman (2006) reported that the cotton-top tamarins (*Saguinus oedipus*) they were observing were moved by the zoo from a naturalistic walk-through exhibit to smaller enclosures pre-cisely because they were becoming habituated to the public. And several studies have shown no change in the activity of felids when visi-tors are present compared to when there are no visitors (e.g. O'Donovan et al. 1993; Margulis et al. 2003). So it does appear to be the case that some animals in zoos disregard visitors; or at least do not show a change in behaviour when visitors are present. Is this due to habituation?

Habituation is usually defined as 'the rela-tively persistent waning of a response as a result of repeated stimulation which is not followed by any kind of reinforcement' (Hinde 1970). Of course, outside of the labo-ratory it is often difficult to know exactly which stimuli the animal is habituating to, so in these situations we sometimes find a somewhat different definition. For example, in the context of habituating wild primates for research purposes, Tutin and Fernandez (1991) define habituation as 'the acceptance by wild animals of a human observer as a natural element in their environment', and this kind of definition by outcome rather than by process is probably more like the sort of thing we mean when we consider that zoo animals might habituate to members of the public. The general idea, though, is that through repeated unreinforced exposure to zoo visitors the animals eventually lose behaviours such as fear responses to people, visual monitoring, avoidance, or startle behaviours. Some of the evidence, then, sup-ports this suggestion, but there are a number of studies which nevertheless show that zoo animals do indeed respond (often adversely) to unfamiliar people (Hosey 2000, 2013). There are at least two explanations that we can put forward to account for this apparent inconsistency. One is that habituation has occurred, but incompletely, so the response to the stimulus of unfamiliar humans is still there, but at a lower intensity. In a study of

cotton-top tamarins, for example, Glatston et al. (1984) found that animals that had previously been off-show still displayed heightened adverse responses to the public a year after being transferred to an on-show enclosure. The other is that unfamiliar humans actually represent an array of stimuli to which animals habituate at different rates. This might help explain why some animals respond more to visitors who, for example, are noisy than those who are quiet (Quadros et al. 2014), if we postulate that habituation to mere presence of people occurs more quickly than habituation to visitor noise. We also sometimes see what appears to be disha-bituation: adverse responses to the public in an unusual situation in animals, which other-wise largely ignore visitors. For example, cheetahs (*Acinonyx jubatus*) at Fota Wildlife Park only showed responses on the few occa-sions when visitors came within the bound-ary rail (O'Donovan et al. 1993).

Habituation in zoo-housed animals has been studied with respect to responses to environmental enrichment (e.g. Anderson et al. 2010) and husbandry and veterinary procedures (e.g. Calle and Bornmann 1988; Phillips et al. 1998), but not apparently as a response to zoo visitors. An analogous situa-tion to zoo animals repeatedly encountering visitors is wild-living animals encountering researchers and tourists, which has appar-ently led to habituation in species as diverse as brown bears (*Ursus arctos*) (Herrero et al. 2005) and Tibetan macaques (*Macaca thi-betana*) (Matheson et al. 2006) at tourist sites. Indeed, wild populations are often deliberately habituated by researchers as an aid to performing the research (Williamson and Feistner 2011), as it increases the visibil-ity of the subjects, permits better identifica-tion of individuals and their relationships with each other, and reduces any effects observers might have on natural behaviour (Goldsmith 2005). We might consider that habituation of zoo-housed animals should be encouraged for much the same reasons, but before making any such recommendations we need more information about the costs of habituation. In wild-living populations the costs to the animals include increased risk of disease transmission from people, and generalisation of the habituation by the animals to other people, rendering them vulnerable to hunting or other human threats (Goldsmith 2005; Williamson and Feistner 2011). It is also unclear how habituation impacts on other aspects of the animals' welfare. Faecal cortisol levels in wild-living capuchins (*Cebus capucinus*) (Jack et al. 2008) and gorillas (Shutt et al. 2014) are more elevated in habituated than non- or less-habituated groups, implying that the process of habituation causes some stress to the animals.

9.3.2 Learning to Use Visitors as a Source of Stimulation

Amongst the studies of how animals in zoos respond to members of the public, a small number show results which can be interpreted as implying that the animals obtain some kind of stimulation from their interactions with visitors. There are also anecdotal accounts of animals inventing ways of interacting with visitors, sometimes to the visitors' detriment. All of these suggest that human contact can in some circumstances be enriching for these animals. In principle there is no reason why HAI could not be an enrichment for captive animals (Hosey 2008; Claxton 2011), and some studies show that increased interactions with familiar people can be viewed this way (e.g. Baker 2004; Carrasco et al. 2009). To what extent the responses to the public can be regarded as enrichment is a moot point, given that these observations are not usually part of a planned enrichment programme, and interpretation of the behavioural change is often a post hoc explanation rather than a predicted change.

Fifty years ago Desmond Morris gave examples of animals at London Zoo incorporating zoo visitors into their activities, and interpreted these as efforts by the animals to add stimulation to an otherwise boring life in the zoo (Morris 1964). These included banging and stamping on the ground, apparently

to attract attention; directing urine and faeces at visitors; and behaviours such as soliciting touch, and then biting the toucher. In the 1960s the chimpanzees at Chester Zoo apparently became proficient at throwing clods of earth at people (Morris and Morris 1966). It is tempting to believe that these apparent efforts to enliven a boring life in unstimulating enclosures are no longer seen in modern zoo enclosures, but we actually have no idea if that is the case; in fact, throwing faeces at visitors increased in a male hamadryas baboon (*Papio hamadryas*) in a Brazilian zoo after transfer from a small traditional cage to a larger more naturalistic enclosure (Bortolini and Bicca-Marques 2011). Additionally a chimpanzee at Furuvik Zoo reportedly stored stones as future missiles to throw at visitors (Osvath 2009).

There are rather more benign ways in which zoo animals learn to interact with visitors. One is to solicit food. At Mexico City Zoo, Fa (1989) found a significant positive correlation between the time that captive green monkeys (*Cercopithecus sabaeus*) spent feeding on supplemental food thrown in by visitors and the number of visitors. Chimpanzees at Chester Zoo interact with visitors, and sequences of interactions can develop between them, in which begging for food is common, and which sometimes culminate in the chimps being offered food (Cook and Hosey 1995). Begging for food is well known in bears, where the behaviour appears to be linked to the display of stereotypies (Van Keulen-Kromhout 1978; Montaudouin and Le Pape 2004).

None of these behaviours would be regarded as enrichment in terms of our modern understanding of the term, and yet the fact that the animals have learned to do these things suggests that there is something rewarding about performing them. To that extent they can be viewed as conditioned responses, although it has also been suggested, at least in the case of food begging by primates, that they may be referential, inasmuch as they indicate an understanding by the animal of how they influence the behaviours of others (Gómez 2005). Nevertheless, the reinforcers for acquiring these behaviours are presumably the stimulating effect of seeing the human response to the behaviour (after, for example, having faeces thrown at them), or else the occasional piece of food. There are a couple of reports, however, where the animal appears to seek interaction with unfamiliar people for its own sake; in other words, the HAI is itself reinforcing.

One example is a long-billed corella (*Cacatua tenuirostris*) named Claude at Adelaide Zoo. On busy days (weekends or public holidays) he spent 90% of his time at the front of the enclosure, and showed behaviours that he didn't show, or rarely showed, on quiet days (weekdays), such as face-to-beak contact, vocalisation in words, and orienting towards human visitors (Nimon and Dalziel 1992). The authors concluded that the presence of people was reinforcing for Claude. Another example is a female gorilla (*Gorilla gorilla graueri*) named Isabelle at Antwerp Zoo, who, unlike the other gorillas in the group, stayed close to the viewing window when visitors were present, and appeared to seek eye contact and mimic the opening and closing of the mouth during speech (Vrancken et al. 1990). In both of these examples the animals were reared in a human-centred environment, with the corella being a former pet and the gorilla having been hand-reared, and we can speculate that learning an attraction to a familiar human has become generalised to include unfamiliar people.

If the opportunity for interactions with people can indeed be reinforcing for animals, we might expect to see the most profound evidence for it when the interactions are direct, rather than indirect. Unfortunately there is very little empirical evidence to inform us on this. Goats and pigs in a petting zoo didn't appear to find grooming by the public enriching (Farrand et al. 2014). Bottlenose dolphins *Tursiops truncatus*, on the other hand, have been reported to increase their play behaviour after interaction sessions in which people get in the water with them and touch them (Trone et al.

2005), and after shows (Miller et al. 2011), which can be interpreted as a positive stimulating effect of HAI on the animals.

9.3.3 Learning to Avoid Visitors

A possible consequence of large naturalistic enclosures is that the animals they house become less visible to the public. Since animals are what the public come to the zoo to see, lack of visibility of those animals is a potential problem, which zoos try to address in a way that doesn't compromise animal welfare (Bashaw and Maple 2001; Kuhar et al. 2010). This lack of visibility might be a result merely of the size and topography of the enclosure, and thus could be independent of animal behaviour, or because the animals prefer certain locations independently of visitor presence. But it might also occur because animals have learned that they can hide from visitors, and thus avoid the more stressful aspects of visitor presence and behaviour. Gorillas at Atlanta Zoo, for example, appear to have a preference for particular types of structure, and if these structures are in less visible parts of the enclosure, then the animals also are less visible (Stoinski et al. 2002). Indeed, alternation of gorillas between a familiar and an unfamiliar enclosure increased their visibility in the unfamiliar enclosure (Lukas et al. 2003). Decreases in adverse responses to zoo visitors do occur when animals are transferred from small 'traditional style' to larger more naturalistic enclosures (Ross et al. 2011), but there appears to be no particular evidence that this occurs because the animals are choosing to be less visible. Furthermore, animals in free-range may be more visible, though this does appear to be dependent upon the species (Sha et al. 2013; Schäfer 2014). They are, however, presumably learning *something* about visitors, or perhaps about how much choice they have about just how visible and interactive they can be.

There is some limited support for the latter interpretation from studies of direct HAI. Undesirable behaviours shown by sheep and goats towards people in a petting area reduced when the animals were provided with a retreat space (Anderson et al. 2002); similarly, use of a retreat space by dolphins *Delphinus delphis* in a swim-with-dolphins programme increased when people were in the pool (Kyngdon et al. 2003). In these cases the animals appear to have learned that the retreat space offers them refuge if they don't want to interact with humans.

9.3.4 The Changing Nature of Visitor Interactions with Animals

Visitors are increasingly interacting with zoo animals at an ever more intimate level with the widespread uptake of interactive educational experience (IEE), often referred to as encounters. Though there is a high degree of variation between IEE offered between zoos and with different animals, they typically include some, if not all of the following characteristics: animals and unfamiliar people are brought into close proximity; often in off-show areas where unfamiliar people are not usually allowed to go; during times when the zoo would otherwise be closed; and often animals are encouraged to interact with these unfamiliar people (Figure 9.2). The advent of these encounters

have required that animals learn a new set of conditions under which they might be expected to interact, be in close proximity with and/or assume otherwise 'normal' behaviour in situations where previously only familiar people would be present. To date, we know very little about the impact of these encounters on the daily behaviour and welfare of the animals taking part. For animals managed within education programmes, these conditions might be considered normal (reviewed Chapter 10); for example ambassador animals are often trained to expect interactions with unfamiliar people in unfamiliar and potentially unexpected circumstances. Data have been published about the methods used to habituate or train animals to engage under these conditions (again see Chapter 10). For example, animals might be provided with food to encourage them into proximity with, or into spaces near, unfamiliar people. These interactive educational experiences provide learning opportunities which might require that the animals either expand their categorisation of people, or the areas of their enclosure and times of day when they might be expected to interact or be active in the presence of unfamiliar people.

(a)

(b)

Figure 9.2 An illustration of a paid interaction available for zoo visitors, with Sumatran tigers (a) and the same animals initiating an interaction with their keeper (b). *Source:* (a) Vicki Melfi; (b) Sheila Roe.

9.4 Discussions/Conclusions

In summary people provide many learning opportunities for animals in zoos, whether these are associations learnt as a consequence of direct or indirect interactions. Unfortunately there are limited published empirical data detailing the diverse and abundant familiar HAI we know to occur in zoos; instead the field is mostly populated with data from visitor studies. It seems likely that zoo visitor–animal interactions are different depending on the context in which they occur. Much of the research in this area has occurred at 'stand and stare' exhibits and has been shown to have mostly negative ramifications. Though the impact of zoo visitor–animal interactions during the large variety of other potential HAI offered by zoos, from feeding events to interactive educational experiences, is less well studied. Most surprisingly, the ubiquitous interactions offered by zoo professionals to the animals in their care have been overlooked completely; with the exception of a handful of studies, which suggest these interactions offer positive learning opportunities.

We have limited our scope in this chapter to the categorisation of familiar and unfamiliar people, but it is likely that some animals are able to conduct much more sophisticated discriminations within these categories. As researchers it is unsurprising that we feel more research is necessary in this area! We feel it would not only provide interesting insights in the field of cognition, but also provide vital information which would facilitate evidence-based zoo animal management. It would be particularly interesting and helpful to have better insight into the degree to which different species are able to discriminate between different categories of people. Cognitive discrimination tasks have been largely dominated by primates, though there is sound research on birds in this area. Zoos of course maintain many different species and so better understanding whether other taxa and species are able to discriminate between familiar and unfamiliar people is crucial. From the limited cognitive research undertaken so far on the forgotten taxa, fish, and reptiles have been observed to show cognitive complexity (see taxa specific boxes by Burghardt and Brown in this book) which might lead us to suspect that these taxa too are sensitive to the different people in their lives. Our working model (Hosey 2008) predicts that familiar people are preferred over unfamiliar people, but we need to remember that these forgotten taxa may perceive the world differently from other taxa. For example, providing scincid lizards *Eulamprus heatwolei* with a novel environment, which is often considered a positive change from a mammal centric view and indeed incorporates many activities performed under the umbrella of environmental enrichment, was observed to elevate cortisol levels and breathing rates denoting stress (Langkilde and Shine 2006). Taking this research to a whole new level might include investigating the impact of dyadic interactions between people and their ramifications for animals; and what animals might learn from these dyadic human relationships and the impacts they might have on zoo animal management. For example, there are numerous anecdotes, which can also be considered as an area rich in research ideas, that zoo animals learn the outcomes of interactions between people in their environment and as a consequence attend to these people differently. For example, animals ignoring cues provided by one person whilst another is present; ignoring a keeper when the curator is present unless the curator supports the keepers cues, either to the keeper or the animal. If indeed the relationships between zoo staff can be understood by the animals and impacts on their management that really would necessitate some changes in our perception of the cognitive abilities of the animals we care for, as well as our conduct and behaviour with our colleagues.

References

Anderson, C., Arun, A.S., and Jensen, P. (2010). Habituation to environmental enrichment in captive sloth bears – effect on stereotypies. *Zoo Biology* 29: 705–714.

Anderson, U.S., Benne, M., Bloomsmith, M.A., and Maple, T.L. (2002). Retreat space and human visitor density moderate undesirable behavior in petting zoo animals. *Journal of Applied Animal Welfare Science* 5: 125–137.

Baker, K.C. (2004). Benefits of positive human interaction for socially housed chimpanzees. *Animal Welfare* 13: 239–245.

Bashaw, M.J. and Maple, T.L. (2001). Signs fail to increase zoo visitors' ability to see tigers. *Curator* 44: 297–304.

Boivin, X., Garel, J.P., Mante, A., and Le Neindre, P. (1998). Beef calves react differently to different handlers according to the test situation and their previous interactions with their caretaker. *Applied Animal Behaviour Science* 55: 245–257.

Bortolini, T.S. and Bicca-Marques, J.C. (2011). The effect of environmental enrichment and visitors on the behaviour and welfare of two captive hamadryas baboons (*Papio hamadryas*). *Animal Welfare* 20: 573–579.

Burrell, A.M. and Altman, J.D. (2006). The effect of the captive environment on activity of captive cotton-top tamarins (*Saguinus oedipus*). *Journal of Applied Animal Welfare Science* 9: 269–276.

Calle, P.P. and Bornmann, J.C. II (1988). Giraffe restraint, habituation and desensitization at the Cheyenne Mountain Zoo. *Zoo Biology* 7: 243–252.

Carrasco, L., Colell, M., Calvo, M. et al. (2009). Benefits of training/playing therapy in a group of captive lowland gorillas (*Gorilla gorilla gorilla*). *Animal Welfare* 18: 9–19.

Choo, Y., Todd, P.A., and Li, D. (2011). Visitor effects on zoo orang-utans in two novel, naturalistic enclosures. *Applied Animal Behaviour Science* 133: 78–86.

Claxton, A.M. (2011). The potential of the human-animal relationship as an environmental enrichment for the welfare of zoo-housed animals. *Applied Animal Behaviour Science* 133: 1–10.

Cook, S. and Hosey, G.R. (1995). Interaction sequences between chimpanzees and human visitors at the zoo. *Zoo Biology* 14 (5): 431–440.

Csatádi, K., Ágnes, B., and Vilmos, A. (2007). Specificity of early handling: are rabbit pups able to distinguish between people? *Applied Animal Behaviour Science* 107: 322–327.

Davis, H. (2002). Prediction and preparation: Pavlovian implications of research animals discriminating among humans. *ILAR Journal* 43: 19–26.

de Passillé, A.M., Rushen, J., Ladewig, J., and Petherick, C. (1996). Dairy calves' discrimination of people based on previous handling. *Journal of Animal Science* 74: 969–974.

Fa, J.E. (1989). Influence of people on the behaviour of display primates. In: *Housing, Care and Psychological Well-being of Captive and Laboratory Primates* (ed. E.F. Segal), 270–290. Park Ridge, USA: Noyes Publications.

Farrand, A., Hosey, G., and Buchanan-Smith, H.M. (2014). The visitor effect in petting zoo-housed animals: aversive or enriching? *Applied Animal Behaviour Science* 151: 117–127.

Glatston, A.R., Geilvoet-Soeteman, E., Hora-Pecek, E., and van Hooff, J.A.R.A.M. (1984). The influence of the zoo environment on social behavior of groups of cotton-topped tamarins, *Saguinus oedipus oedipus*. *Zoo Biology* 3: 241–253.

Goldsmith, M.L. (2005). Habituating primates for field study: ethical considerations for African great apes. In: *Biological Anthropology and Ethics: Form Repatriation to Genetic Identity* (ed. T.R. Turner), 49–64. Albany, NY: State University of New York Press.

Gómez, J.C. (2005). Requesting gestures in captive monkeys and apes: conditioned responses or referential behaviours? *Gesture* 5: 91–105.

Hediger, H. (1970). *Man and Animal in the Zoo*. London: Routledge & Kegan Paul.

Herrero, S., Smith, T., DeBruyn, T.D. et al. (2005). From the field: brown bear habituation to people – safety, risks and benefits. *Wildlife Society Bulletin* 33: 362–373.

Hinde, R.A. (1970). *Animal Behaviour: A Synthesis of Ethology and Comparative Psychology*. New York: McGraw Hill.

Hosey, G. (2008). A preliminary model of human-animal relationships in the zoo. *Applied Animal Behaviour Science* 109: 105–127.

Hosey, G. (2013). Hediger revisited: how do zoo animals see us? *Journal of Applied Animal Welfare Science* 16: 338–359.

Hosey, G. and Melfi, V. (2012). Human-animal bonds between zoo professionals and the animals in their care. *Zoo Biology* 31: 13–26.

Hosey, G.R. (2000). Zoo animals and their human audiences: what is the visitor effect? *Animal Welfare* 9: 343–357.

Jack, K.M., Lenz, B.B., Healan, E. et al. (2008). The effects of observer presence on the behaviour of Cebus capucinus in Costa Rica. *American Journal of Primatology* 70: 490–494.

Koba, Y. and Tanida, H. (2001). How do miniature pigs discriminate between people? Discrimination between people wearing overalls of the same colour. *Applied Animal Behaviour Science* 73: 45–56.

Kuhar, C.W., Miller, L.J., Lehnhardt, J. et al. (2010). A system for monitoring and improving animal visibility and its implications for zoological parks. *Zoo Biology* 29: 68–79.

Kyngdon, D.J., Minot, E.O., and Stafford, K.J. (2003). Behavioural responses of captive common dolphins *Delphinus delphis* to a 'swim-with-dolphin' programme. *Applied Animal Behaviour Science* 81: 163–170.

Langkilde, T. and Shine, R. (2006). How much stress do researchers inflict on their study animals? A case study using a scincid lizard, *Eulamprus heatwolei*. *Journal of Experimental Biology* 209 (6): 1035–1043.

Lee, W.Y., Lee, S., Choe, J.C., and Jablonski, P.G. (2011). Wild birds recognise individual humans: experiments on magpies, *Pica pica*. *Animal Cognition* 14: 817–825.

Lukas, K.E., Hoff, M.P., and Maple, T.L. (2003). Gorilla behaviour in response to systematic alternation between zoo enclosures. *Applied Animal Behaviour Science* 81: 367–386.

Margulis, S.W., Hoyos, C., and Anderson, M. (2003). Effect of felid activity on zoo visitor interest. *Zoo Biology* 22: 587–599.

Martin, R.A. and Melfi, V. (2016). A comparison of zoo animal behavior in the presence of familiar and unfamiliar people. *Journal of Applied Animal Welfare Science* 19 (3): 234–244.

Marzluff, J.M., Walls, J., Cornell, H.N. et al. (2010). Lasting recognition of threatening people by wild American crows. *Animal Behaviour* 79 (3): 699–707.

Masson, J.M. and McCarthy, S. (1996). *When Elephants Weep: The Emotional Lives of Animals*. New York: Delta Publishing.

Matheson, M.D., Sheeran, L.K., Li, J.-H., and Wagner, R.S. (2006). Tourist impact on Tibetan macaques. *Anthrozoös* 19: 158–168.

Melfi, V.A. and Thomas, S. (2005). Can training zoo-housed primates compromise their conservation? A case study using Abyssinian colobus monkeys (*Colobus guereza*). *Anthrozoös* 18 (3): 304–317.

Miller, L.J., Mellen, J., Greer, T., and Kuczaj, S.A. II (2011). The effects of education programs on Atlantic bottlenose dolphin (*Tursiops truncatus*) behaviour. *Animal Welfare* 20: 159–172.

Mitchell, G., Herring, F., and Obradovich, S. (1992). Like threaten like in mangabeys and people? *Anthrozoös* 5: 106–112.

Mitchell, G., Obradovich, S.D., Herring, F.H. et al. (1991). Threats to observers, keepers, visitors, and others by zoo mangabeys (*Cercocebus galeritus chrysogaster*). *Primates* 32: 515–522.

Montaudouin, S. and Le Pape, G. (2004). Comparison of the behaviour of European brown bears (Ursus arctos arctos) in six different parks, with particular attention to stereotypies. *Behavioural Processes* 67: 235–244.

Morris, D. (1964). The response of animals to a restricted environment. *Symposium of the Zoological Society of London* 13: 99–118.

Morris, D. and Morris, R. (1966). *Men and Apes*. London: Hutchinson.

Munksgaard, L., de Passillé, A.M., Rushen, J. et al. (1997). Discrimination of people by dairy cows based on handling. *Journal of Dairy Science* 80: 1106–1112.

Nimon, A.J. and Dalziel, F.R. (1992). Cross-species interaction and communication: a study method applied to captive siamang (*Hylobates syndactylus*) and long-billed corella (*Cacatua tenuirostris*) contacts with humans. *Applied Animal Behaviour Science* 33: 261–272.

O'Donovan, D., Hindle, J.E., McKeown, S., and O'Donovan, S. (1993). Effect of visitors on the behaviour of female cheetahs Acinonyx jubatus. *International Zoo Yearbook* 32: 238–244.

Osvath, M. (2009). Spontaneous planning for future stone throwing by a male chimpanzee. *Current Biology* 19 (5): R190–R191.

Parker, E.N., Bramley, L., Scott, L. et al. (2018). An exploration into the efficacy of public warning signs: a zoo case study. *PLoS One* 13 (11): e0207246. https://doi.org/10.1371/journal.pone.0207246.

Phillips, M., Grandin, T., Graffam, W. et al. (1998). Crate conditioning of bongo (*Tragelaphus eurycerus*) for veterinary and husbandry procedures at the Denver Zoological Gardens. *Zoo Biology* 17: 25–32.

Quadros, S., Goulart, V.D.L., Passos, L. et al. (2014). Zoo visitor effect on mammal behaviour: does noise matter? *Applied Animal Behaviour Science* 156: 78–84.

Ross, S.R., Wagner, K.E., Schapiro, S.J. et al. (2011). Transfer and acclimatization effects on the behaviour of two species of African great ape (*Pan troglodytes* and *Gorilla gorilla gorilla*) moved to a novel and naturalistic zoo environment. *International Journal of Primatology* 32: 99–117.

Rousing, T., Ibsen, B., and Sørensen, J.T. (2005). A note on: on-farm testing of the behavioural response of group-housed calves towards humans; test-retest and inter-observer reliability and effect of familiarity of test person. *Applied Animal Behaviour Science* 94: 237–243.

Rybarczyk, P., Koba, Y., Rushen, J. et al. (2001). Can cows discriminate people by their faces? *Applied Animal Behaviour Science* 74: 175–189.

Schäfer, F. (2014). To see or not to see: animal-visibility in Apenheul Primate Park. *International Zoo News* 61: 5–20.

Sha, J.C.M., Kabilan, B., Alagappasamy, S., and Guha, B. (2013). Benefits of naturalistic freeranging primate displays and implications for increased human-primate interactions. *Anthrozoös* 26: 13–26.

Sherwen, S.L., Magrath, S.J.L., Butler, K.L. et al. (2014). A multi-enclosure study investigating the behavioural response of meerkats to zoo visitors. *Applied Animal Behaviour Science* 156: 70–77.

Shutt, K., Heistermann, M., Kasim, A. et al. (2014). Effects of habituation, research and ecotourism on faecal glucocorticoid metabolites in wild western lowland gorillas: implications for conservation management. *Biological Conservation* 172: 72–79.

Sinnot, J.M., Speaker, H.A., Powell, L.A., and Mosteller, K.W. (2012). Perception of scary Halloween masks by zoo animals and humans. *International Journal of Comparative Psychology* 25: 83–96.

Snyder, R.L. (1975). Behavioral stress in captive animals. In: *Research in Zoos and Aquariums*, 41–76. Washington DC: National Academy of Sciences.

Stemmler-Morath, C. (1968). Ultimate responsibility rests with the keeper. In: *The World of Zoos* (ed. R. Kirchshofer), 171–183. London: Batsford Books.

Stoinski, T.S., Hoff, M.P., and Maple, T.L. (2002). The effect of structural preferences, temperature, and social factors on visibility in Western lowland gorillas (*Gorilla g. gorilla*). *Environment & Behavior* 34: 493–507.

Stone, S.M. (2010). Human facial discrimination in horses: can they tell us apart? *Animal Cognition* 13: 51–61.

Tanida, H. and Nagano, Y. (1998). The ability of miniature pigs to discriminate between a stranger and their familiar handler. *Applied Animal Behaviour Science* 56 (2–4): 149–159.

Trone, M., Kuczaj, S., and Solangi, M. (2005). Does participation in dolphin-human interaction programs affect bottlenose dolphin behaviour? *Applied Animal Behaviour Science* 93: 363–374.

Tutin, C.E.G. and Fernandez, M. (1991). Responses of wild chimpanzees and gorillas to the arrival of primatologists: behaviour observed during habituation. In: *Primate Responses to Environmental Change* (ed. H.O. Box), 187–197. London: Chapman & Hall.

Van Keulen-Kromhout, G. (1978). Zoo enclosures for bears. *International Zoo Yearbook* 18: 177–186.

Vrancken, A., Van Elsacker, L., and Verheyen, R.F. (1990). Preliminary study on the influence of the visiting public on the spatial distribution in captive eastern lowland gorillas (*Gorilla gorilla graueri*). *Acta Zoologica et Pathologica Antverpiensia* 81: 9–15.

Waiblinger, S. (2019). Chapter 3. Agriculture animals. In: *Anthrozoology Human–Animal Interactions in Domesticated and Wild Animals* (eds. G. Hosey and V. Melfi), 32–58. UK: Oxford University Press.

Ward, S.J. and Melfi, V. (2013). The implications of husbandry training on zoo animal response rates. *Applied Animal Behaviour Science* 147: 179–185.

Ward, S.J. and Sherwen, S. (2019). Chapter 5. Zoo Animals. In: *Anthrozoology Human–Animal Interactions in Domesticated and Wild Animals* (eds. G. Hosey and V. Melfi), 81–103. UK: Oxford University Press.

Williamson, E.A. and Feistner, A.T.C. (2011). Habituating primates: processes, techniques, variables and ethics. In: *Field and Laboratory Methods in Primatology: A Practical Guide* (eds. J.M. Setchell and D.J. Curtis), 33–49. Cambridge: Cambridge University Press.

Box B1

Elephant Training in Zoos

Greg A. Vicino

Training is deeply embedded in the culture of zoo elephant care. It would be unthinkable, at least in the United States, to house elephants in zoos without relying heavily on training to care for them. In fact elephants remain the only zoo animals in the United States for which accreditation standards specifically mention and evaluate training methods and outcomes (Association of Zoos and Aquaria 2018). Although the training of zoo animals is now commonplace, it is the specific details involved in elephant training that make it perhaps one of the more unique, and arguably, necessary paradigms in the captive management of any species.

The size and power of an adult elephant in a zoo setting is quite simply the most basic explanation as to why training is crucial for the effective management of the species. From the most basic need of shifting an elephant (moving it from one location to another), to the more complex need of medical management such as wound treatment and pregnancy monitoring, in which anaesthesia carries a high risk (Fowler et al. 2000). Daily husbandry routines with zoo elephants often involve a high level of training, and as long-lived species, this can serve not only to provide mental stimulation, but also to aid in the medical management of ailments associated with old age. Training zoo elephants can serve as a proxy for the problem solving challenges they would face in the wild by providing them with

the ability to control the outcome of a situation through behaviours that are beneficial to them. Perhaps the most logical argument is simply that elephants in zoos often face a myriad of conditions that would not exist in the wild, some of which the outcomes must be facilitated by humans. Elephant training in zoos continues to evolve and some validated measures of positive welfare have been associated with training. For example, Greco et al. (2016) found a negative correlation between the time an elephant spent demonstrating stereotypic behaviour and the time they spent in the presence of human caretakers practicing positive reinforcement training; essentially suggesting that the more time spent training the elephants, the less stereotypies they performed. Training zoo housed elephants, remains a fundamental part of elephant husbandry and will continue to advance in line with our knowledge of how best to care for this species.

The past two decades have seen a monumental shift in zoo elephant care, most notably characterised by the influence of positive reinforcement training and the concept of protected-contact. Within the basic framework of training methods, positive reinforcement is a method of operant conditioning in which desired behaviours (as a response to a trainers cue) are rewarded when successful (Daugette et al. 2012). That definition in this context serves to differentiate from the

Zoo Animal Learning and Training, First Edition. Edited by Vicky A. Melfi, Nicole R. Dorey, and Samantha J. Ward.
© 2020 John Wiley & Sons Ltd. Published 2020 by John Wiley & Sons Ltd.

former method of training most commonly associated with elephant management, positive punishment. Positive punishment dominated elephant management in the first few centuries of human–elephant relationships and relies heavily on applying aversive stimuli to the animal following the failure to execute a behaviour on cue (Hockenhull and Creighton 2013). The comparable efficacy, reliability, and practicality of these opposing methods are well documented (Ramirez 1999), and simply defining their role in the history of elephant training should suffice in this context. The term 'protected-contact' (see Figure B1.1) is the opposite of 'free-contact' (see Figure B1.2) and they both serve as a descriptor of the degree to which humans and elephants share, or do not, the same space. Protected-contact requires that human and elephants are always 'protected' from one another by a barrier (bars, or bollards), and has become assumed as a part of positive reinforcement due to the nature of the physical arrangement. Free-contact is however, the contrary, with elephants and humans sharing the same space, and typically requires a positive punishment training method. The argument for shifting paradigms is that protected-contact is based on trust and motivation (as an elephant can simply walk away without punishment or reward), whereas free-contact requires a dominant/submissive relationship. Proponents of free-contact argue that they would be unable to care for an elephant properly if it was up to the elephant to choose to participate in a treatment that may be uncomfortable. Proponents of protected-contact and positive reinforcement training consider the efficacy of treatment improved by not only adding choice, but also by establishing a trusting relationship that is reliant on reward for

Figure B1.1 An example of how keepers can provide foot care for elephants *Elephas maximus* within a protected management regime. The degree to which the contact between animal and keeper is restricted in protected-contact can vary, in this illustration, physical contact between elephants and keepers is still available to both parties. *Source:* Jeroen Stevens.

Figure B1.2 An example of how keepers can perform husbandry tasks within a free-contract management regime; free-contact is denoted by the animal and keeper sharing the same physical space as one another. In this situation, a young elephant is being washed by their keeper. *Source:* Jeroen Stevens.

participation (Ramirez 1999). The fundamental principles of these training methods are based in deep-rooted training doctrine and certainly have results-based arguments on both sides. Moreover, it is becoming decreasingly favoured by the public to accept a method that is based on punishment rather than reward.

Another, more noticeable, shift in zoo elephant training has been the slow and deliberate decline in elephant shows or presentations of trained elephants demonstrating behaviour unrelated to medical management or husbandry. Twenty years ago, most American zoos that housed elephants had at least one scheduled presentation a day. The presentations remained fairly consistent, highlighting the strength (for example, hauling logs or lifting weights), agility (for example, standing on two legs or sitting), balance (for example, standing on small platforms), and the relationship with the trainer. Often these shows shared information about natural history, the elephant's unique physiology, and emphasised the intelligence and tractability of the species. For many guests this was the first time they were exposed to elephant training in a zoo environment, however, on

the surface they were not much different from the elephant demonstrations in circuses. Once positive reinforcement and protected-contact became the norm these demonstrations either disappeared from zoos or were replaced with conservation oriented presentations highlighting the plight of wild elephants. Many elephant training programmes have gone as far as to remove the word 'No' from the training jargon. This is not to avoid the anthropomorphism of the communication system, but to remind the trainer that elephants don't need to understand what they can't do, but what they can do. This paradigm shift has us looking more towards the decisions elephants in zoos can make for themselves and less about how we dictate those decisions, with an increasing focus on allowing these animals some choice within their environment.

Of course as with other zoo animal training programmes, elephant management takes an almost unprecedented amount of collaboration and cooperation between trainers, managers, medical staff, and researchers. Robust and effective training programmes open up the possibilities for elephants to participate in their own care at a level that could someday surpass the objectives and benefits outlined above. Further understanding of disease progress and transmission in elephants will no doubt advance in line with advancements in elephant training (Magnuson et al. 2017). As will the ability of caretakers to probe deeper into the cognitive capacities of one of the world's most socially complex animals, with an eye towards educational outreach (Carey 2018) and the proliferation of a conservation oriented culture amongst zoo visitors. In the future, zoo elephants may be trained to rec-

ognise symbols that represent choices the animal would like to make regarding its daily routine, not unlike some great apes (Savage-Rumbaugh et al. 2001). Even activities that an elephant would like to participate in, could be presented as choices he or she makes by communicating with a trainer. As the zoos of the world continue to focus efforts towards conservation and the appreciation of nature, the effective training of zoo elephants dedicated to increased animal welfare will serve a critical role.

References

Association of Zoos and Aquariums (2018). Accreditation Standards and Related Policies. https://www.speakcdn.com/assets/2332/aza-accreditation-standards.pdf. p. 64–68.

Carey, J. (2018). Animal cognition research offers outreach opportunity. *Proceedings of the National Academy of Sciences of the United States of America* 115 (18): 4522.

Daugette, K.F., Hoppes, S., Tizard, I., and Brightsmith, D. (2012). Positive reinforcement training facilitates the voluntary participation of laboratory macaws with veterinary procedures. *Journal of Avian Medicine and Surgery* 26 (4): 248.

Fowler, M., Steffey, E., Galuppo, L., and Pasoe, J. (2000). Facilitation of Asian elephant (*Elephas maximus*) standing immobilization and anesthesia with a sling. *Journal of Zoo and Wildlife Medicine* 31 (1): 118–123.

Greco, B.J., Meehan, C.L., Miller, L.J. et al. (2016). Elephant management in North

American Zoos: environmental enrichment, feeding, exercise, and training. *PLoS One* 11 (7): e0152490. https://doi.org/10.1371/journal.pone.0152490.

Hockenhull, J. and Creighton, E. (2013). Training horses: positive reinforcement, positive punishment, and ridden behavior problems. *Journal of Veterinary Behavior* 8 (4): 245.

Magnuson, R.J., Linke, L.M., Isaza, R., and Salman, M.D. (2017). Rapid screening for mycobacterium tuberculosis complex in clinical elephant trunk wash samples. *Research in Veterinary Science* 112: 52–58. https://doi.org/10.1016/j.rvsc.2016.12.008.

Ramirez, K. *Animal Training : Successful Animal Management Through Positive Reinforcement*, vol. 1999. Chicago, IL: Shedd Aquarium.

Savage-Rumbaugh, Shanker, S.G., and Taylor, T.J. (2001). *Apes, Language, and the Human Mind*. Oxford University Press.

Box B2

Human–Elephant Interactions in Semi-captive Asian Elephants of Myanmar

Khyne U. Mar

Due to the centuries-old tradition of close proximity with humans, captive elephants in the range states of Asia are now generally recognised as 'domesticated' or 'tamed' elephants. In Myanmar, elephants have been used in logging practises since around 1800 where elephants were of great use to British imperialists (Saha 2015), especially in the construction of roads and railway tracks, and transporting goods and people in difficult terrain (Bryant 1993). In earlier days, the captive stock of timber elephants were replenished by capturing wild elephants and as a combination of this and breeding from them, the numbers of working elephants slowly increased in captivity. Alongside this rise in the captive elephant population was an increase in veterinary knowledge, training, and husbandry to help maintain the elephants' health and productivity. Most traditional elephant trainers are ethnic minorities of Karen and Khamti-Shan with an in-depth knowledge and understanding of the management, taming, and husbandry of elephants (Saha 2015).

Myanmar is home to the world's largest captive population of about 5500 Asian elephants (*Elephas maximus*); the majority of which are still working in the timber industry. Elephants are also used to work in forest patrols, tourism, transport, or as kunkies (specially trained elephants that are typically used to drive wild elephants). Traditionally, each elephant has its own mahout (head rider; see Figure B2.1) assigned from a young age, who cares for the elephant through to retirement, who they interact with on a daily basis. Timber elephants have been recorded to have a lifespan twice that of their zoo living counterparts with an average of 56 years in the wild compared to 16.9 years for zoo elephants (Clubb et al. 2008). Furthermore, captive-born timber elephants have been found to have higher levels of fitness and survival compared to their wild-caught counterparts (Mar 2007).

Management of Working Elephants in Myanmar

Almost 47% of Myanmar is covered in forest, representing one of the largest forested countries in mainland Southeast Asia. Since 1856, Myanmar has applied selective logging, called the Myanmar Selection System (MSS), where mature trees are selectively harvested with a cutting cycle of 30 years. MSS uses elephants for this process, which is considered to contribute to sustainable forest management and to maintain biodiversity at the logging site (Khai et al. 2016).

Figure B2.1 Historically in Myanmar, and other elephant habitat countries, elephants have been managed by pairing them when young with a mahout, who is often a young man, with the intention that the pair will remain together in perpetuity. *Source:* Jeroen Stevens.

At the age of four years, all calves born in captivity, and sub-adult elephants measuring 1.40 m (4.6 ft) at shoulder height if their age is unknown, are weaned and tamed/trained during the cool season (November to January) (Gale 1974). Typically, trained elephants between 5 and 17 years are used as baggage elephants and their training is continued until they get used to the verbal cue, logging/baggage harnesses, and fettering chains. Elephants are classified as fully grown at the age of 17 years. Two mahouts generally handle each individual elephant in the work force. Any bull in *musth* and some elephants with aggressive or unreliable temperaments are assigned an extra mahout who is armed with a spear (www.myanmatimber.com.mm). At 55 years old working elephants will retire, and spend most of their time roaming and foraging with one mahout who is assigned to take care of their well-being. Some bull elephants sire calves during retirement.

At night working elephants may forage unsupervised in the adjacent forests in their family groups, where they encounter tame and wild conspecifics. The working elephants are maintained as mixed herds consisting of adult males and females and calves of various ages, thus mimicking the social structure of wild elephant herds.

The main characteristic of traditional elephant management in Asia, including Myanmar, is close contact between man and elephant; which has some similarities to free-contact in zoos (Mar 2007; Kurt et al. 2008). The basic principal is that the animal is controlled by a handler/mahout through domination (Montesso 2010) using negative and positive reinforcement to modify behaviour.

Taming Training

Historically most elephants used by people that were captured from the wild were tamed and used as draught animals for various purposes. The only method that existed to tame the freshly-caught animal was to break its spirit, so that they were submissive to humans. Wild-caught elephants take more time to tame than those born in captivity, which is thought to be because of prior negative experience with humans. The breaking procedure, therefore, undoubtedly incorporates stress and compromises the welfare of the animal, especially during the first few days of taming. In modern times, elephants are treasured as flagship species with the highest level of international legal protection. Many countries now ban wild capture of both Asian and African elephants (*Loxodonta africana*) and thus the majority of captive elephants in most Asian countries are of captive-born stock. Captive-born calves grow up in a human-dominated landscape, so are half-tamed from birth, therefore traditional methods of taming are not necessary. As trust and mutual affection should be the basis of a relationship between an elephant and its trainer, the use of hooks or sharpened spikes that may traumatise the elephant are discouraged.

The first step in elephant training in Myanmar is for the mahout to intentionally take the role of an alpha animal within the herd. Once the elephant is considered to accept the mahout as dominant, evidenced by allowing him to ride on the elephant's neck, a trusting and positive relationship can then be orchestrated through positive interactions. This may take 8–10 years, and will include activities such as taking the elephant to foraging sites, hand-feeding, playing, and bathing.

About 70–100 four-year old captive born calves are trained annually in Myanma Timber Enterprise (MTE). Taming training is conducted using half-tamed young calves, and uses one of three methods:

1) The crush method (two walls) or one-sided crush method (one wall): this has a wide safety margin for trainers, but can cause injuries to elephants when they push against the walls.
2) The cradle only method: this is considered safer for elephants and trainers so is preferred. The trainee calf can be moved freely in the cradle without injuries,

except for some skin abrasions due to ropes (Oo 2010).
3) A combination of both crush and cradle.

The main aim of taming is to train the calf to understand and obey common commands such as stop, come/move, kneel, laydown, and back. It is also intended that they will develop a close relationship with their mahout. Whilst training of zoo elephants aims to support a range of activities, including feeding, exercise, training, and environmental enrichment (details in Stevenson and Walter 2006; Greco et al. 2016), Myanmar calf training focuses on desired behaviour that is useful in logging in a free-contact situation. Based on the calf's behavioural responses to cues a combination of positive and negative reinforcement strategies are used to coerce them into complying with the requested cue, after which, when the desired behaviour is performed, they are rewarded. Taming can take between 21 and 30 days.

Adult Myanmar elephants need to be trained to accept fetters being put on and released, being ridden on the neck and back, walking with fetters, kneeling down for mounting and to complete requested tasks, such as lifting, pushing, and dragging. Most of this training is achieved through desensitisation, where the trainers lessen the negative experience to a variety of procedures over time. The elephants are trained with serial (highly repetitive) exercises. During the course of training, the trainer mahouts should consider the animals' instincts and biology, but also individual variations in these, as they are likely to affect their learning abilities (McGreevy and Boakes 2007).

Conclusion

The majority of captive elephants in Asia undergo a taming procedure before they reach adulthood. Advances in animal welfare science, along with a deeper understanding of the basic needs of elephants in captivity, have enabled methods for taming elephants to be fine-tuned over the past decade. Myanmar captive elephants are not exposed to confinement-specific stressors such as restricted movement, reduced retreat space, forced proximity to humans, reduced feeding opportunities, or maintenance in abnormal social groups as seen in zoo elephants but they do tend to suffer from work- and weather-related stress. More research is needed to explore the most efficient way to tame elephants in captivity so that these elephants can survive and breed well without sustained effect of trauma they suffer in taming operations.

References

Bryant, R.L. (1993). Forest problems in colonial Burma: historical variations on contemporary themes. *Global Ecology and Biogeography Letters* 3 (4–6): 122–137.

Clubb, R., Rowcliffe, M., Lee, P. et al. (2008). Compromised survivorship in zoo elephants. *Science* 322 (5908): 1649–1649.

Gale, U.T. (1974). *Burmese Timber Elephant*. Trade Corporation.

Greco, B.J., Meehan, C.L., Miller, L.J. et al. (2016). Elephant management in North American Zoos: environmental enrichment, feeding, exercise, and training. *PLoS One* 11 (7): 26.

Khai, T.C., Mizoue, N., Kajisa, T. et al. (2016). Effects of directional felling, elephant skidding and road construction on damage to residual trees and soil in Myanmar selection system. *International Forestry Review* 18 (3): 296–305.

Kurt, F., Mar, K.U., and Garai, M. (2008). Giants in chains: history, biology and preservation of Asian elephants in Asia. In: *Elephants and Ethics: Toward a Morality of Coexistence* (eds. C. Wemmer and C.A. Christen), 327–345. Marryland, USA: The John Hopkins University Press.

Mar, K. (2007). The demography and life history of timber elephants of myanmar. PhD thesis. University College London.

McGreevy, P.D. and Boakes, R.A. (2007). *Carrots and Sticks: Principles of Animal Training*. Cambridge, UK: Cambridge University Press.

Montesso, F. (2010). The horse. In: *The UFAW Handbook on the Care and Management of Laboratory and Other Research Animals* (eds. R.C. Hubrecht and J. Kirkwood), 525–542. Wiley Blackwell.

Oo, Z.M. (2010). The training methods used in Myanma Timber Enterprise. *Gajah* 33 (33): 58–61.

Saha, J. (2015). Among the beasts of Burma: animals and the politics of colonial sensibilities, c. 1840–1940. *Journal of Social History* 48 (4): 910–932.

Stevenson, M. and Walter, O. (2006). *Management Guidelines for the Welfare of Zoo Animals. Elephants*, 2e, 1–219. British and Irish Association of Zoos and Aquariums.

Box B3

Elephant Cognition: An Overview

Sarah L. Jacobson and Joshua M. Plotnik

Although elephant behaviour has been studied extensively through decades of ethological field research (Douglas-Hamilton and Douglas-Hamilton 1975; Poole 1996; Moss et al. 2011), far less is known about elephant cognition, especially when compared to our knowledge of widely studied non-human primate cognition (Byrne et al. 2009; Irie and Hasegawa 2009; Byrne and Bates 2011). However, recently, researchers studying wild and captive elephants have made strides in better understanding the social and physical cognition of the family Elephantidae, primarily focusing on African savanna elephants (*Loxodonta africana*) and Asian elephants (*Elephas maximas*).

Sensory Modalities of Elephant Cognition

Investigating the cognition of species that are evolutionarily distinct from humans is a challenge, as the variability in sensory perspectives across such species must be accounted for in experimental design (Plotnik et al. 2013). Much emphasis within the field of animal cognition has been on the design of experiments that rely on an animal's capacity for using vision in the decision-making process, probably due to the field's focus on largely visual animals such as primates (Tomasello and Call 1997) and birds (Emery 2006). Two of the first experimental studies on elephants addressed their capacity to make visual discriminations (Rensch 1957; Nissani et al. 2005), but the results indicated that their capacity for visual discrimination varied across subjects. More recently, a group of wild elephants demonstrated their ability to classify ethnic groups visually by garment colour (Bates et al. 2007). It appears elephants can use vision to make decisions, but probably not nearly as much as they use olfaction and audition (Plotnik et al. 2014).

Elephants produce many vocalisations across a broad acoustic range and have sensitive systems for detecting these vocalisations (Poole et al. 1988; Langbauer 2000; O'Connell-Rodwell 2007). Specifically, African elephants have demonstrated that they can discriminate between familiar and unfamiliar infrasonic vocalisations (O'Connell-Rodwell et al. 2007) and recognise the vocalisations of individuals over distances greater than 1 km (McComb et al. 2003). African elephants are also able to discriminate between ethnicities, genders, and ages of humans from vocal cues (McComb et al. 2014). Elephants possess sophisticated olfactory systems as well (Shoshani et al. 2006). Studies of elephant olfactory discrimination abilities have demonstrated their ability to distinguish between ethnic groups (Bates et al. 2007) and discrete chemical compounds (Rizvanovic et al. 2013; Schmitt et al. 2018), to match different odours in a

Zoo Animal Learning and Training, First Edition. Edited by Vicky A. Melfi, Nicole R. Dorey, and Samantha J. Ward.
© 2020 John Wiley & Sons Ltd. Published 2020 by John Wiley & Sons Ltd.

match-to-sample paradigm (von Dürckheim et al. 2018), and to detect TNT (Miller et al. 2015). Plotnik et al. (2014) found that Asian elephants could use olfactory cues to find food and exclude non-rewarding choices, but not acoustic cues to find the same food. This evidence implies that elephants are using a complement of multiple senses to navigate their worlds, an important consideration for the design of future cognitive experiments.

Social Cognition

Elephants are social animals whose complex fission–fusion social structure suggests that their cognitive abilities evolved to maintain strong social relationships (Payne 2003). Ethological experiments conducted in populations of African savanna elephants have greatly informed our understanding of the social cognition of this species. From these studies, we know that elephants recognise their social companions' vocalisations (McComb et al. 2003) and odour cues, as well as track the location of other individuals in relation to themselves (Bates et al. 2008). Long-term ethological research has also described the level of information exchange between individual elephants (Lee and Moss

1999), and it is assumed that due to their long lives and slow maturation, social transfer of knowledge is likely, although this has yet to be explicitly tested. Distinct socio-cognitive abilities have also been tested with elephants in more controlled, captive settings. Plotnik et al. (2006) demonstrated that elephants are capable of mirror self-recognition (Figure B3.1), an ability linked to self-awareness and which develops in human children concurrently with empathy and sympathetic concern (Zahn-Waxler et al. 1992). This evidence, combined with a study revealing that elephants reassure others in distress (Plotnik and de Waal 2014), suggests a level of social understanding in elephants that may be comparable to that of the great apes. Elephants are also able to cooperate in a complex task in which they must recognise their partner's role to coordinate and retrieve food (Figure B3.2; Plotnik et al. 2011).

Physical Cognition

Elephants' physical cognition abilities – their knowledge of space, objects, and causal relationships – have also been investigated. Their spatial knowledge has, for instance, been investigated in studies tracking wild elephants'

Figure B3.1 TangMo the elephant inspects her mirror image at the Golden Triangle Asian Elephant Foundation in Chiang Rai, Thailand. *Source:* Joshua Plotnik.

Figure B3.2 Two elephants cooperate to pull in a table with food rewards at the Thai Elephant Conservation Center in Lampang, Thailand. *Source:* Plotnik et al. 2011; photograph: Joshua Plotnik.

movements across large distances to permanent sources of food and water during drought (Moss et al. 2011). Several experiments have reported elephants' abilities to discriminate quantities of food (Irie-Sugimoto et al. 2009; Perdue et al. 2012; Irie et al. 2018; Plotnik et al. 2019) make summations about such quantities (Irie and Hasegawa 2012), and understand means–end relationships (Irie and Hasegawa 2012). Unlike many primate species, elephants are not widely recognised for their tool-use abilities, a characteristic thought to be an important component of the evolution of physical cognition (Byrne 1997). It is possible that elephants' feeding ecology does not require the extractive foraging techniques leading to tool use in primates and perhaps their highly flexible trunk acts as a tool instead. Wild elephants have been seen using branches for body care such as scratching (Chevalier-Skolnikoff and Liska 1993), which prompted a controlled experiment with captive Asian elephants. The subjects were provided with branches and were observed modifying them to use as tools to swat flies (Hart et al. 2001). Another experimental study observed elephants blowing food to within their reach, which the authors argued could indicate the elephants' use of air as a tool (Mizuno et al. 2016). In a demonstration of insightful problem solving, one captive elephant was also observed using a plastic cube as a tool to stand on to obtain out-of-reach food (Foerder et al. 2011).

Implications and Future Directions of Elephant Research

Research on the cognitive abilities of elephants is important from a theoretical and applied perspective. Since elephants have evolved similar cognitive abilities to primates but have evolved in evolutionarily distant taxa, further comparative research could provide important insight on the convergent evolution of physical and socio-cognitive traits in a variety of intelligent species. Future research on elephant cognition, however, should focus on experimental designs that recognise the elephants' strength in acoustic and olfactory tasks so that comparisons can be made without neglecting the differences in sensory perspectives across species. Information about elephant cognition can also aid in the development of appropriate management and conservation methods for the three endangered elephant species (Mumby and Plotnik 2018). Lastly, information about elephants' advanced physical and socio-cognitive abilities, especially those that are similar to humans, can be an important educational tool to encourage the public to become more engaged in conservation efforts.

References

Bates, L.A., Sayialel, K.N., Njiraini, N.W. et al. (2007). Elephants classify human ethnic groups by odor and garment color. *Current Biology* 17 (22): 1938–1942. https://doi.org/10.1016/j.cub.2007.09.060.

Bates, L.A., Sayialel, K.N., Njiraini, N.W. et al. (2008). African elephants have expectations about the locations of out-of-sight family members. *Biology Letters* 4 (1): 34–36. https://doi.org/10.1098/rsbl.2007.0529.

Byrne, R.W. (1997). The technical intelligence hypothesis: an additional evolutionary stimulus to intelligence. In: *Machiavellian Intelligence II*, 2e (eds. A. Whiten and R. Byrne), 289–311. Cambridge University Press.

Byrne, R.W. and Bates, L.A. (2011). Elephant cognition: what we know about what elephants know. In: *The Amboseli Elephants: A Long-Term Perspective on a Long-Lived Mammal* (eds. C.J. Moss, H. Croze and P.C. Lee), 174–182. University of Chicago Press.

Byrne, R.W., Bates, L.A., and Moss, C.J. (2009). Elephant cognition in primate perspective. *Comparative Cognition and Behavior Reviews* 4: 65–79. https://doi.org/10.3819/ccbr.2009.40009.

Chevalier-Skolnikoff, S. and Liska, J. (1993). Tool use by wild and captive elephants. *Animal Behaviour* 46 (2): 209–219. https://doi.org/10.1006/anbe.1993.1183.

Douglas-Hamilton, I. and Douglas-Hamilton, O. (1975). *Among the Elephants*, 285. London: Collins.

von Dürckheim, K.E.M., Hoffman, L.C., Leslie, A. et al. (2018). African elephants (*Loxodonta africana*) display remarkable olfactory acuity in human scent matching to sample performance. *Applied Animal Behaviour Science* 200: 123–129. https://doi.org/10.1016/j.applanim.2017.12.004.

Emery, N.J. (2006). Cognitive ornithology: the evolution of avian intelligence. *Philosophical Transactions of the Royal Society, B: Biological Sciences* 361 (1465): 23–43. https://doi.org/10.1098/rstb.2005.1736.

Foerder, P., Galloway, M., Barthel, T. et al. (2011). Insightful problem solving in an Asian elephant. *PLoS ONE* 6 (8): e23251. https://doi.org/10.1371/journal.pone.0023251.

Hart, B.L., Hart, L.A., McCoy, M., and Sarath, C.R. (2001). Cognitive behaviour in Asian elephants: use and modification of branches for fly switching. *Animal Behaviour* 62 (5): 839–847. https://doi.org/10.1006/anbe.2001.1815.

Irie, N. and Hasegawa, T. (2009). Elephant psychology: what we know and what we would like to know: elephant cognition. *Japanese Psychological Research* 51 (3): 177–181. https://doi.org/10.1111/j.1468-5884.2009.00404.x.

Irie, N. and Hasegawa, T. (2012). Summation by Asian elephants (*Elephas maximus*). *Behavioral Science* 2 (2): 50–56. https://doi.org/10.3390/bs2020050.

Irie, N., Hiraiwa-Hasegawa, M., and Kutsukake, N. (2018). Unique numerical competence of Asian elephants on the relative numerosity judgment task.

Journal of Ethology https://doi.org/10.1007/s10164-018-0563-y.

Irie-Sugimoto, N., Kobayashi, T., Sato, T., and Hasegawa, T. (2009). Relative quantity judgment by Asian elephants (*Elephas maximus*). *Animal Cognition* 12 (1): 193–199. https://doi.org/10.1007/s10071-008-0185-9.

Langbauer, W.R. (2000). Elephant communication. *Zoo Biology* 19 (5): 425–445. https://doi.org/10.1002/1098-2361 (2000)19:5<425::AID-ZOO11>3.0.CO;2-A.

Lee, P.C. and Moss, C.J. (1999). The social context for learning and behavioural development among wild African elephants. *Mammalian Social Learning: Comparative and Ecological Perspectives* 72: 102.

McComb, K., Reby, D., Baker, L. et al. (2003). Long-distance communication of acoustic cues to social identity in African elephants. *Animal Behaviour* 65 (2): 317–329. https://doi.org/10.1006/anbe.2003.2047.

McComb, K., Shannon, G., Sayialel, K.N., and Moss, C. (2014). Elephants can determine ethnicity, gender, and age from acoustic cues in human voices. *Proceedings of the National Academy of Sciences* 111 (14): 5433–5438. https://doi.org/10.1073/pnas.1321543111.

Miller, A.K., Hensman, M.C., Hensman, S. et al. (2015). African elephants (*Loxodonta africana*) can detect TNT using olfaction: implications for biosensor application. *Applied Animal Behaviour Science* 171: 177–183. https://doi.org/10.1016/j.applanim.2015.08.003.

Mizuno, K., Irie, N., Hiraiwa-Hasegawa, M., and Kutsukake, N. (2016). Asian elephants acquire inaccessible food by blowing. *Animal Cognition* 19 (1): 215–222. https://doi.org/10.1007/s10071-015-0929-2.

Moss, C.J., Croze, H., and Lee, P.C. (eds.) (2011). *The Amboseli Elephants: A Long-Term Perspective on a Long-Lived Mammal*. University of Chicago Press.

Mumby, H.S. and Plotnik, J.M. (2018). Taking the elephants' perspective: remembering elephant behavior, cognition and ecology in human-elephant conflict mitigation.

Frontiers in Ecology and Evolution 6 https://doi.org/10.3389/fevo.2018.00122.

Nissani, M., Hoefler-Nissani, D., Lay, U.T., and Htun, U.W. (2005). Simultaneous visual discrimination in Asian elephants. *Journal of the Experimental Analysis of Behavior* 83 (1): 15–29. https://doi.org/10.1901/jeab.2005.34-04.

O'Connell-Rodwell, C.E. (2007). Keeping an "ear" to the ground: seismic communication in elephants. *Physiology* 22 (4): 287–294. https://doi.org/10.1152/physiol.00008.2007.

O'Connell-Rodwell, C.E., Wood, J.D., Kinzley, C. et al. (2007). Wild African elephants (*Loxodonta africana*) discriminate between familiar and unfamiliar conspecific seismic alarm calls. *The Journal of the Acoustical Society of America* 122 (2): 823–830. https://doi.org/10.1121/1.2747161.

Payne, K. (2003). Sources of Social Complexity in the Three Elephant Species. In: *Animal Social Complexity: Intelligence, Culture, and Individualized Societies* (eds. W. de FBM and P.L. Tyack), 57–86. Harvard University Press.

Perdue, B.M., Talbot, C.F., Stone, A.M., and Beran, M.J. (2012). Putting the elephant back in the herd: elephant relative quantity judgments match those of other species. *Animal Cognition* 15 (5): 955–961. https://doi.org/10.1007/s10071-012-0521-y.

Plotnik, J.M., Brubaker, D.L., Dale, R., Tiller, L.N., Mumby, H.S., and Clayton, N.S. (2019). Elephants have a nose for quantity. *Proceedings of the National Academy of Sciences*, 116 (25): 12566–12571. https://doi.org/10.1073/pnas.1818284116.

Plotnik, J.M. and de Waal, F.B.M. (2014). Asian elephants (*Elephas maximus*) reassure others in distress. *Peer J* 2: e278. https://doi.org/10.7717/peerj.278.

Plotnik, J.M., de Waal, F.B.M., and Reiss, D. (2006). Self-recognition in an Asian elephant. *Proceedings of the National Academy of Sciences of the United States of America* 103 (45): 17053–17057. https://doi.org/10.1073/pnas.0608062103.

Plotnik, J.M., Lair, R., Suphachoksahakun, W., and de Waal, F.B.M. (2011). Elephants know when they need a helping trunk in a cooperative task. *Proceedings of the National Academy of Sciences of the United States of America* 108 (12): 5116–5121. https://doi.org/10.1073/pnas.1101765108.

Plotnik, J.M., Pokorny, J.J., Keratimanochaya, T. et al. (2013). Visual cues given by humans are not sufficient for Asian elephants (*Elephas maximus*) to find hidden food. *PLoS ONE* 8 (4): e61174. https://doi.org/10.1371/journal.pone.0061174.

Plotnik, J.M., Shaw, R.C., Brubaker, D.L. et al. (2014). Thinking with their trunks: elephants use smell but not sound to locate food and exclude nonrewarding alternatives. *Animal Behaviour* 88: 91–98. https://doi.org/10.1016/j.anbehav.2013.11.011.

Poole, J. (1996). The African elephant. In: *Studying Elephants* (ed. K. Kangwana), 1–8. African Wildlife Foundation.

Poole, J.H., Payne, K., Langbauer, W.R., and Moss, C.J. (1988). The social contexts of some very low frequency calls of African elephants. *Behavioral Ecology and Sociobiology* 22 (6): 385–392. https://doi.org/10.1007/BF00294975.

Rensch, B. (1957). The intelligence of elephants. *Scientific American* 196 (2): 44–49.

Rizvanovic, A., Amundin, M., and Laska, M. (2013). Olfactory discrimination ability of Asian elephants (*Elephas maximus*) for structurally related odorants. *Chemical Senses* 38 (2): 107–118. https://doi.org/10.1093/chemse/bjs097.

Schmitt, M.H., Shuttleworth, A., Ward, D., and Shrader, A.M. (2018). African elephants use plant odours to make foraging decisions across multiple spatial scales. *Animal Behaviour* 141: 17–27. https://doi.org/10.1016/j.anbehav.2018.04.016.

Shoshani, J., Kupsky, W.J., and Marchant, G.H. (2006). Elephant brain. *Brain Research Bulletin* 70 (2): 124–157. https://doi.org/10.1016/j.brainresbull.2006.03.016.

Tomasello, M. and Call, J. (1997). *Primate Cognition*. Oxford University Press.

Zahn-Waxler, C., Radke-Yarrow, M., Wagner, E., and Chapman, M. (1992). Development of concern for others. *Developmental Psychology* 28 (1): 126–136. https://doi.org/10.1037/0012-1649.28.1.126.

Box B4

Marine Mammal Training

Sabrina Brando

Training has many benefits for captive marine mammals and should be part of a professional animal care programme. It facilitates daily and regular management procedures, and allows for participation in research, education and conservation programmes. Training and animal learning in a broader sense, also offers animals more complexity in their environment. For example choice can be provided to the animals, by providing them with different stimuli, like photos which represent different toys or activities, which they can then choose between (see Figure B4.1). Furthermore, instead of the trainer making all the decisions for the animal, the animal can be trained to indicate their preferences. For example, animals can be trained to press a lever that has been associated with different people or animals to indicate who they want to spend time with (Adams and MacDonald 2018). This technique can also be used for allowing the animal to choose a reinforcer (see Figure B4.2) (Fernandez et al. 2004; Gaalema et al. 2011), enrichment items (Bashaw et al. 2016; Fay and Miller 2015; Mehrkam and Dorey 2014, 2015), or even between enrichment and training (Dorey et al. 2015). This box will give a short overview of marine mammal training examples and opportunities.

Today many different species of marine mammals are trained using positive reinforcement techniques to enable the animals to collaborate in their own daily care, (for example the treatment of health issues such as sea lions and seals trained to accept eye drops to treat cataracts: Colitz et al. 2010; Gage 2011), in research, education and conservation programmes (for a review see Kuczaj and Xitco 2002; Brando 2010). Initially the training of marine mammals was motivated by the entertainment industry to present animals performing and participating in movies and TV series. Marine Studios (now Marineland) in Florida was one of the first 'oceanarium' facilities housing cetaceans and pinnipeds, starting with bottlenose dolphins (*Tursiops truncates*) in 1938 (Marineland 2018). The purpose of the facility at the time was to film underwater footage for TV series and movies. Thus dolphins were trained to preform many behaviours that aren't seen in the wild, such as leaping through a hoop suspended above the water or jumping to touch a ball or behaviours performed together with the trainers, like rides on the back of the animal, and jumping over ropes together. In addition to these behaviours, the facility also displayed natural behaviours such as aerial performances including jumps, spins, fast surface swims, and summersaults. All of these behaviours, whether natural or unnatural, are trained behaviours shaped by the trainers who work with the animals on a regular basis.

Zoo Animal Learning and Training, First Edition. Edited by Vicky A. Melfi, Nicole R. Dorey, and Samantha J. Ward.
© 2020 John Wiley & Sons Ltd. Published 2020 by John Wiley & Sons Ltd.

Figure B4.1 Choice can be given to marine mammals in captivity; in this instance the animals had learned that they could have a choice between two different objects of activities, which were represented by different photos. *Source:* Sabrina Brando.

Figure B4.2 Training programmes can be set-up whereby the animal chooses which reinforcer it will receive after the successful performance of a behaviour. In this photograph the dolphin is choosing between two different 'buoys'. *Source:* Sabrina Brando.

Training Logistics

Some individual animals and/or species such as Steller sealions (*Eumetopias jubatus*) and polar bears (*Ursus maritimus*) can be dangerous, and so for the safety of all those involved can be trained using protected-contact techniques. For example, Steller sea lions could be trained using a partially open board (see Figure B4.3) that the animal can back onto when providing a blood sample

Figure B4.3 In situations where animals might be dangerous it can be prudent (and necessary) to train in a protected-contact situation; where the animal and trainer are separated in space by a barrier. *Source:* Robert van Schie.

from their hind flippers. The board gives the trainer more time to back away if the animal leaves the station or if they turn towards the trainer. Some species may require multiple trainers that work together during non-pro-tected contact with large marine mammals such as walruses (*Odobenus rosmarus*) or orca (*Orcinus orca*) in a way to maintain safety. Trainers can therefore watch each other whilst training to make sure the han-dler does not place themselves between the animal and water, which can be a hazardous position to be in and potentially provide feedback on the training session.

Acoustic, visual, and tactile cues can be used both above and below the water level to facilitate a variety of behaviours. The stimu-lus chosen should fit with the need of the behaviour being trained. For example, acous-tic and some visual stimuli can be perceived over long distances. Underwater tones such as electronic sounds, words and even musical instruments can also cue behaviours in pres-entations without trainers being visible to the spectators. Interactive screens with symbols work well, some of which can be activated by

echolocation, to give an individual the oppor-tunity to request preferred reinforcers from their trainer (Starkhammar et al. 2007).

The provision of food for most marine mammals is considerably different compared to many other zoo-housed species. Most species housed in zoos, e.g. bears, elephants, primates, parrot, and other species, might be given multiple foraging opportunities throughout the day. Various means of food provisions and distributions are used, like scatter feeds, puzzles, or other feeding opportunities in the enclosures. Most marine mammal species however, such as sea lions, seals, and dolphins, obtain almost all their food from the trainer's hand, in training ses-sions and presentations, and are therefore dependent on successful participation. A very small amount of food is given in envi-ronment enrichment devices or as frozen treats. A 'free feed' is a session where no behaviours are trained, and the animal is just fed a part of the diet. During a 'free feed' the animals are mostly stationed in front of the trainer whilst they are fed the food and these sessions are a regular practice in marine mammals care and management. Instead of these 'free feed' types of sessions, it should be considered to have more environmental enrichment activities through which food can be obtained, through, e.g. foraging or cognitive tasks.

Trainer–Animal Relationship

Hosey (2008) highlighted the importance of positive human–animal interactions within zoos, and data from terrestrial mammal (zoo) research indicates that positive reinforcement training acts as a means of positive interactions between zoo keepers and their animals (Ward and Melfi 2013) and that these positive interactions lead to positive human–animal relationships (Ward and Melfi 2015).

Marine mammal trainers not only focus on building relationships based on positive interactions and positive reinforcement, but will carefully observe animal behaviour and preferences, and pay attention to the effect of human body language, posture and communication on the animals in their care (Davis and Harris 2006). The use of fish, play, toys, high energy exercises, stroking, and playing games such as 'hide and seek', may all contribute to building positive relationships based on positive interactions. Clegg et al. (2015) identified 11 critical components to ensure good dolphin welfare, one of these being good human-animal relationships and was measured using the animal's response to the trainer whilst not under stimulus control and non-food tactile interactions.

Conclusion

As in all animal training, when marine mammals do not want to collaborate in training it is important to evaluate the situation and consider why the animal chooses not to participate in order to find a solution to the situation. Marine mammals partaking in presentations and/or interactive programmes should be offered the opportunity to participate, or to opt out. When participating in a session an area of the pool or land should be set aside to provide a retreat away from visitors and trainers.

Whilst training is used and relied upon a lot in marine mammal care programmes, training is only one of many tools to promote positive welfare and will only constitute a small part of an animal's day and as with other species housed in captivity, marine mammal welfare requires adopting a 24/7 approach. Zoo professionals spend only a limited amount of time in the zoo training animals and so it is important to consider the time when the animal is not being trained. Providing a complex environment with various features such as different pools, vegetation, underwater activities, and haul out areas (depending on the species), as well as having other individuals to interact with is crucial.

References

Adams, L.C. and MacDonald, S.E. (2018). Spontaneous preference for primate photographs in Sumatran orangutans (*Pongo abelii*). *International Journal of Comparative Psychology* 31: 1–16. https://escholarship.org/uc/item/08t203bk.

Bashaw, M.J., Gibson, M.D., Schowe, D.M., and Kucher, A.S. (2016). Does enrichment improve reptile welfare? Leopard geckos (*Eublepharis macularius*) respond to five types of environmental enrichment. *Applied Animal Behaviour Science* 184: 150–160.

Brando, S.I.C.A. (2010). Advances in husbandry training in marine mammal care programs. *International Journal of Comparative Psychology* 23: 777–791.

Clegg, I.L.K., Borger-Turner, J.L., and Eskelinen, H.C. (2015). C-well: the development of a welfare assessment index for captive bottlenose dolphins (*Tursiops truncatus*). *Animal Welfare* 24 (3): 267–282.

Colitz, C.M., Saville, W.J., Renner, M.S. et al. (2010). Risk factors associated with cataracts and lens luxations in captive pinnipeds in the United States and the Bahamas. *Journal of the American Veterinary Medical Association* 237 (4): 429–436.

Davis, C. and Harris, G. (2006). Redefining our relationships with the animals we train – leadership and posture. *Soundings* 31 (4): 6–8.

Dorey, N.R., Mehrkam, L.R., and Tacey, J. (2015). A method to assess relative preference for training and environmental enrichment in captive wolves (*Canis lupus and Canis lupus arctos*). *Zoo Biology* 34 (6): 513–517.

Fay, C. and Miller, L.J. (2015). Utilizing scents as environmental enrichment: preference assessment and application with Rothschild giraffe. *Animal Behavior and Cognition* 2: 285–291. https://doi.org/10.12966/abc.08.07.2015

Fernandez, E.J., Dorey, N., and Rosales-Ruiz, J. (2004). A two choice preference assessment with five cotton-top tamarins (*Saguinus oedipus*). *Journal of Applied Animal Welfare Science* 7 (3): 163–169. https://doi.org/10.1207/s15327604jaws0703_2

Gaalema, D.E., Perdue, B.M., and Kelling, A.S. (2011). Food preference, keeper ratings, and reinforcer effectiveness in exotic animals: the value of systematic testing. *Journal of Applied Animal Welfare Science* 14: 33–41.

Gage, L.J. (2011). Captive pinniped eye problems: we can do better. *The Journal of Marine Animals and Their Ecology* 4: 25–28.

Hosey, G. (2008). A preliminary model of human–animal relationships in the zoo. *Applied Animal Behaviour Science* 109 (2): 105–127.

Kuczaj, S.A. II and Xitco, M.J. Jr. (2002). It takes more than fish: the psychology of marine mammal training. *International Journal of Comparative Psychology* 15: 186–200.

Marineland (2018). A history of adventure. https://marineland.net/our-history (accessed 7 October 2018).

Mehrkam, L.R. and Dorey, N.R. (2014). Is preference a predictor of enrichment efficacy in Galapagos tortoises (*Chelonoidis nigra*)? *Zoo Biology* 33: 275–284.

Mehrkam, L.R. and Dorey, N.R. (2015). Preference assessments in the zoo: keeper and staff predictions of enrichment preference. *Zoo Biology* 34: 418–430.

Starkhammar, J., Amundin, M., Olsén, H., et al. (2007). Acoustic touch screen for dolphins, first application of ELVIS – an echo-location visualization and interface system. *Proceedings of the 4th International Conference on Bio-Acoustics,* Loughborough, UK (10–12 April 2007). Institute of Acoustics, 29(3): 63–68.

Ward, S.J. and Melfi, V. (2013). The implications of husbandry training on zoo animal response rates. *Applied Animal Behaviour Science* 147: 179–185.

Ward, S.J. and Melfi, V. (2015). Keeper-animal interactions: differences between the behaviour of zoo animals affect stockmanship. *PLoS ONE* 10 (10): e0140237. https://doi.org/10.1371/journal.pone.0140237.

Box B5

Cognitive Abilities of Marine Mammals

Gordon B. Bauer

Scientific studies of dolphins (order: Cetacea) and sea lions (order: Pinnipedia) reveal a constellation of sophisticated cognitive abilities, including imitation, learning set acquisition, sequence learning, problem solving, and concept learning. Manatees (order: Sirenia) are the only marine mammal herbivores and reveal an interesting set of sensory processes, although little is known about their broader cognitive abilities. These three orders of marine mammals consist of many species, but cognitive/perceptual studies are dominated by only a few, bottlenose dolphins (*Tursiops truncatus*) amongst the cetaceans, California sea lions (*Zalophus californianus*), amongst the pinnipeds, and West Indian manatees (*Trichechus manatus*), amongst the sirenians. These species are commonly found in captivity and therefore are readily available for behavioural study. However, it needs to be recognised that these orders contain diverse species from a wide range of habitats, which may be associated with different cognitive characteristics. There is a paucity of investigations of cognition in several other marine mammal taxa, including sea otters (*Enhydra lutris*) and polar bears (*Ursis maritimus*), so they are not discussed in this review.

Sensory perception is important for understanding cognition because it constrains the type of information an animal has available for processing. Also, discrimination learning is inherent in the methodology involved in behaviourally testing sensory processes, so psychophysical reports of sensory detection and discrimination imply discrimination learning abilities, an important cognitive trait. Bottlenose dolphins, hereafter referred to simply as dolphins, have exquisite hearing abilities including high frequency hearing extending well into the ultrasound range (Johnson 1967) auditory temporal processing abilities (Supin et al. 2001), and sensitive discrimination of sound amplitude (Au and Hastings 2008). Their hearing is integrated with sound production to generate highly sensitive echolocation abilities (Au 1993). By typical mammalian standards, they have good visual acuity, although modest compared to primates. Their underwater and in-air vision brackets the United States of America's criterion for legal blindness (10 arc minutes); somewhat better than legally blind underwater and somewhat worse in air (Herman et al. 1975). They are monochromats (Ahnelt and Kolb 2000) that most likely see their world in shades of grey (Madsen and Herman 1980) (but see Griebel and Peichel 2003 for an alternative view; colour vision based on rod-cone interaction). Dolphin sense of touch has been minimally investigated (Dehnhardt and Mauck 2008), although morphological and physiological studies suggest that they are sensitive in the facial, cranial, and genital regions (Ridgway and Carder 1990). The small size or absence of olfactory bulbs suggest that toothed

Zoo Animal Learning and Training, First Edition. Edited by Vicky A. Melfi, Nicole R. Dorey, and Samantha J. Ward.
© 2020 John Wiley & Sons Ltd. Published 2020 by John Wiley & Sons Ltd.

whales, the odontocetes, have minimal or no sense of smell, an important path to cognition in terrestrial mammals (e.g. Otto and Eichenbaum 1992), whilst the presence of innervated taste buds and results of psychophysical testing suggest that dolphins may have a satisfactory sensitivity for some compounds, including citric acid (sour), quinine sulphate (bitter), and sodium chloride (salty) (Kusnetsov 1990; Nachtigall and Hall 1984). The baleen whales have better developed olfactory bulbs but as yet little is known about their chemical detection capabilities (Pihlström 2008).

Dolphins have demonstrated the ability to categorise objects in same/different and match-to-sample paradigms. They can develop learning sets for acoustic stimuli, measured by second trial performance on two choice discriminations, with appropriate training (Herman 1980). They form representations of objects, not just their perceptual characteristics. Object representation is illustrated by excellent cross-modal matching abilities, for example, in which dolphins discriminate objects presented visually using a different modality, echolocation, and vice versa (Harley et al. 1996, 2003; Pack et al. 2004).

Dolphins can respond appropriately to chains of commands delivered by arm gestures or acoustic signals representing objects, actions, or modifiers to guide behaviour. For example, 'surface-pipe-fetch-bottom-hoop', led the dolphin to take the pipe on the surface of the water to a hoop on the bottom of the tank when there were a variety of other objects in diverse locations (Herman 1986). Although there is some controversy over whether to describe this behaviour as understanding language, demonstrating semantic and syntactic abilities, or more basic rule learning, the fact is that the resulting behaviour is complex (Herman 1988, 1989; Schusterman and Gisiner 1988, 1989).

Dolphins can be observed responding to hand signals in commercial aquaria, as well as in scientific studies, but they can also respond to signals delivered on TV or video monitors. They can even respond correctly to signs when the only stimuli observable on the screen are bright balls against a black background tracking the pattern of hand signals (Herman et al. 1990) and can mimic behaviours viewed on a video monitor (Herman et al. 1993; Pack 2015).

Dolphins have excellent mimetic skills in both behavioural and vocal domains, an unusual ability amongst mammals outside of humans, although also seen in some birds. Dolphins imitate other dolphin phonations, trainer whistles, and computer-generated sounds (Richards et al. 1984). They mimic the behaviour of conspecifics and other species such as humans and seals. They mimic behaviour and sounds spontaneously as well as under stimulus control by a trainer (Herman 2002).

Bottlenose and rough-toothed dolphins (*Steno bredanensis*) innovate by responding correctly to commands to produce new behaviours (Pryor et al. 1969). This ability has been cleverly used in conjunction with imitation to demonstrate the ability to synchronously perform an innovated behaviour with another dolphin (Herman 2002). Furthermore, when embedded in a set of diverse commands a dolphin can accurately do a directed behaviour, imitate the behaviour of another dolphin, or perform a behaviour that it had executed in a previous trial (Mercado et al. 1999). The last ability indicates memory for specific past actions.

A dolphin can initiate and respond to pointing gestures (Xitco et al. 2004) and identify objects by eavesdropping on the echolocation of another dolphin (Xitco and Roitblat 1996). The ability to mimic behaviour, initiate and respond to pointing, and eavesdrop all suggest ways that dolphins can share information about their environments with conspecifics. They can also use tools for foraging in natural environments (e.g. sponge fishing) (Mann et al. 2008; Smolker et al. 1997) and solve problems to gain food rewards using tools in captive settings. For example, Gory (reviewed in Kuczaj and Walker 2006) reported that dolphins learned observationally to drop four weights (tools) into an experimental apparatus to release a fish reward. The dolphins also exhibited planning as demonstrated by the fact that when the weights were moved a long distance away from the apparatus, they collected up to four or five weights at a time in contrast to picking up only one weight at a time and dropping it into the apparatus when the weights were close.

Overall, bottlenose dolphins display a complex array of cognitive abilities that would be important to investigate in other toothed-whale species. Pinnipeds, primarily the California sea lion have been studied on many similar tasks with similar results, as described below (see Figure B5.1).

Pinnipeds have good underwater and in-air visual acuity (Schusterman 1972), hearing sensitivity extending into the ultrasonic range (Cunningham et al. 2014), and modest sound localisation (Supin et al. 2001). Although they do not have echolocation, they have exquisite active touch (discrimination of shapes) and passive touch (detection and following of hydrodynamic stimuli) abilities, using the sensory hairs (vibrissae) on their faces (Dehnhardt and Mauck 2008), which probably helps them navigate. Pinnipeds can also discriminate saline gradients, which could serve as an effective navigation tool. They demonstrate abilities to categorise objects in match-to-sample (MTS) and oddity paradigms and acquire learning sets (learning to learn) (Schusterman and Kastak 2002). They have been trained successfully in artificial 'language' sign sequences, similar to those demonstrated by dolphins (Schusterman and Kastak 2002). They are also flexible vocal learners with at least some capacity for vocal imitation (Reichmuth and Casey 2014). Schusterman and colleagues have engaged in comprehensive studies that demonstrate that sea lions can be trained to develop a concept of stimulus equivalence (Schusterman et al. 2000), i.e. categorisation of dissimilar stimuli, which is important in understanding cognitive and social concepts. These concepts can be developed in MTS procedures, which pinnipeds readily acquire and maintain. For example, a sea lion has been shown to remember an MTS task with no loss of accuracy after 10 years (Reichmuth Kastak and Schusterman 2002).

The primary focus of captive behavioural research of the Sirenia has been the sensory processes of manatees. Manatees have poor

Figure B5.1 Most of our knowledge about pinniped cognitive ability has come from our understanding of Californian sealions, *Zalophus californianus*; as a consequence this species is frequently featured in shows which demonstrate their various cognitive skills. *Source:* Jeroen Stevens.

visual acuity, about 20 arc minutes (Bauer et al. 2003; Mass et al. 1997). Although augmented with dichromatic colour vision, apparently unique amongst marine mammals (Ahnelt and Kolb 2000; Griebel and Schmid 1996; Newman and Robinson 2005). Their excellent hearing sensitivity extends into the ultrasonic range (Gaspard et al. 2012; Gerstein et al. 1999), they have high temporal processing rates (Mann et al. 2005), good sound localisation for broadband sounds (Colbert-Luke et al. 2015), excellent tactile discrimination of textures (Bachteler and Dehnhardt 1999; Bauer et al. 2012), and highly sensitive detection of hydrodynamic stimuli, aided by the vibrissae that cover their bodies (Gaspard et al. 2013, 2017; Reep et al. 2011). Although formal tests of memory have not been conducted, memory of experimental procedures over a year and memory for navigation routes and destinations in the wild (Marsh et al. 2011; Reep and Bonde 2006), suggest ability to retain information over long time periods.

The marine mammals comprise diverse taxa with members only distantly related evolutionarily. Their large size makes them difficult in many cases to bring into captive settings, however, those that are, are frequently studied in small samples. Caution should be used in generalising the described attributes of the few species studied to other marine mammals, which may vary substantially in a wide range of morphological, sensory, and ecological parameters.

More thorough reviews of the cognitive and/or sensory characteristics of cetaceans, pinnipeds, and sirenians can be found in the following sources:

Marine mammals in general: Clark (2013), Dehnhardt (2002), Supin et al. (2001), and Thewissen and Nummela (2008). Kuczaj also edited two special issues of the *International Journal of Comparative Psychology*, (2010), on captive research, most of which addressed cognition and sensory perception of marine mammals.

Cetaceans: Clark (2013), Hanke and Erdsack (2015), Harley and Bauer (2017), and Pack (2015).

Pinnipeds: Schusterman and Kastak (2002).

Sirenians: Bauer et al. (2010), Bauer and Reep (2018), and Reep and Bonde (2006).

References

Ahnelt, P.K. and Kolb, H. (2000). The mammalian photoreceptor mosaic-adaptive design. *Progress in Retinal and Eye Research* 19: 711–777.

Au, W.W. and Hastings, M.C. (2008). *Principles of Marine Bioacoustics*. New York: Springer.

Au, W.W.L. (1993). *The Sonar of Dolphins*. New York: Springer.

Bachteler, D. and Dehnhardt, G. (1999). Active touch performance in the Antillean manatee: evidence for a functional differentiation of the facial tactile hairs. *Zoology* 102: 61–69.

Bauer, G.B., Colbert, D., Fellner, W. et al. (2003). Underwater visual acuity of Florida manatees, *Trichechus manatus latirostris*. *International Journal of Comparative Psychology* 16: 130–142.

Bauer, G.B., Colbert, D.E., and Gaspard, J.C. (2010). Learning about manatees: a collaborative program between New College of Florida and Mote Marine Laboratory to conduct laboratory research for manatee conservation. *International Journal of Comparative Psychology* 23: 811–825.

Bauer, G.B., Gaspard, J.C. III, Colbert, D.E. et al. (2012). Tactile discrimination of textures by Florida manatees (*Trichechus manatus latirostris*). *Marine Mammal Science* 28: 456–471.

Bauer, G.B. and Reep, R.L. (2018). Sirenian sensory systems. In: *Encyclopedia of Animal Cognition and Behavior* (eds. J. Vonk and T.K. Shackelford). Springer https://doi.org/10.1007/978-3-319-47829-6_1318-1.

Clark, F.E. (2013). Marine mammal cognition and captive care: a proposal for cognitive enrichment in zoos and aquariums. *Journal of Zoo and Aquarium Research* 1: 1–6.

Colbert-Luke, D.E., Gaspard, J.C. III, Reep, R. et al. (2015). Eight-choice sound localization by manatees: performance abilities and head related transfer functions. *Journal of Comparative Physiology A* 201: 249–259.

Cunningham, K., Southall, B.L., and Reichmuth, C. (2014). Auditory sensitivity of seals and sea lions in complex listening scenarios. *Journal of the Acoustical Society of America* 136: 3410–3421.

Dehnhardt, G. (2002). Sensory systems, 1. In: *Marine Mammal Biology: An Evolutionary Approach* (ed. A.R. Hoelzel), 116–141. Oxford: Blackwell.

Dehnhardt, G. and Mauck, B. (2008). Mechanoreception in secondarily aquatic vertebrates. In: *Sensory Evolution on the Threshold: Adaptations in Secondarily Aquatic Vertebrates* (eds. J.G.M. Thewissen and S. Nummela), 295–314. Berkeley: University of California Press.

Gaspard, J.C. III, Bauer, G.B., Mann, D.A. et al. (2017). Detection of hydrodynamic stimuli by the postcranial sensory body of Florida manatees (*Trichechus manatus latirostris*). *Journal of Comparative Physiology A* 203: 111–120. https://doi.org/10.1007/s00359-016-1142-8.

Gaspard, J.C. III, Bauer, G.B., Reep, R.L. et al. (2012). Audiogram and auditory critical ratios of two Florida manatees (*Trichechus manatus latirostris*). *Journal of Experimental Biology* 215: 1442–1447.

Gerstein, E.R., Gerstein, L., Forsythe, S.E., and Blue, J.E. (1999). The underwater audiogram of the West Indian manatee (*Trichechus manatus*). *Journal of the Acoustical Society of America* 105: 3575–3583.

Griebel, U. and Peichel, L. (2003). Colour vision in aquatic mammal – facts and open questions. *Aquatic Mammals* 29: 18–30.

Griebel, U. and Schmid, A. (1996). Color vision in the manatee (*Trichechus manatus*). *Vision Research* 36: 2747–2757.

Hanke, W. and Erdsack, N. (2015). Ecology and evolution of dolphin sensory systems. In: *Dolphin Communication and Cognition: Past, Present, and Future* (eds. D.L. Herzing and C.M. Johnson), 49–74. Cambridge: MIT Press.

Harley, H.E. and Bauer, G.B. (2017). Cetacean cognition. In: *Encyclopedia of Animal Cognition and Behavior* (eds. J. Vonk and T.K. Shackelford). Springer https://doi.org/10.1007/978-3-319-47829-6_997-1.

Harley, H.E., Putnam, E.A., and Roitblat, H.L. (2003). Bottlenose dolphins perceive object features through echolocation. *Nature* 424: 667–669.

Harley, H.E., Roitblat, H.L., and Nachtigall, P.E. (1996). Object recognition in the bottlenose dolphin (*Tursiops truncatus*): integration of visual and echoic information. *Journal of Experimental Psychology: Animal Behavior Processes* 22: 164–174.

Herman, L.M. (1980). Cognitive characteristics of dolphins. In: *Cetacean Behavior: Mechanisms and Functions* (ed. L.M. Herman), 363–429. Malabar: Krieger Publishing.

Herman, L.M. (1986). Cognition and language competencies of bottlenose dolphins. In: *Dolphin Cognition and Behavior: A Comparative Approach* (eds. R.J. Schusterman, J. Thomas and F.G. Wood), 221–251. Hillsdale, NJ: Lawrence Erlbaum Associates.

Herman, L.M. (1988). The language of animal language research. *Psychological Record* 38: 349–362.

Herman, L.M. (1989). In which procrustean bed does the sea lion sleep tonight? *Psychological Record* 39: 19–49.

Herman, L.M. (2002). Vocal, social, and self-imitation by bottlenosed dolphins. In: *Imitation in Animals and Artifacts* (eds. C. Nehaniv and K. Dautenhahn), 63–108. New York: Academic Press.

Herman, L.M., Morrel-Samuels, P., and Pack, A.A. (1990). Bottlenosed dolphin and human recognition of veridical and degraded video displays of an artificial gestural language. *Journal of Experimental Psychology: General* 119: 215–230.

Herman, L.M., Pack, A.A., and Morrel-Samuels, P. (1993). Representational and conceptual skills of dolphins. In: *Language and*

Communication: Comparative Perspectives (eds. H.L. Roitblat, L.M. Herman and P.E. Nachtigall), 403–442. Hillsdale: Erlbaum.

Herman, L.M., Peacock, M.F., Yunker, M.P., and Madsen, C.J. (1975). Double-slit pupil yields equivalent aerial and underwater diurnal acuity. *Science* 189: 650–652.

Gaspard III, J.C., Bauer, G.B., Reep, R.L. et al. (2013). Detection of hydrodynamic stimuli by the Florida manatee (*Trichechus manatus latirostris*). *Journal of Comparative Physiology A* 199: 441–450.

Johnson, C.S. (1967). Sound detection thresholds in marine mammals. In: *Marine Bio-Acoustics II* (ed. W.N. Tavolga), 247–260. New York: Pergamon.

Kuczaj, S.A. II (2010). Research with captive marine mammals is important, part I/part II. *International Journal of Comparative Psychology* 23 (3,4): 225–226.

Kuczaj, S.A. II and Walker, R.T. (2006). How do dolphins solve problems? In: *Comparative Cognition: Experimental Explorations of Animal Intelligence* (eds. E.A. Wasserman and T.R. Zentall), 580–601. New York: Oxford University Press.

Kusnetsov, V.B. (1990). Chemical sense in the dolphin (*Tursiops truncatus*): quasi olfaction. In: *Sensory Abilities of Cetaceans: Laboratory and Field Evidence* (eds. J.A. Thomas and R.A. Kastelein), 481–503. New York: Plenum.

Madsen, C.M. and Herman, L.M. (1980). Social and ecological correlates of cetacean vision and visual appearance. In: *Cetacean Behavior: Mechanisms and Functions* (ed. L.M. Herman), 101–147. Malabar: Krieger Publishing.

Mann, D., Hill, M., Casper, B. et al. (2005). Temporal resolution of the Florida manatee (*Trichechus manatus latirostris*) auditory system. *Journal of Comparative Physiology* 191: 903–908.

Mann, J., Sargeant, B.L., Watson-Capps, J.J. et al. (2008). Why do dolphins carry sponges? *PLoS One* 3: e3868. https://doi.org/10.1371/journal.pone.0003868.

Marsh, H., O'Shea, T.J., and Reynolds, J.E. III (2011). *Ecology and Conservation of Sirenia: Dugongs and Manatees.* Cambridge: Cambridge University Press.

Mass, A.M., Odell, D.K., Ketten, D.R., and Supin, A.Y. (1997). Ganglion layer topography and retinal resolution of the Caribbean manatee *Trichechus manatus latirostris*. *Doklady Biological Sciences* 355: 392–394.

Mercado, E., Uyeyama, R.K., Pack, A.A., and Herman, L.M. (1999). Memory for action events in the bottlenose dolphin. *Animal Cognition* 2: 17–25.

Nachtigall, P.E. and Hall, R.W. (1984). Taste reception in the bottlenosed dolphin. *Acta Zoologica Fennica* 172 (Supplement): 147–148.

Newman, L.A. and Robinson, P.R. (2005). Cone visual pigments of aquatic mammals. *Visual Neuroscience* 22: 873–879.

Otto, T. and Eichenbaum, H. (1992). Olfactory learning and memory in the rat: a "model system" for studies of the neurobiology of memory. In: *The Science of Olfaction* (eds. M. Serby and K. Chobor), 213–244. New York: Springer-Verlag.

Pack, A.A. (2015). Experimental studies of dolphin cognitive abilities. In: *Dolphin Communication and Cognition: Past, Present, and Future* (eds. D.L. Herzing and C.M. Johnson), 175–200. Cambridge: MIT Press.

Pack, A.A., Herman, L.M., and Hoffmann-Kuhnt, M. (2004). Dolphin echolocation shape perception: from sound to object. In: *Echolocation in Bats and Dolphins* (eds. J.A. Thomas, C. Moss and M. Vater), 288–308. Chicago: University of Chicago Press.

Pihlström, H. (2008). Comparative anatomy and physiology of chemical senses in aquatic mammals. In: *Sensory Evolution on the Threshold: Adaptations in Secondarily Aquatic Vertebrates* (eds. J.G.M. Thewissen and S. Nummela), 95–109. Berkeley: University of California Press.

Pryor, K.W., Haag, R., and O'Reilly, J. (1969). The creative porpoise: training for novel behavior. *Journal of the Experimental Analysis of Behavior* 12: 653–661.

Reep, R.L. and Bonde, R.K. (2006). *The Florida Manatee: Biology and Conservation.* University Press of Florida.

Reep, R.L., Gaspard, J.C., Sarko, D.K. et al. (2011). Manatee vibrissae: evidence for a "lateral line" function. *Annals of the New York Academy of Sciences* 1225: 101–109.

Reichmuth, C. and Casey, C. (2014). Vocal learning in seals, sea lions, and walruses. *Current Opinion in Neurobiology* 28: 66–71.

Reichmuth Kastak, C. and Schusterman, R.J. (2002). Long-term memory for concepts in a California sea lion (*Zalophus californianus*). *Animal Cognition* 5: 225–232. https://doi.org/10.1007/s10071-002-0153-8.

Richards, D.G., Wolz, J.P., and Herman, L.M. (1984). Vocal mimicry of computer generated sounds and vocal labeling of objects by a bottlenosed dolphin (*Tursiops truncatus*). *Journal of Comparative Psychology* 98: 10–28.

Ridgway, S.H. and Carder, D.A. (1990). Tactile sensitivity, somatosensory responses, skin vibrations, and the skin surface ridges of the bottlenose dolphin, *Tursiops truncatus*. In: *Sensory Abilities of Cetaceans* (eds. J. Thomas and R. Kastelein), 163–179. New York: Plenum Press.

Schusterman, R.J. (1972). Visual acuity in pinnipeds. In: *Behavior of Marine Animals*, vol. 2, vertebrates (eds. H.E. Winn and B.L. Olla), 469–492. New York: Plenum Press.

Schusterman, R.J. and Gisiner, R. (1988). Artificial language comprehension in dolphins and sea lions: the essential cognitive skills. *The Psychological Record* 38: 311–348.

Schusterman, R.J. and Gisiner, R. (1989). Please parse the sentence: animal cognition in the procrustean bed of linguistics. *The Psychological Record* 39: 3–18.

Schusterman, R.J. and Kastak, D. (2002). Problem solving and memory. In: *Marine Mammal Biology: An Evolutionary Approach* (ed. A.R. Hoelzel), 371–387. Oxford: Blackwell.

Schusterman, R.J., Reichmuth, C.J., and Kastak, D. (2000). How animals classify friends and foes. *Current Directions in Psychological Science* 9: 1–6.

Smolker, R.A., Richards, A., Connor, R. et al. (1997). Sponge-carrying by Indian Ocean bottlenose dolphins: possible tool-use by a delphinid. *Ethology* 103: 454–465.

Supin, A.Y., Popov, V.V., and Mass, A.M. (2001). *The Sensory Physiology of Aquatic Mammals*. Boston, MA: Kluwer Academic Publishers.

Thewissen, J.G.M. and Nummela, S. (2008). *Sensory Evolution on the Threshold: Adaptations in Secondarily Aquatic Vertebrates*. Berkeley: University of California Press.

Xitco, M.J., Gory, J.D., and Kuczaj, S.A.I. (2004). Dolphin pointing is linked to the attentional behavior of a receiver. *Animal Cognition* 7: 231–238.

Xitco, M.J. and Roitblat, H.L. (1996). Object recognition through eavesdropping: passive echolocation in bottlenose dolphins. *Animal Learning and Behavior* 24: 355–365.

Box B6

The Application of Positive Reinforcement Training to Enhance Welfare of Primates in Zoological Collections

Jim Mackie

The first record of a primate in a modern, scientific, zoo is at the Zoological Gardens at Regents Park, now known as ZSL London Zoo, when in 1835, Tommy the chimpanzee arrived and caused a sensation in Victorian London. Our fascination with primates in zoos has continued ever since, with the first chimp's tea party being held at London Zoo in 1926. The popularity of these events, and others like them in various zoos, are an indication of how much we humans love primates but how little we used to understand about the cognitive ability and emotional capacity of our closest living animal relatives.

More recently, zoos have been able to provide scientists with unprecedented access to observe primate behaviour, resulting in research projects, especially with great apes, that have led to a radical change in the way we think about primates, and subsequently care for them in captivity. In 1994, one of the first studies into the use of operant conditioning with wild animals was conducted with 37 chimpanzees that were trained to voluntarily enter crates (Kessell-Davenport and Gutierrez 1994). This research, and a growing understanding of the potential welfare benefits of using positive reinforcement techniques highlighted by such organisations as the International Primatological Society (Prescott and Buchanan-Smith 2003), has led to major refinement in the management of primates in zoos, with an emphasis on using

cooperative trained behaviours to replace more coercive husbandry techniques. Nowadays primate training in zoos encompasses a huge repertoire of behaviours as diverse as voluntary blood draws, ultrasound exams, X-rays and dental care. One recent example of this is with orangutan (*Pongo pygmaeus*) at Waco Zoo Texas (Franklin 2005), which incorporated a protected-contact sleeve and positive reinforcement training to achieve the goal of voluntary blood withdrawal (e.g. Figure B6.1).

The use of positive reinforcement training in primate management has moved on from just the great apes to incorporate all types of primates. At ZSL London Zoo, keepers have trained their aye aye's *(Daubentonia madagascariensis)* to accept a hand injection for anaesthetic. The keepers in this instance also trained the animals to enter a travel crate to go to sleep in comfort before being moved to the veterinary theatre for the blood withdrawal required to test vitamin D consumption *(Training of nocturnal primates to enhance welfare, BIAZA Awards winner 2018, Christina Stender)*. Simple medical procedures such as X-ray exams, which in the past would have been conducted under anaesthetic, likely following a manual restraint, can now be completed consciously and voluntarily using simple operant training techniques. A good example of this was with a male ring tailed lemur *(Lemur catta)* at ZSL

Zoo Animal Learning and Training, First Edition. Edited by Vicky A. Melfi, Nicole R. Dorey, and Samantha J. Ward.
© 2020 John Wiley & Sons Ltd. Published 2020 by John Wiley & Sons Ltd.

(a)

(b)

Figure B6.1 Examples of how husbandry training can be used to support the provision of veterinary care include training for (a) blood sampling and (b) to use an inhaler. *Source:* Steve Martin.

Whipsnade Zoo who was trained to present his hand onto an X-ray pad for regular examinations to monitor the healing progress of a dislocated finger. At the Smithsonian Institute in Washington, USA, not only had the adult female orangutan *(Pongo spp.)* been trained to present her body for an X-ray exam, but the procedure also included a stationing behaviour for her young baby too.

An early example of the use of training to assist with a complex management scenario was at the Brookfield Zoo in the USA, where positive reinforcement training was successfully used to help reintegrate an orangutan mother and infant after a period of hand rearing by its human carers. This situation arose when the infant and mother had to be separated, initially for medical reasons and subsequently an absence of nursing. To facilitate early reintroduction a simultaneous training programme was initiated for both mother and infant (e.g. see Figure B6.2). The idea was to train the infant to take a bottle feed from keepers through a mesh panel and for the mother to accept this process after they were reintroduced. The training was a success even resulting in the mother presenting the baby to the keepers for bottle feeding. The authors of a paper recounting these

events attribute one of the key reasons for the success of the process being due to the 'close, trusting relationship' that developed between the mother orangutan and her keepers (Sodaro and Webber 2000). This was something that was not necessarily encouraged in zoos in the late 1980s and perhaps should be considered more keenly when approaching any primate – keeper interactions in zoos today, not just during formal training sessions.

Training for husbandry and medical procedures play an important role in the management of smaller primate groups, with Old and New World monkeys and lemurs particularly well represented in this field. Many collections train simple foundation behaviours like targeting, stationing, recall, and crate training for transport, which allow more invasive procedures such as hand injections for general anaesthesia and regular medication such as insulin for diabetic animals. At ZSL London Zoo a diabetic Sulawesi crested macaque *(Macaca nigra)* was routinely injected with insulin and urine collected on a daily basis to test ketones. This individual maintained good health into old age.

Another group where training is becoming a more established component of daily

Figure B6.2 Training young animals can help during times of transition, especially when hand reared infants need to be reintroduced to their natal group or a surrogate mother; here an infant Orangutan has been trained to stay in a location and open mouth on cue. *Source:* Steve Martin.

management is with callitrichids. In Europe there has been workshops focusing solely on their training and the EAZA (European Association for Zoos and Aquariums) callitrichid advisory group has a database of information on various successful training programmes. Perhaps the most influential work on this subject was conducted by the Bronx Zoo displaying a remarkable, 17 different species of callitrichid, with a wide range of behaviours trained, including ultrasound and stethoscope examinations (Savastano et al. 2003).

The one group of primates in which the use of training could be described as developing, is with nocturnal primates. With the exception of aye aye, of which there are several examples of training in the zoo community, this group is arguably the most under-represented with regards to training programmes, and research generally. They are however, very well provided for with other areas of behaviour-based husbandry, especially regarding lighting and behavioural enrichment. There are obvious reasons for this relative lack of inclusion in zoo husbandry

training. Firstly, the enclosures provided for these species tend to be very controlled environmentally and operate on a reverse light cycle to enable visitors to observe active animals. These environmental controls often result in relatively small enclosures, which mean that the presence of a zoo keeper can result in the animal demonstrating the species appropriate natural behaviour of … hiding! At ZSL London Zoo's Nightlife exhibit, the keepers overcame this issue by conducting crate training with slender loris (*loris lydekkerianus*) in protected-contact conditions. A specially designed crate attached to the outside of the enclosure meant the animal could choose whether to participate and allowed the trainer to add reinforcement when the correct behaviour criteria was reached. Other nocturnal prosimian species trained include Moholi bushbaby (*Galago moholi*), at London Zoo, trained for weighing and transportation.

Of all the groups of animals housed in zoological collections, primates are surely the best represented of all in terms of training for cooperative medical and husbandry

Figure B6.3 Incorporation of training into primate husbandry regimes can lead to various benefits, especially those which lead to maintaining and supporting good health. *Source:* Jim Mackie.

procedures (e.g. see Figure B6.3). However, there is still a considerable way to go with certain groups such as the nocturnal species, where the provision of larger enclosures would enable more comprehensive training programmes. Additionally, there is increasingly a greater appreciation in the zoo community that training does not automatically equate to improved animal welfare. Zoo professionals must consider ways of giving the animal greater choice and control during training sessions, continually assessing the welfare potential of each training programme. Adding a greater variety of behaviours and reinforcers, offering the animal greater power to decide when and where the training happens, and even asking them to choose who conducts the training; will help maintain motivation and enhance welfare. According to legislation in the UK, each training programme should 'provide a net welfare benefit to the animal' (Defra 2012), which is a crucial directive as it takes away the human agenda from training. Additionally it is important to use training appropriately. As zoo keepers become better at training animals we must avoid the temptation to use it to solve problems which should be achieved by other means such as enrichment or enclosure design. Perhaps the most important criteria for training is that it must be engaging, challenging but not stressful, and above all ... **fun!**

References

DEFRA (2012). *Secretary of State's Standards of Modern Zoo Practice*. Department for Environment Food and Rural Affairs https://assets.publishing.service.gov.uk/government/uploads/system/uploads/attachment_data/file/69596/standards-of-zoo-practice.pdf.

Franklin, J.A. (2005). Orangutan husbandry manual, positive reinforcement training. https://www.czs.org/custom.czs/media/CenterAnimalWelfare/Orangutan-Husbandry-Manual/Positive-Reinforcement-Training.pdf (accessed 30 October 2018).

Kessell-Davenport, A. and Gutierrez, T. (1994). Training captive chimpanzees for movement in a transfer box. https://awionline.org/content/training-captive-chimpanzees-movement-transfer-box (accessed 30 October 2018).

Prescott, M.J. and Buchanan-Smith, H.M. (2003). Training non-human primates using positive reinforcement techniques.

Journal of Applied Animal Welfare Science 6 (3): 157–161.

Savastano, G., Hanson, A., and McCann, C. (2003). The development of an operant conditioning training program for New World primates at the Bronx zoo. *Journal of Applied Animal Welfare Science* 6 (3): 247–261.

Sodaro, C. and Webber, B. (2000). Hand-rearing and early reintroduction of a Sumatran orang-utan. *International Zoo Yearbook* 37: 374–380.

Box B7

Species-specific Considerations: Primate Learning

Betsy Herrelko

A chimpanzee *(Pan troglodytes)* sees a banana, a high-value piece of food, high above the ground and out of reach. Jumping up to reach the banana does not get the chimpanzee close enough to grab it, nor does trying to hit it with a pole. Over time, the chimpanzee gathers items around the enclosure to stack as a way to get closer to the coveted item. After much trial and error, boxes and a stick were combined to be able to reach the food reward. In Wolfgang Köhler's (1925) study of problem solving and tool use, we see many instances of learning: an animal changing its behaviour as a result of an experience.

Primates represent the quintessential example of animal learning, this is in large part because we (humans) are primates and are drawn to anthropocentric concepts that appear to be similar to us (e.g. Rees 2001). We expect them to be smart because we are smart. There are risks and benefits of attributing human characteristics to animals (anthropomorphism), but when it comes to the world of cognition and how animals process information, scientists work hard to help animal abilities shine through.

An area of cognition that is regularly debated is the concept of building upon information learned by those who came before them, the 'ratchet effect' (Tomasello 1999). Also discussed as cumulative transmission, it is presumed to be the reason why humans have developed such a complex world of processes and industry and other primate species have not. Detailed texts outlining contributions of primate learning literature are available from iconic primatologists such as Tomasello and Call (1997), Matsuzawa (2001), Lonsdorf et al. (2010), amongst others.

Learning is a part of everything we do; at least it should be if we have opportunities to experience new things. Mentally stimulating challenges need to be appropriately challenging (Meehan and Mench 2007), animals need to have the tools to solve the problem, be it physical tools (as in the Köhler study) or mental abilities (e.g. being able to understand the rules of a task or what a complicated training cue means). We do not want to overwhelm or expect too much of an animal when we have not done the groundwork to prepare them for the task at hand. An important concept in designing opportunities for learning success is factoring in animals' natural history, particularly with respect to their physical characteristics, sensory abilities, and life history.

Physical Characteristics

How an animal physically functions in the world is key to understanding the way they learn. Failed learning may be the result of not ensuring the animal has the 'tools' to solve the problem or because we are simply not asking the right question or using the right methods. This is where the physical characteristics of different primates come into play. For example, gibbons (Hylobatidae) were thought to be less intelligent than other primate species because they were not successful in performing cognitive tasks in which others excelled (e.g. Spence 1937). That is, until a scientist modified the task to accommodate the gibbons' elongated hand (Beck 1967). When the items were elevated off a flat surface, the gibbons were able to learn how to participate in the task and correctly solve the problem.

Sensory Abilities

The ability to respond to sensory signals is crucial to the success of any living species (e.g. Krebs and Dawkins 1984). For example, vervet monkey *(Chlorocebus pygerythrus)* vocalisations can be referential and represent objects or events in the world (Cheney and Seyfarth 1982); gorillas *(Gorilla gorilla*

gorilla) produce distinct, identifying odours (Hepper and Wells 2010); and male gelada baboons *(Theropithecus gelada)* indicate social status with chest colour (Bergman et al. 2009). Vision, in particular, is important for primates, but not all primates see the world in the same way (Jacobs 1993). Colour vision, whilst helpful for detecting ripe fruit (Regan et al. 1998) or palatable leaves (Lucas et al. 1998), is different amongst primates. Although humans, other apes, and Old World monkeys (Catarrhines) are trichromatic, colour vision in New World monkeys (Platyrrhines) varies with many species exhibiting allelic trichomacy, meaning only females are trichromatic, and is limited in prosimians (Strepsirrhines), which are mono and dichromatic (Surridge et al. 2003). The sensory ability of each species is linked to the capacity to learn. When we consider the strength of an animal's sense of smell and what they can see, or even the natural cues they are drawn towards, we are designing training methodology to their

Figure B7.1 Batang and Redd orangutan *Pongo pygmaeus* travelling from one habitat to another via the O-Line, orangutan transit system, at Smithsonian's National Zoo. *Source:* Smithsonian Institution.

strengths and increase the likelihood of successful learning.

Life Experiences

Primatologists are often asked who are the smartest amongst the non-human great ape species, but it is an apples-to-oranges comparison. Scientifically, we know more about chimpanzees than any other non-human great ape species largely because we have a long history of working with them both in the field and in captivity. Additionally, the natural history of their social structures may impact willingness or motivation to participate in learning-based activities. Chimpanzees (Kummer 1971), bonobos (Nishida and Hiraiwa-Hasegawa 1987), and orangutans (Schaik 1999) live in fission–fusion societies where individuals come together (fusion) and break apart in different groups (fission). All three species are familiar with movement in group structure (see Figure B7.1). Gorillas, on the other hand, live in fairly stable groups, predominantly with a single adult male, the silverback (Robbins et al. 2004). This aspect of great ape natural history, i.e. sociality, may explain the motivation behind the willingness (or lack thereof) to participate in individually-based cognitive activities and it might be that these individuals may be more likely to participate in activities if alone or away from their group.

Individual history also plays a role; we know that certain personality characteristics self-select for willingness to participate in research sessions (Herrelko et al. 2012; Morton et al. 2013), and it would be reasonable to assume that those who have positive experiences participating in cognitive activities would be more likely to continue participation in the future and vice versa. There is also a difference in those with atypical life histories. Enculturated apes (e.g. those reared in an environment rich with social and communication opportunities with humans) appear to exhibit better cognitive skills compared to their non-enculturated counterparts (Call and Tomasello 1996), though the reasoning behind it is debated (Bering 2004; Tomasello and Call 2004).

Conclusion

Learning is something that happens every day and can be shaped to improve animals' lives (e.g. captive management) and teach us about their cognitive abilities. Adapting training and research methodologies used in learning activities should be appropriate for the species as well as the individual. The information provided in this textbox focuses on overarching categories and only scratches at the surface of primate uniqueness. Learning concepts are stable, but can become meaningful with additional information. All we need to do to catch a glimpse of how they view their world is to pay attention to how they experience life.

References

Beck, B. (1967). A study of problem solving by gibbons. *Behaviour* 28 (1/2): 95–109.

Bergman, T.J., Ho, L., and Beehner, J.C. (2009). Chest color and social status in male geladas (*Theropithecus gelada*). *International Journal of Primatology* 30: 791–806.

Bering, J.M. (2004). A critical review of the "enculturation hypothesis": the effects of human rearing on great ape social cognition. *Animal Cognition* 7: 201–212.

Call, J. and Tomasello, M. (1996). The effect of humans on the cognitive development of apes. In: *Reaching into Thought* (eds. A.E. Russon, K.A. Bard and S.T. Parker), 371–403. New York: Cambridge University Press.

Cheney, D.L. and Seyfarth, R.M. (1982). How Vervet monkeys perceive their grunts: field playback experiments. *Animal Behaviour* 30 (2): 739–751.

Hepper, P.G. and Wells, D.L. (2010). Individually identifiable body odors are produced by the gorilla and discriminated by humans. *Chemical Senses* 35: 263–268.

Herrelko, E.S., Vick, S.-J., and Buchanan-Smith, H.M. (2012). Cognitive research in zoo-housed chimpanzees: influence of personality and impact on welfare. *American Journal of Primatology* 74: 828–840.

Jacobs, G.H. (1993). The distribution and nature of color vision among the mammals. *Biological Reviews* 68: 413–471.

Köhler, W. (1925). *The Mentality of Apes*. London: Routledge and Kegan Paul.

Krebs, J.R. and Dawkins, R. (1984). Animal signals: mind-reading and manipulation. In: *Behavioural Ecology: An Evolutionary Approach*, 2e (eds. J.R. Krebs and N.B. Davies), 390–402. Oxford UK: Blackwell Scientific Publications.

Kummer, H. (1971). *Primate Societies: Group Techniques of Ecological Adaptation*. Chicago: Aldine.

Lonsdorf, E.V., Ross, S.R., and Matsuzawa, T. (2010). *The Mind of the Chimpanzee*. Chicago: University of Chicago Press.

Lucas, P.W., Darvell, B.W., Lee, P.K.D. et al. (1998). Colour cues for leaf food selection by long-tailed macaques (*Macaca fascicularis*) with a new suggestion for the evolution of Trichomatic colour vision. *Folia Primatologica* 69: 139–152.

Matsuzawa, T. (2001). *Primate Origins of Human Cognition and Behavior*. Tokyo: Springer.

Meehan, C. and Mench, J. (2007). The challenge of challenge: can problem solving opportunities enhance animal welfare? *Applied Animal Behaviour Science* 102: 246–261.

Morton, F.B., Lee, P.C., and Buchanan-Smith, H.M. (2013). Taking personality selection bias seriously in animal cognition research: a case study in capuchin monkeys (*Sapajus apella*). *Animal Cognition* 16: 677–684.

Nishida, T. and Hiraiwa-Hasegawa, M. (1987). Chimpanzees and bonobos: cooperative relationships among males. In: *Primate Societies* (eds. B.B. Smuts, D.L. Cheney, R.M. Seyfarth, et al.), 165–177. Chicago: University of Chicago Press.

Rees, A. (2001). Anthropomorphism, anthropocentrism, and anecdote: primatologists on primatology. *Science, Technology & Human Values* 26 (2): 227–247.

Regan, B.C., Julliot, C., Simmen, B. et al. (1998). Frugivory and colour vision in *Alouatta seniculus*, a trichomatic platyrrhine monkey. *Vision Research* 38: 3321–3327.

Robbins, M.M., Bermejo, M., Cipolletta, C. et al. (2004). Social structure and life-history patterns in Western gorillas (*Gorilla gorilla gorilla*). *American Journal of Primatology* 64: 145–159.

van Schaik, C.P. (1999). The Socioecology of fission-fusion sociality in orangutans. *Primates* 40 (1): 69–86.

Spence, K.W. (1937). Experimental studies of learning and the higher mental processes of infra-human primates. *Psychological Bulletin* 34: 906–850.

Surridge, A.K., Osorio, D., and Mundy, N.I. (2003). Evolution and selection of trichromatic vision in primates. *Trends in Ecology and Evolution* 18 (4): 198–205.

Tomasello, M. (1999). *The Cultural Origins of Human Cognition*. Cambridge, MA: Harvard University Press.

Tomasello, M. and Call, J. (1997). *Primate Cognition*. Oxford: Oxford University Press.

Tomasello, M. and Call, J. (2004). The role of humans in the cognitive development of apes revisited. *Animal Cognition* 7 (4): 213–215.

Box B8

Training Reptiles in Zoos: A Professional Perspective

Richard Gibson

Until the turn of the century if you'd asked the average reptile keeper in a zoo whether they did any training with their animals you'd have been lucky not to be laughed out of the building. But times are changing. As our understanding and appreciation of the intellectual abilities of reptiles has grown (see Burghardt 2013), so too has the incidence and breadth of training programmes exploiting a 'new found' (or at least newly recognised) reptilian aptitude for training.

That reptiles may quickly habituate to normally aversive stimuli has been long appreciated by many a reptile lover who enjoys handling their charges and indeed by the less-than-well-behaved visitor to the reptile house who discovers, perhaps to their dismay, that banging on the window has little if any effect upon the 'lifeless' reptile within. Such a propensity for desensitisation can be put to good use in the presentation of suitably prepared animals for public displays and even contact, and for delivery of veterinary procedures including visual inspection, claw-clipping, ultrasound examination, and even injections and blood draw (e.g. Bryant et al. 2016; Davis 2006). The addition of operant conditioning, employing simple associative learning through positive reinforcement (most often, but not always, food), greatly expands the training possibilities to include targeting for transfer between enclosures, crating, safe feeding, exercise, and behavioural enrichment; stationing for veterinary work; and trained behaviours for public presentations.

Several zoos and aquariums now routinely train tortoises and turtles, crocodilians, and many lizards (especially monitor lizards, which are widely believed, if not proven, to be the most 'intelligent' of the lizards) for all of the above reasons (see website links https://www.cabq.gov/culturalservices/biopark/news/training-with-reptiles, https://maddiekduhon.wordpress.com/2014/08/07/did-you-know-reptiles-can-be-trained, and https://nationalzoo.si.edu/animals/exhibits/reptile-discovery-center). The species benefitting most commonly from this new found herpetological husbandry tool are probably: the giant Galapagos and Aldabra tortoises (*Chelonoidis niger* and *Aldabrachelys gigantea* respectively) which, owing to their enormous weight and strength, are managed and manipulated far more effectively through training than against their will (see Weiss and Wilson 2003; Gaalema and Benboe 2008); crocodilians, whose size and potential to do harm, require a 'hands-off' approach to management wherever possible (see Augustine 2009; Hellmuth and Gerrits 2008); and the world's largest and arguably most charismatic lizard, the Komodo dragon (*Varanus komodoensis*), which has perhaps seen the greatest attention to training – indeed, few zoos maintaining this impressive saurian do so without implementing some degree of target and/or stationing training – and as a result, boasts amongst the most notable achievements in this field (see Figures B8.1 and B8.2). For example, several zoos report carrying out veterinary procedures on unrestrained dragons including nail-clipping,

Figure B8.1 A conditioned adult Komodo dragon *Varanus komodoensis* receives an unrestrained ultrasound scan to assess reproductive condition. *Source:* ZSL.

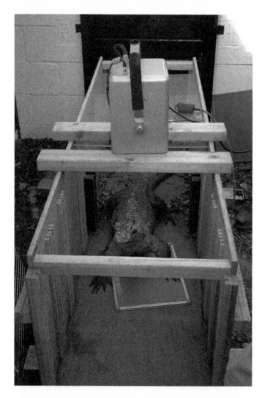

Figure B8.2 A sub-adult Komodo dragon *Varanus komodoensis* in a training crate modified to allow conscious, unrestrained X-rays and other procedures. *Source:* ZSL.

injections, and blood-draw from the caudal vein, ultrasound examination, 'stationing' on X-ray plates and weighing scales, and even application of a mask for anaesthesia (Hellmuth et al. 2012).

Targeting and desensitisation have been put to further use in shifting dragons and crocodilians between enclosures, crate and crush acceptance, directing feeding, increasing exercise and enrichment, positioning and controlling animals for filming and photography, presentations to zoo visitors, and acceptance of a harness to allow walks outside of their enclosure in suitable weather (see Figure B8.3).

Snakes are certainly the most neglected reptile group (excluding the tuatara [*Sphenodon punctatus*], the sole survivor of the order Rhynchocephalia) in terms of active conditioning (though habituation to handling is common). Hellmuth et al. (2012) mention only two snakes in their appendix of examples of reptile training in zoos; the green mamba (*Dendroapsis angusticeps*) and false water cobra (*Hydrodynastes gigas*). However, Kleinginna (1970) reported that indigo snakes (*Drymarchon corias*) show a similar propensity to operant conditioning as that of rats or pigeons tested in similar studies so certainly it is not a lack of aptitude for training that leaves snakes clinging to the bottom rung of the training ladder.

Nevertheless, there are some inherent complications in training reptiles, which relate largely to their physiology, one of which applies most pointedly to snakes, the infrequency of feeding. Most people know that most snakes don't eat every day. Indeed some don't eat for weeks or months at a time and when they do it tends to be single large meals. Using small titbits to repeatedly reinforce behaviour is therefore impractical with most species. Training snakes however, needn't rely on food rewards and many keepers report deliberately or incidentally training their charges to retreat to the opposite end of the enclosure or enter trap-boxes (where they are undisturbed) in response to a signal, a gentle touch (with a hook in the case

Figure B8.3 An adult female alligator focused on her training target. This posture/behaviour allows the chin and teeth to be observed for damage. *Source:* Auckland Zoo.

of venomous species) or even just the rattle of the keepers keys in the door.

Snakes are not alone in offering food-reward related challenges in training. Reptiles as a group, owing to their ectothermic physiology, are highly prone to obesity in captivity, either through the wrong quality of food or too high a quantity of food. Dieting overweight reptiles is very challenging since their highly efficient metabolism uses so little energy, and should be avoided at all costs. Extra care must therefore be taken to ensure that food-based rewards for training are carefully calculated as a part of the animals overall nutritional intake.

One way to mitigate the disadvantage imposed by irregularity of feeding in some reptiles is to consider alternate forms of reward. As described above, some snakes quickly teach themselves a behaviour which is rewarded by access to security and safety – that of entering a hide box – and lizards can be trained to find environmental shelters on cue (Zuri and Bull 2000). Giant tortoises and other testudines, may respond well to touch, the erect posture of giant tortoises in response to a scratch in the right place is well known

(Bryant et al. 2016; see Figure B8.4), and lizards and even crocodilians have been reported to apparently take pleasure from and seek out, human contact; itself a possible form of enrichment (Melfi 2013).

A further challenge in training reptiles, intimately linked to their physiology and natural history and influencing their diet and food intake, is the importance of meeting their circadian, seasonal and reproductive activity cycles. Being ectothermic, reptiles are greatly influenced by climate and their interest in training and indeed their ability to respond, are dependent upon them being in an appropriate physiological state. Correctly maintained reptiles experience temperature (and other climatic component) variation on circadian and annual seasonal cycles. These not only impact upon a reptile's body temperature and resultant motivation to react to stimuli such as training and food, but also stimulate a variety of behavioural and physiological conditions including brumation and aestivation, and reproductive cycling (during which many species will cease to eat and may behave very differently during part or all of the reproductive phase).

Figure B8.4 Following a period of patting and scratching to encourage the 'finching' posture, this adult male Galapagos tortoise *Chelonoidis nigra* is held in position with a target during desensitisation of the neck for future conscious blood draws. *Source:* Auckland Zoo.

Training reptiles therefore, may not always be as straight forward as with endothermic 'gas-guzzling' birds and mammals. Keepers should take into account alternate forms of communication with their animals and cuing beyond traditional auditory and visual. Exploiting the extraordinary olfactory abilities of most reptiles and the sensitivity of snakes especially, to non-airborne vibration, may provide alternate means of cuing animals to particular behaviours. Whilst rewards might include access to preferred resting/hiding places and favourable basking/thermal localities, positive human contact and, of course, food.

Training reptiles poses some significant challenges but the rewards are well worth the effort. Being able to work safely with dangerously large crocodilians and lizards because food has been disassociated from keepers and the animals conditioned to allow close proximity, even contact, and daily husbandry makes life less stressful for keeper and animal alike. Veterinary procedures are usually quicker, cheaper (no anaesthesia) and far less traumatic, and the opportunities to include reptiles in the traditionally bird and mammal-dominated world of public presentations and trained-animal displays are gradually being realised. These benefits though, are yet to be widely appreciated by the broader zoological community. Reptile training is not yet the 'norm' and traditional ways of thinking and keeping take a long time to change. Whilst anecdotes of personal experiences with reptile training are increasingly common, and indeed YouTube and other online video sharing websites are peppered with films of all manner of scaly animals demonstrating variably authentic trained behaviours, publications in the field are hard to come by. The zoo reptile keeping industry would benefit greatly from a concerted effort to generate wider interest and develop reptile-specific skills in training through dedicated workshops and publications in relevant literature.

References

Augustine, L. (2009). Husbandry training with an exceptional South African crocodile. *ABMA Wellspring* 10 (3): 2–3.

Bryant, Z., Harding, L., Grant, S., and Rendle, M. (2016). A method for sampling the Galapagos tortoises, *Chelonoidis nigra*, using operant conditioning for voluntary blood draws. *Herp. Bull.* 135 (Spring): 7–10.

Burghardt, G. (2013). Environmental enrichment and cognitive complexity in reptiles and amphibians: concepts, review, and implications for captive populations. *Appl. Anim. Behav. Sci.* 174 (3–4): 286–298.

Davis, A. (2006). Target Training and Voluntary Blood Drawing of the Aldabra Tortoise (*Geochelone gigantea*). Proc. AAZK Conf. 156–164.

Gaalema, D.E. and Benboe, D. (2008). Positive reinforcement training of Aldabra tortoises (*Geochelone gigantea*) at Zoo Atlanta. *Herpetol. Rev.* 39: 331–334.

Hellmuth, H. and Gerrits, J. (2008). Croc in a box: crate training an adult gharial. *ABMA Conference Proceedings,* Phoenix, AZ. ABMA.

Hellmuth, H., Augustine, L., Watkins, B.A., and Hope, K. (2012). Using operant conditioning and desensitization to facilitate veterinary care with captive reptiles. *Vet. Clin. Exot. Anim.* 15: 425–443.

Kleinginna, P. (1970). Operant conditioning of the Indigo snake. *Psychon. Sci.* 18: 53–55.

Melfi, V. (2013). Is training zoo animals enriching? *Appl. Anim. Behav. Sci.* 147 (3–4): 299–305.

Weiss, E. and Wilson, S. (2003). The use of classical and operant conditioning in training Aldabra tortoises (*Geochelone gigantea*) for venipuncture and other husbandry issues. *J. Appl. Anim. Welf. Sci.* 6 (1): 33–38.

Zuri, I. and Bull, C. (2000). The use of visual cues for spatial orientation in the sleepy lizard (*Tiliqua rugosa*). *Can. J. Zool.* 79 (4): 515–520.

Box B9

The Learning Repertoire of Reptiles

Gordon M. Burghardt

Reptiles have been neglected in research on cognition, emotions, sociality, and needs for psychological welfare. Yet the picture of them as sedentary, impassive creatures operating as instinctive machines is rapidly being replaced by one that views them as having behavioural complexity and plasticity that approach, if not rival, what has been found in many mammals and birds. Whilst the traditional view is partly due to an anthropomorphic tendency to view them through a human lens (Rivas and Burghardt 2002), it is also true that the scientific community did not look closely at their behavioural plasticity until recent decades (Burghardt 1977). With the advent of ethology and its emphasis on the study of the diversity of behaviour and its control, a focus on species typical behaviour expanded to the behavioural plasticity needed to survive in an often changing ecology. We now know that all groups of non-avian reptiles behave in fascinating ways, and have many traits in common with birds and mammals including sophisticated communication, problem solving, parental care, play, complex sociality, individual recognition, and even social learning and tool use (see Manrod et al. 2008 for examples). Furthermore, turtles, crocodilians, and lizards have all been shown to be quite adept at most traditional learning tasks (Burghardt 2013; Wilkinson and Huber 2012). The secret is to make sure the problems tested

accommodate their sensory abilities and behavioural repertoires. Here I will just mention a few examples.

Turtles have been used in studies of spatial learning (review in Wilkinson and Huber 2012). Painted turtles (*Chrysemys*), sliders (*Trachemys*), and cooters (*Pseudemys*) have been used in many (as far as reptiles go) learning studies due to their hardiness in captivity, use of visual cues, and trainability (Burghardt 1977). Red-bellied cooters (*Pseudemys nelsoni*) can be readily trained to climb out of the water and knock over a bottle for a food pellet (Davis and Burghardt 2007), and they can retain both the behaviour and discrimination for at least two years without any training (Davis and Burghardt 2012). Given the annual return of females to specific nest sites over many years, such retained memory is something we suspected could occur, but to demonstrate such a skill in captivity is an advance and opens up the possibility of more refined studies. But exciting field studies are also ongoing, as in the finding that mother aquatic turtles return from long distances to where they laid eggs on nesting beaches as the eggs near hatching, vocally communicate with them, and then guide them hundreds of kilometres to feeding areas (Ferrara et al. 2014). Turtles were the first reptiles to be experimentally documented as being able to learn from other animals solving a problem and thus learn it by

Zoo Animal Learning and Training, First Edition. Edited by Vicky A. Melfi, Nicole R. Dorey, and Samantha J. Ward.
© 2020 John Wiley & Sons Ltd. Published 2020 by John Wiley & Sons Ltd.

observation, if not actual imitation. For example, the earliest studies showed that they could learn which way to go around a barrier to obtain food, or which coloured bottle to dislodge to obtain food, by merely observing a trained animal (Davis and Burghardt 2011; Wilkinson et al. 2010). Turtles have also been shown to show concept formation (respond to darker or lighter stimuli, for example, not to the hue of the training stimulus) (Leighty et al. 2013).

Hatchling crocodilians stay with their mothers, and perhaps fathers, for months to years in 'crèches' where they are protected from predators and follow her on migrations to favourable habitats, especially important where dry seasons are severe. The agonistic behaviour of hatchlings develops early in some species and later in others and seems tuned to the nature of their territorial and social system. They have been observed playing with objects, coordinating hunting for fish, and decorating themselves with sticks to attract (and capture) birds, looking for such sticks to build their nests (Dinets 2015; Dinets et al. 2015). They have the most complex vocal repertoire of any group of reptiles. Their extensive parental care and communication skills (visual, chemical, tactile, auditory) perhaps reflect their close relationship with both dinosaurs and birds (Dinets 2013). Although difficult to study in laboratory settings as adults, they have been shown to be quite proficient in traditional learning tasks including reversal learning (Burghardt 1977).

Lizards are the largest group of reptiles and have diversified in fantastic ways to exploit marine, freshwater, arboreal, subterranean, rainforest, mountain, and desert habitats. For example, lizards learn many kinds of visual discriminations, eavesdrop on the warning sounds other species give to predators in their proximity, show spatial and reversal learning (e.g. Leal and Powell 2012), and some monitor lizards seem to be able to count as well as solve problem boxes with one trial learning (review in Burghardt 2013). Social learning has been demonstrated in lizards (Kis et al. 2015). Recently it has been documented that incubation temperature

may impact their learning abilities (Clark et al. 2014). Many lizards have complex social lives and social organisation including some that are monogamous and live in family groups (review in Doody et al. 2013). The great diversity in lizard ecology, reproduction, and sociality suggest that similar cognitive diversity will be shown.

Snakes, closely related to lizards, are very chemosensory oriented and, lacking limbs and, virtually, the ability to hear, cannot be tested in many traditional apparatus. Still, learning studies with pythons, ratsnakes, and gartersnakes, amongst others, show they can learn discriminations, modify antipredator behaviour, learn better ways of handling prey, etc. Hognose snakes, famous for playing 'dead' when threatened, monitor the gaze direction of 'predators' when death-feigning, focusing on the eyes themselves, not just head orientation (Burghardt and Greene 1988). However, there are few traditional learning tasks, with the exception of habituation (Herzog et al. 1989), in which snakes have been successfully trained (Burghardt 1977). The work of Holtzman and colleagues (e.g. Holtzman et al. 1999) on escape learning is a rare exception. Work is ongoing in several zoos on using bridging and targeting in controlling the behaviour of potentially dangerous snakes. Behaviours being studied include individualised feeding and movements into switch boxes. Snakes do rely on chemical cues as well as vision in feeding in rather complex ways that suggest interesting cognitive abilities that need more exploration. For example, multimodal matching may be involved in learning about noxious prey in plains gartersnakes (Terrick et al. 1995).

It is clear that learning and cognition in non-avian reptiles is in a phase of great research interest (Matsubara et al. 2017), and this box could only mention some of the diverse avenues being pursued. Besides, species differences, there are also individual differences in temperament, sociality, preferences, decision making and even personality, which also opens up many research questions (Waters et al. 2017).

References

Burghardt, G.M. (1977). Learning processes in reptiles. In: *The Biology of the Reptilia*, vol. 7 (ecology and behavior) (eds. C. Gans and D. Tinkle), 555–681. New York, NY: Academic Press.

Burghardt, G.M. (2013). Environmental enrichment and cognitive complexity in reptiles and amphibians: concepts, review and implications for captive populations. *Applied Animal Behaviour Science* 147: 286–298.

Burghardt, G.M. and Greene, H.W. (1988). Predator simulation and duration of death feigning in neonate hognose snakes. *Animal Behaviour* 36 (6): 1842–1844.

Clark, B.F., Amiel, J.J., Shine, R. et al. (2014). Colour discrimination and associative learning in hatchling lizards incubated at 'hot' and 'cold' temperatures. *Behavioral Ecology and Sociobiology* 68: 239–247.

Davis, K.M. and Burghardt, G.M. (2007). Training and long-term memory of a novel food acquisition task in a turtle (*Pseudemys nelsoni*). *Behavioural Processes* 75 (2): 225–230.

Davis, K.M. and Burghardt, G.M. (2011). Turtles (*Pseudemys nelsoni*) learn about visual cues indicating food from experienced turtles. *Journal of Comparative Psychology* 125: 404–410.

Davis, K.M. and Burghardt, G.M. (2012). Long-term retention of visual tasks by two species of emydid turtles, *Pseudemys nelsoni* and *Trachemys scripta*. *Journal of Comparative Psychology* 126: 213–223.

Dinets, V. (2013). Long-distance signalling in Crocodylia. *Copeia* 2113: 517–526.

Dinets, V. (2015). Play behavior in crocodilians. *Animal Behavior and Cognition* 2: 49–55.

Dinets, V., Brueggen, J.C., and Brueggen, J.D. (2015). Crocodilians use tools for hunting. *Ethology Ecology and Evolution* 27: 74–78.

Doody, J.S., Burghardt, G.M., and Dinets, V. (2013). Breaking the social-nonsocial dichotomy: a role for reptiles in vertebrate social behaviour research? *Ethology* 119: 95–103.

Ferrara, C.R., Vogt, R.C., Sousa-Lima, R.S. et al. (2014). Sound communication and social behavior in an Amazonian river turtle (*Podocnemis expansa*). *Herpetologica* 70: 149–156.

Herzog, H.A. Jr., Bowers, B.B., and Burghardt, G.M. (1989). Development of antipredator responses in snakes: IV. Interspecific and intraspecific differences in habituation of defensive behavior. *Developmental Psychobiology* 22 (5): 489–508.

Holtzman, D.A., Harris, T.W., Aranguren, G., and Bostock, E. (1999). Spatial learning of an escape task by young corn snakes, *Elaphe guttata guttata*. *Animal Behaviour* 57 (1): 51–60.

Kis, A., Huber, L., and Wilkinson, A. (2015). Social learning by imitation in a reptile (*Pogona vitticeps*). *Animal Cognition* 18: 325–331.

Leal, M. and Powell, B.J. (2012). Behavioural flexibility and problem solving in a tropical lizard. *Biology Letters* 8: 28–30.

Leighty, K.A., Grand, A.P., Pittman Courte, V.L. et al. (2013). Relational responding by eastern box turtles (*Terrepene Carolina*) in a series of color discrimination tasks. *Journal of Comparative Psychology* 127: 256–264.

Manrod, J.D., Hartdegen, R., and Burghardt, G.M. (2008). Rapid solving of a problem apparatus by juvenile black-throated monitor lizards *(Varanus albigularis albigularis)*. *Animal Cognition* 11: 267–273.

Matsubara, S., Deeming, D.C., and Wilkinson, A. (2017). Cold-blooded cognition: new directions in reptile cognition. *Current Opinion in Behavioral Sciences* 16: 126–130.

Rivas, J. and Burghardt, G.M. (2002). Crotalomorphism: a metaphor for understanding anthropomorphism by omission. In: *The Cognitive Animal: Empirical and Theoretical Perspectives on Animal Cognition* (eds. M. Bekoff, C. Allen and G.M. Burghardt), 9–17. Cambridge, MA: MIT Press.

Terrick, T.D., Mumme, R.L., and Burghardt, G.M. (1995). Aposematic coloration enhances chemosensory recognition of noxious prey in the garter snake, *Thamnophis radix*. *Animal Behaviour* 49: 857–866.

Waters, R.M., Bowers, B.B., and Burghardt, G.M. (2017). Personality and individuality in reptile behavior. In: *Personality in Non-human Animals* (eds. J. Vonk, A. Weiss and S. Kuczaj), 153–184. New York, NY: Springer.

Wilkinson, A. and Huber, L. (2012). Cold-blooded cognition: reptilian cognitive abilities. In: *The Oxford Handbook of Comparative Evolutionary Psychology* (eds. J. Vonk and T.K. Shackelford), 129–143. New York, NY: Oxford University Press.

Wilkinson, A., Kuenstner, K., Mueller, J., and Huber, L. (2010). Social learning in a non-social reptile. *Biology Letters* https://doi.org/10.1098/rsbl.2010.0092.

Box B10

Training Birds from a Zoo Professional's Perspective

Heidi Hellmuth

Birds have been actively trained for bird shows in zoos for many decades, but ironically the avian taxa lags behind many of their animal counterparts in husbandry training in zoos. With such a long history of training, why are there challenges to training birds for veterinary and husbandry needs? There are a few possible reasons for this discrepancy. Birds in shows tend to be housed individually, allowing more control of diet and other stimulation, which facilitates training. Whereas birds housed within the zoo collection (exhibit birds) tend to be housed in pairs or groups and frequently in mixed-species exhibits; social and larger more complex housing can all provide distractions during training. Some bird show training, especially historically, has been based, at least in part, on weight management; a birds' diet (calorific content) was dependent on the behaviour it performed. An underlying concept was that hungry birds would be more motivated to take part in training; this technique is no longer favoured for most current show bird training (Heidenreich 2014). Most exhibit birds in zoos have fixed daily diets, determined by nutritional content.

Importantly, the primary consideration that dictates the overall management strategy of birds, and other animals, in shows, is their behaviour, whereas the focus for exhibit birds tends to be breeding or visibility to the public. Another difference that sets birds apart in zoos, and possibly one of the most important,

are the staff that look after them. Bird show staff are hired on the basis of their knowledge of animal learning theory and experience of training, but historically this has been less of a focus for keepers looking after exhibit birds. Even now, these skills are not as widely recognised and focused upon for non-mammalian taxa in many zoos. Training birds in zoos might seem to be a challenge, due to a (mis) perception that training birds is difficult and requires special techniques, compared to training other species. Importantly, training is training; regardless of the taxa, species, or individual involved. The science and techniques are the same, but as with all forms of animal management, their application is guided by natural history, behavioural biology, and individual animal needs. Since 'training is training', why is there a separate box talking about tips for training birds (and other taxa) in zoos? Good question! The reason is to share some of the general training strategies that have proven most helpful and effective in working with birds in zoos, to lessen the training 'learning curve' and maximise chances for success. So now, on with the ideas!

Watch the Birds

The first step in almost any animal management endeavour is to observe the animals. See what areas of their exhibits they prefer, including heights, perch types, distance from

Zoo Animal Learning and Training, First Edition. Edited by Vicky A. Melfi, Nicole R. Dorey, and Samantha J. Ward.
© 2020 John Wiley & Sons Ltd. Published 2020 by John Wiley & Sons Ltd.

the public etc. See what activity patterns and behaviours they perform; when do they tend to be active versus rest? The key is to use your birds' behaviour as a guide when setting up your training programme. Pick times of day that the birds are generally active, and choose areas where they are the most comfortable and spend most of their time in. Too often in zoos, training times are dictated by the keeper's schedule, and training locations are selected for convenience and ease of access. Turning this around and having the birds' behaviour dictate the training parameters will go a long way towards setting the stage for a more successful programme. As most bird species are prey animals, asking them to learn new behaviours in an area that they are not comfortable in, is beginning the process with one hand (or wing) tied behind your back. If the bird is uncomfortable, there is a strong risk of associating training in general, with nervousness or fear. So watch and listen to the birds and enable this to shape how you set up their training programme.

Good Keeper–Animal Relationships

Keepers are an inevitable, inescapable daily presence, whose behaviour has a significant role in animal training and welfare. Great keepers have 'animal sense' that refers to their ability to read animals' behaviour and modify their own behaviour to make the animal feel comfortable. Animal sense is difficult to teach as it is guided by experience, intuition, and possibly an innate natural ability but everyone can work to build a positive keeper animal relationship (KAR; Ward and Melfi 2013, 2015) that is vital when working around birds, or other skittish or flighty (pun intended) animals.

Though the behaviour performed is determined by its consequences in operant conditioning, success in bird training can be greatly facilitated with a strong focus on the antecedents; setting up the circumstances which make the desired behaviours more

likely to occur. Despite being critical it is sometimes underappreciated in training zoo birds. One simple, yet powerful, strategy to working successfully with birds is to consistently give them a verbal warning, or cue, before taking *any* action. This is important for multiple reasons:

1) It reduces the chances that they will be startled by an unexpected movement or action, which will lessen their trust of you and their comfort in their environment.
2) It makes the keeper more aware of their own behaviour and its potential effect on the birds in their care.
3) It enhances the KAR in daily care and interactions.
4) It provides the bird with a predictable environment, which conveys to them a perception of control in their lives.

So how does it work? It is a very simple concept, and once practiced can become second nature. To earn and keep the birds' trust, the verbal cues must be used consistently (regularly and reliably), and they must be delivered before the action. For example, the keeper says 'door', then opens the door into the exhibit; 'entering', then steps inside the exhibit; 'bowls', then picks up or puts down food/water bowls; 'moving', then walks around the perimeter to get to the other side of the exhibit and so on and so forth. Some keepers feel silly using this level of verbal interaction/cueing, but I have personally observed birds, including burrowing owls (*Athene cunicularia*) and toco toucans (*Ramphastos toco*) that after implementation of this strategy, became calmer and relaxed in a few days where they had previously flown around in a panic during routine care. The birds' reactions and behaviours to these verbal cues should also guide the keeper's actions. For example, if a bird reacts to a verbal cue by moving or changing its behaviour, the keeper should wait until the bird is still, then recue before taking further action.

Another strategy is the 'freeze and go' technique where the keeper's behaviour is dictated by the bird's behaviour. This technique is also thought to promote a positive KAR. Simply, if the bird moves, the keeper freezes and waits for the bird to settle and become stationary, before resuming any movement (preceded by a verbal cue, of course). Movement and unpredictability can mean danger to the bird. If a bird is moving to create a flight distance between itself and the keeper, or otherwise try to find a more comfortable situation when a keeper is in their enclosure, a keeper that continues to move or remain active may elicit a panic or fear

Figure B10.1 Training show birds in zoos can raise some special challenges, but when successful can inspire awe: (a) and (b) Steve and Andean condor, c. 1989; (c) Steve and cinereous vulture, c. 1978; (d) Steve and golden eagle, c. 1978; (e) Steve and Joanna, Minnesota Zoo, c. 1987; (f) Steve and white-backed vulture, c. 1989. *Source:* Heidi Hellmuth.

response in the bird. If birds are in a nervous or tense state when keepers are in their environment, it will likely hinder the learning process during training, so this simple technique is not just good bird care but effective bird training.

A combination of both active and passive training can also be used, to give the birds more opportunities to engage in and practice the desired behaviour(s) that can aid in achieving training goals. Active training refers to formal training sessions where the bird is cued to perform specific behaviours. Passive training occurs when the bird is given the opportunity to engage in a behaviour when the keeper is absent. For example, crate training is likely to progress more quickly with a combination of active training sessions with the keeper approximating the bird closer to and then into the crate, and passive sessions where reinforcers are left near/in the crate, allowing the bird to explore and gain the reinforcement on its own.

Finally, knowing the most powerful reinforcer for birds is a critical component of an effective programme. Many zoo animals are hand or target fed parts of their diet by keepers as the reinforcer. Most exhibit birds tend to be fed primarily out of bowls in specific feeding areas. To effectively use portions of a birds' diet as primary reinforcement for training, it is important to know what items the birds favour, and to begin associating these items more directly with the keeper versus the food bowl. This can be easily done, starting with a food preference test. Separate each food type into a separate pile within the bowl and observe which the bird(s) choose to eat first, second, and third. Do this for several days or more to get an idea of which food types they prefer. Remove the most preferred food types from their daily feed and begin feeding them separately. This allows the keeper to find and use the birds' favoured foods as training reinforcement, and associates these preferred foods more directly with the keeper, which again can help set a stronger foundation for training.

All of these suggestions and tips have a common theme: listen to the birds and let their natural history and behaviour guide your actions. This will help you have a stronger relationship, reduce the potential stress associated with required daily husbandry and implement a more successful training programme (Figure B10.1a–f). Good luck!

References

Heidenreich, B. 2014. Weight management in animal training: pitfalls, ethical considerations and alternative options. http://www.goodbirdinc.com/pdf/Heidenreich_%20Weight%20Management_ABMA_2014.pdf.

Ward, S.J. and Melfi, V. (2013). The implications of husbandry training on zoo animal response rates. *Applied Animal Behaviour Science* 147: 179–185.

Ward, S.J. and Melfi, V. (2015). Keeper-animal interactions: differences between the behaviour of zoo animals affect Stockmanship. *PLoS ONE* 10 (10): e0140237. https://doi.org/10.1371/journal.pone.0140237.

Box B11

Learning and Cognition in Birds

Jackie Chappell

The derogatory expression 'bird brain' has a lot to answer for. Historically, birds have been considered less intelligent than mammals, but an increasing number of studies over the past few decades have established birds as equals to many mammals in their capacity for learning and the sophistication of their cognitive capacities. Learning in birds can be broadly categorised into two main forms. At one extreme, learning is relatively inflexible, and individuals learn about specific relationships between particular stimuli and events. Nevertheless, such learning can be powerful and long lasting, and the performance of birds is equivalent to that of many mammals. For example, pigeons (*Columba livia*) trained on a set of 160 photographic slides to respond to one of each pair, recalled these associations with little loss of performance two years later (Vaughan and Greene 1984). At the other extreme, learning can be highly flexible, and birds can apply relatively abstract rules to new situations, rather than learning associations by rote. Some show the ability to reason about relations between objects (such as whether two items in a pair match), applying those relations correctly to novel situations by analogy (Smirnova et al. 2015).

There is substantial variation in learning and cognition between different taxonomic groups of birds, depending upon the perceptual abilities, social environment, and ecological niche of the species. It is therefore helpful to categorise cognition into different 'domains', each allowing the animal to process and use information of a particular kind. Three such domains will be outlined below: spatio-temporal, social, and physical cognition.

Spatio-temporal Cognition

Using information about space is important for many bird species. For example, migratory species, or those which forage some distance from a home location, use a variety of navigational mechanisms and must process and recall information about spatial locations, paths or directions in order to find their way (Wiltschko and Wiltschko 2003). Recent research employing GPS-equipped route recorders carried by homing pigeons (*C. livia*) has shown that when repeatedly released from a site, they fly on an idiosyncratic path to which they become more and more faithful on each release (Biro et al. 2004). This and other evidence strongly suggests that the pigeons use visual landmarks to memorise and recognise their path, which may help to explain why pigeons' visual memory capacity is so extensive and enduring.

Spatial cognition is also important for food-storing or caching birds, which make use of seasonal gluts by storing food for later consumption. Many food-storing bird species remember where they have placed their stored food, and recall those locations with some accuracy many months later. Experiments suggest that visual cues surrounding the caching site are used to remember the locations (Gould-Beierle and Kamil 1996), and since retrieval performance in some species is better when retrieving items they have stored themselves, it suggests that something about the personal experience of storing the food item facilitates storage and recall of its location (Shettleworth 1990). These 'episodic-like' memories have been probed in elegant experiments using food-storing Western scrub jays (*Aphelocoma californica*). These have exploited the fact that they prefer wax worms to peanuts, but only when they are fresh and not decayed. Scrub jays were allowed to store peanuts in one tray and wax worms in another, whilst manipulating the elapsed time between storing wax worms and the opportunity to retrieve them. As predicted, scrub jays preferred the tray in which they had stored wax worms, but only if they had stored the worms recently, so the worms were still fresh (Clayton and Dickinson 1998). This suggests that the jays remembered what they had stored, when they had stored it, and where (which tray) they stored it in.

Social Cognition

Living in a complex social group can provide opportunities to learn by observing others (Reader and Laland 2003). However, it also introduces the problem of maintaining relationships and keeping track of interactions, which has been argued as a cognitive constraint on group size in primates (Dunbar 1992). Social learning can take a variety of forms. In emulation, observers copy the outcome of another individual's behaviour, which contrasts with imitation, in which they learn about and copy the actions of the demonstrator (Auersperg et al. 2014; Huber et al. 2001). These mechanisms can result in novel behaviours spreading rapidly through a local population, as individuals learn from an innovator (Morand-Ferron and Quinn 2011). True imitation (in which the action itself is copied) is rare in non-human animals (even in apes), and whilst vocal imitation occurs in several species, there has been only one demonstration of motor imitation in a bird, an African grey parrot *Psittacus erithacus* (Moore 1992).

Theory of Mind (the ability to reason about the mental states of another individual) has been argued as an important capability for individuals living in complex social groups. In animals lacking language, it is difficult to determine whether they are inferring mental states of other individuals, or simply responding to their behaviour. However, experiments with scrub jays have suggested that personal experience of having pilfered another scrub jay's cache of food determined whether or not an individual would choose to relocate their caches to a new tray, if another scrub jay had seen them cache the food and could thus potentially steal their cache (Emery and Clayton 2001).

Physical Cognition

Interacting with physical objects in the environment is an important aspect of many birds' lives, from nest-building (Healy et al. 2008) to extractive foraging. Physical cognition may involve an understanding of the functional properties of objects, reasoning about causal relations, and predicting the behaviour of objects. Much of the focus of research in this area has been on tool-using or tool-making birds. In the wild, the list of species that habitually make tools is quite short (New Caledonian crows *Corvus moneduloides*, and woodpecker finches *Camarhynchus pallidus*), but several non-tool using species have been shown to be capable of tool use in captivity, including other corvids (e.g. rooks, *Corvus frugilegus*; Bird and Emery 2009) and parrots (Goffin's cockatoo, *Cacatua goffini*; Auersperg et al. 2012, and kea, *Nestor notabilis*; Auersperg et al. 2011). Tool use is thought to be potentially cognitively demanding because it requires the animal to control an object (rather than part of their own body) dynamically to act on a goal (Bentley-Condit and Smith 2010). However, it is possible that birds using or even making tools habitually do so using standard processes of learning, whereby making or using a tool of a particular type in a particular way results in a reward. Nevertheless, experiments suggest that habitual and non-habitual tool users alike are able to make a novel tool to solve a problem (e.g. Auersperg et al. 2011, 2012; Bird and Emery 2009; Morand-Ferron and Quinn 2011; Weir et al. 2002), suggesting that the process may be much more flexible than would be possible through associative learning alone.

It should be clear from the discussion above, that on-going research demonstrates that many species of bird have advanced learning and cognitive abilities, rivalling that of many mammals. As with any animal, the mixture of 'expertise' is shaped by the ecological niche, but if provided with the right circumstances in which to demonstrate their prowess, birds are anything but 'bird-brained'.

References

Auersperg, A.M.I., von Bayern, A.M.P., Gajdon, G.K. et al. (2011). Flexibility in problem solving and tool use of kea and New Caledonian crows in a multi access box paradigm. *PLoS One* 6: e20231 EP.

Auersperg, A.M.I., Szabo, B., von Bayern, A.M.P., and Kacelnik, A. (2012). Spontaneous innovation in tool manufacture and use in a Goffin's cockatoo. *Current Biology* 22: R903–R904.

Auersperg, A.M.I., von Bayern, A.M.I., Weber, S. et al. (2014). Social transmission of tool use and tool manufacture in Goffin cockatoos (*Cacatua goffini*). *Proceedings of the Biological Sciences* 281: 20140972–20140972.

Bentley-Condit, V. and Smith, E.O. (2010). Animal tool use: current definitions and an updated comprehensive catalog. *Behaviour* 147: 185–221.

Bird, C.D. and Emery, N.J. (2009). Insightful problem solving and creative tool modification by captive nontool-using rooks. *Proceedings of the National Academy of Sciences of the United States of America* 106: 10370–10375.

Biro, D., Meade, J., and Guilford, T. (2004). Familiar route loyalty implies visual pilotage in the homing pigeon. *Proceedings of the National Academy of Sciences of the United States of America* 101: 17440–17443.

Clayton, N.S. and Dickinson, A. (1998). Episodic-like memory during cache recovery by scrub jays. *Nature* 395: 272–274.

Dunbar, R.I.M. (1992). Neocortex size as a constraint on group-size in primates. *Journal of Human Evolution* 22: 469–493.

Emery, N.J. and Clayton, N.S. (2001). Effects of experience and social context on prospective caching strategies by scrub jays. *Nature* 414: 443–446.

Gould-Beierle, K.L. and Kamil, A.C. (1996). The use of local and global cues by Clarks nutcrackers, *Nucifraga columbiana*. *Animal Behaviour* 52: 519–528.

Healy, S., Walsh, P., and Hansell, M. (2008). Nest building by birds. *Current Biology* 18: R271–R273.

Huber, L., Rechberger, S., and Taborsky, M. (2001). Social learning affects object exploration and manipulation in keas, *Nestor notabilis*. *Animal Behaviour* 62: 945–954.

Moore, B.R. (1992). Avian movement imitation and a new form of mimicry: tracing the evolution of a complex from of learning. *Behaviour* 122: 231–263.

Morand-Ferron, J. and Quinn, J.L. (2011). Larger groups of passerines are more efficient problem solvers in the wild. *Proceedings of the National Academy of Sciences of the United States of America* 108: 15898–15903.

Reader, S.M. and Laland, K.N. (2003). *Animal Innovation*. Oxford: Oxford University Press.

Shettleworth, S.J. (1990). Spatial memory in food-storing birds. *Philosophical Transactions: Biological Sciences* 329: 143–151.

Smirnova, A., Zorina, Z., Obozova, T., and Wasserman, E. (2015). Crows spontaneously exhibit analogical reasoning. *Current Biology* 25: 256–260.

Vaughan, W. and Greene, S.L. (1984). Pigeon visual memory capacity. *Journal of Experimental Psychology: Animal Behavior Processes* 10: 256–271.

Weir, A.A., Chappell, J., and Kacelnik, A. (2002). Shaping of hooks in New Caledonian crows. *Science* 297: 981.

Wiltschko, R. and Wiltschko, W. (2003). Avian navigation: from historical to modern concepts. *Animal Behaviour* 65: 257–272.

Box B12

Species-specific Considerations when Planning and Implementing Training with Aquatics

Heather Williams

Training has been happening with different taxa in zoos, and some aquariums, for many years. Although, in general terms, aquariums are lagging behind zoos in the adoption of training and behavioural management. Why? The widely held false perception of goldfish having only a three-second memory may have something to do with it. However, Gee et al. (1994) showed goldfish can learn to push a lever for food, and are able to anticipate a feeding window as small as 1 hour in 24. There is also the view of fish being less intelligent than mammals, which, may be true in some circumstances, however many species are certainly capable of a spectrum of cognitive abilities (see Box B13) and can be trained complicated behaviours. These include cownose rays (*Rhinoptera bonasus*) swimming with a camera mounted on a hoop over their head (Phoenix Zoo), to various species, trained to enter boxes or slings for weighing, or movements of zebra sharks (*Stegastoma fasciatum*) allowing voluntary blood draws (Aquarium of the Pacific). The good news is that more and more aquarists are learning the basics of animal learning theory, how to apply it to train behaviours and are getting involved. In the last few years, there has been a massive increase in the amount of training carried out within aquariums, with a heavy focus on elasmobranchs. In 1993, 34.1% of surveyed aquaria were carrying out some form of training; increasing to 88.8% in 2013 (Janssen et al.

2017). There is still a lot of catching up to do and the training techniques used in aquariums mostly focus on the use of primary reinforcement alone, for the time being; but we're getting there!

The biggest obstacle in terms of setting up a training programme is probably pretty obvious to most readers – time!!! Although, in the long run this can be viewed as an investment, which will pay dividends in the future. It is also important to think about what is needed and whether it is worth spending the time and effort for the desired outcome.

In general terms, training is training and the basic premise is the same regardless of the species involved. In practise, however, things can change drastically depending upon the cognitive ability of the species involved, and when using food as a primary reinforcer, their appetite levels. Aquariums can present a particular challenge with regards to training as it is relatively rare to find a single species exhibit. Accessing the area can be fraught with difficulty as not every aquarist is lucky enough to be able to access their tank from all angles. However, the most obvious barrier is the water. In order to have good access to our animals, we either need to go to them (with the added complication of extra equipment, whether that is as simple as rubber boots, or as complicated as full diving equipment), or we need them to come to us at the surface of the water.

Zoo Animal Learning and Training, First Edition. Edited by Vicky A. Melfi, Nicole R. Dorey, and Samantha J. Ward.
© 2020 John Wiley & Sons Ltd. Published 2020 by John Wiley & Sons Ltd.

Many fish are trained to some degree without the aquarists involved in their care even realising it. This happens simply through feeding a particular individual or species at a certain area, and over time, they learn to associate food with a certain area of their enclosure. Some even appear to react when they see the colour of staff uniform, despite being fed from a public area where there are many different people around; so likely learning to discriminate between familiar and unfamiliar people.

As aquarium tanks generally house a few different species, with differing behaviours and appetite levels, training can help with general daily husbandry. For example, in a tank with larger predatory (faster) fish such as golden trevally (*Gnathanodon speciosus*) or crevalle jacks (*Caranx hippos*), they can locate and eat food more quickly than their tankmates. An aquarist may decide to train the predatory fish to come to a particular area of their tank for their food, freeing up an area of the tank for the slower fish to feed without competition. Alternatively, the slower fish can be trained to approach the aquarist for hand feeding (see Figure B12.1).

Benthic (or bottom dwelling) creatures can also prove challenging to training attempts, particularly in deeper tanks. For example, I worked with a honeycomb whiptailed ray (*Himantura undulata*) which was used to being scatter fed on the bottom of a relatively shallow (2 m) tank during her time in quarantine (see Figure B12.2). However, she was being moved into a mixed species exhibit that was 4.5 m deep so scatter feeding was no longer an option. In order to keep control over her diet, the staff fashioned a feeding pole for her to come to. Unfortunately, for one reason or another, she did not feed well in the main exhibit, so she was moved into an adjoining acclimation tank (~1.5 m deep). In this acclimation tank, the ray was fed next to a target which was gradually moved up the side of the tank wall, so that she learnt to surface for her food. The second time she was moved into the mixed species tank, the aquarists were able to successfully feed the ray and a much more positive outcome was accomplished.

It is possible in many cases to carry out a training session using a proportion of the fish's daily food allowance and then feed the remaining food as normal. However, in my personal experience, training in an aquarium generally involves the use of all the fish's diet for that day. The training session and feeding time becomes

Figure B12.1 Many different fish species can be trained to support husbandry systems, for example this stonebass *Polyprion americanus* was trained by offering food rewards on the surface of the water at the National Marine Aquarium. *Source:* Oliver Reed.

Figure B12.2 Even fish species which do not necessarily come to the surface to feed can be trained effectively, for example this honeycomb whiptailed ray *Himantura undulata* was fed using feeding tongs which facilitated target training in 1.5 m deep acclimation tank at the National Marine Aquarium. *Source:* Oliver Reed.

one. This does mean, although you still need the training to happen, it may move more slowly as it is important to ensure the animal receives enough food. Having said that, keeping track of the amount of food eaten by the fish, is a focal part of our training.

Many aquariums will not weigh their fish on a regular basis. However, it may be necessary to keep track of the health of some specific individuals such as elasmobranchs or young fish. This is where forward planning which incorporates training can help no end as without it, weighing can cause stress leading to fish struggling and the process of weighing can take longer than necessary.

If the fish are desensitised to a sling or a net, firstly, they will be more relaxed leading to a more accurate weight, the process will also be less dangerous for all involved and can mean less staff involvement.

'Crate training' is not simple within tanks, as most don't have an off-display section. Because fish are generally not moved from tank to tank on a regular basis, crate training may not in fact be worth the effort. There are certain circumstances when it can be useful to carry out a different version, i.e. moving fish for housekeeping reasons (problematic anemones), or safety (venomous creatures or sharp teeth!). With crate training and training more generally, the public can see what is going on, and so it can help to have staff explaining the situation or put up some simple signs with explanatory information. It may be best in some circumstances to have the crate in the exhibit for the training session and then remove it again, thus removing any chance of entanglement, but also to keep the exhibit clutter free for the public – after all they pay our wages!

It can be very important to provide training in zoos and aquaria for ethical reasons. It is our job to ensure our animals are kept healthy and well fed, and training relating to feeding can ensure this is more likely than just simply scatter feed in a large mixed exhibit. I have experienced in the past a group of a nervous species, tarpon (*Megalops atlanticus*) that were not feeding well following a move into a 2.5 million litre, 10.5 m deep tank. A simple training programme was introduced which involved a target placed into the water and the species was only fed in the vicinity of this target. This made a huge difference and the individuals gained the confidence needed to feed well. It was also possible to remove the target when one of the sand tiger sharks (*Carcharhias taurus*) swam by, as the tarpon would get a little too focused on the target sometimes!

People are probably even less inclined to think of training invertebrates than they are fish, however, octopus are extremely intelligent and it can really improve their welfare to carry out some training but also general enrichment. Giant Pacific octopus (*Enteroctopus dofleini*) are particularly receptive to play and secondary reinforcement. Many aquariums will interact with their octopus on a daily basis and the aquarists will provide various enrichment items to the octopus. The octopus will generally interact with people and can show different reactions to different staff members carrying out tactile enrichment (personal observation and discussions with colleagues). They will also work for their food and when offered food contained within one of many containers, the octopus generally does not take long to figure it out!

One of my most important tips is to have fun! If you make the sessions enjoyable both for yourself and the fish involved, the outcomes are much more likely to be positive.

References

Gee, P., Stephenson, D., and Wright, D.E. (1994). Temporal discrimination learning of operant feeding in goldfish (*Carassius auratus*). *Journal of Experimental Animal Behaviour* 62 (1): 1–13.

Janssen, J.D., Kidd, A., Ferreira, A., and Snowden, S. (2017). Training and conditioning of elasmobranchs in aquaria. In: *The Elasmobranch Husbandry Manual II: Recent Advances in the Care of Sharks, Rays and their Relatives* (eds. M. Smith, D. Warmolts, D. Thoney, et al.), 209–221. Ohio Biological Survey.

Box B13

The Cognitive Abilities of Fish

Culum Brown

Fish are the forgotten majority. They comprise more than 50% of the total vertebrate diversity but are seldom represented in zoos. Naturally they are the showcase of public aquariums the world over. While most people don't think of fish as being intelligent, there is a huge body of research that suggest that they match the rest of the vertebrates in just about every facet (Bshary and Brown 2014; Brown 2015). Indeed, evolutionary theory tells us that all vertebrates are simply modified fish, thus from this perspective perhaps the finding that fish are smart is less surprising. Fish cognition has a pronounced influence on their behaviour, which is of direct relevance to housing them in captivity.

In order to understand fish behaviour it is vital to gain a glimpse into how they view the world around them (their umwelt). Let us begin with vision since most people can relate to the visual world. The standard fish is tetrachromatic although many species have more visual pigments than this and some, such as deep sea fish, have none. Most fish can see colours more vividly than we can. For the majority of shallow water species this means colour is very important to them. Colour is used during courtship and foraging for instance. In cichlid fish, for example, variation in colour and preferences for certain colours has led to massive species diversification (Seehausen et al. 2008). Many fish have innate colour preferences, which means it is easier to train them when using these preferred col-

ours. Red, for instance, seems to be universally attractive perhaps because red coloured prey are rich in keratin which is a limiting resource. It has been theorised that some species have incorporated red into their mating displays to make the most of this pre-existing sensory bias (Seehausen et al. 2008). Classic examples include guppies *(Poeciliilia reticulate)* and sticklebacks *(Gasterosteus aculeatus)*. When given a choice between two foraging patches, one indicated by a red marker and another by a neutral coloured marker (e.g. green), many species of fish will show a natural tendency to forage near the red marker (e.g. Laland and Williams 1997). Studies examining spatial learning in guppies where fish are required to swim through a tunnel to access a foraging site show that a pre-existing bias for red makes it much easier to train them to swim through the red tunnel as opposed to some other colour (Laland and Williams 1997). Fish also have a degree of unconscious visual processing and thus fall for the same sorts of visual illusions as humans. For example, both sharks and bony fish can discriminate between Kanizsa figures (Fuss et al. 2014).

To date much of the research on spatial learning in fish has tended to focus on visual cues. Fish are capable of using a single cue (a beacon) to locate a given location. They can also use integration by relying on the relative position of a number of beacons to locate a specific spot. If you test their spatial learning

Zoo Animal Learning and Training, First Edition. Edited by Vicky A. Melfi, Nicole R. Dorey, and Samantha J. Ward.
© 2020 John Wiley & Sons Ltd. Published 2020 by John Wiley & Sons Ltd.

in a rectangular arena and then remove the beacons, they can use the geometry of the arena to locate a reward, or at least narrow it down to one of two alternative locations (Warburton 1990). Once they have learnt the location they can remember it for a staggering amount of time. For example, it took just five trials for rainbowfish to reliably locate a hole in an approaching trawl net out and they retained the information for more than a year (Brown and Warburton 1999). At the top end of the spatial learning continuum, fish are also capable of developing cognitive maps. A classic example is intertidal gobies (*Bathygobis soporator*) which explore the surrounding rock platform at high tide and return to their home pool to sit out the low tide (Aronson 1951). Studies have shown that these fish are nearly always found in their home pools. Even if they are displaced they quickly return. When these fish are threatened during low tide, they can jump into neighbouring pools. They cannot see these pools but they know their location based on the spatial map they built up during their high tide excursions. Fish can also do time–place learning where they must keep track of time and location simultaneously. A classic example is when convict cichlids (*Cichlasoma nigrofasciatum*) gather at a feeding spot in anticipation of food arriving (Reebs 1993).

Guppies trained to locate a specific foraging patch can pass that information on to naive individuals through a process of social learning. Experiments show that even after the trained individuals are removed from the school, the naive fish retain the information. Information can pass horizontally (i.e. with a generation) or vertically (between generations; Brown and Laland 2011). When social information passes between generations there is the possibility to establish population specific cultural traditions as is the case with many song birds. Field studies have shown that many migration routes are maintained by cultural transmission (Brown and Laland 2011). For example, the daily migration movements of French grunts (*Haemulon flavolineatum*) from their daily hiding spots to

their night time foraging patches is passed on via social learning. It is believed that the breeding migration routes of Atlantic cod (*Gadus morhua*) are also based on cultural traditions (Brown and Laland 2011).

While vision is important to many fish, chemosensory cues are arguably more important. The reason for this is simply that fish live in an aquatic medium, which is highly suited to the transmission of chemical information (see also Box A3). The sense of smell in fish is far more powerful than most terrestrial animals and fish are capable of detecting chemicals at concentrations lower than one in a billion (Brown and Chivers 2006). Chemosensory recognition plays a vital role in just about every aspect of a fish's life. Obviously they can recognise various food items by smell, but they can also recognise each other and potential predators (Brown and Chivers 2006). If given a choice between the smell of a familiar and an unfamiliar conspecific they will move towards the smell of a familiar fish (Ward and Hart 2003). They can also detect the difference between related and unrelated individuals based on smell alone. While predator recognition can be partly innate in some species of fish, many can learn to associate the smell of alarm pheromones with the appearance of a predator. Naive fish can be rapidly trained in effective predator avoidance in this manner (e.g. Brown and Laland 2011). Fish can use smell for orientation. For example, many salmonids return to their natal streams as breeding adults using chemosensory cues (Dittman and Quinn 1996).

Water is also a very good medium for sound (see also Box A2). Fish are mostly made of water and they are surrounded by water which means sound waves can travel through the fish to activate the sensory cells directly. Moreover sound travels faster and attenuates less quickly under water than in air, which means there is ample opportunity for fish to use hearing to sample their world. Research has shown that fish communicate using sound in a variety of contexts including mating and aggressive interactions (e.g. damsel fish; Myrberg et al. 1986). There is a dawn and dusk chorus underwater as there is in terrestrial ecosystems. Experiments using benthic sharks have shown that they can be trained to associate the sound of bubbles and the arrival of food, but they cannot associate flashing lights and food (Guttridge and Brown 2014). This is undoubtedly because these sharks do not readily rely on vision while foraging so light does not make a very good conditioning stimulus.

While fish can also use electroreception to sample the world, and even navigate, there is very little by way of research in this area. We do know, however, that many species of knife fish find prey, navigate, and communicate with one another by sending out pulses of electricity into the water much in the way that dolphins use sonar.

The question of whether fish feel pain and respond to stress as we do has been an issue of heated debate (Brown 2015). The main reason for the opposition is largely because of the huge commercial interests in harvest-ing fish from the wild. There is no doubt, however, that fish are highly intelligent animals and their behaviour suggests they are sentient. To gain an unbiased account of pain perception it is wise to turn to evolutionary theory (Brown 2017). It is clear from comparative physiology and molecular studies that the pain receptors in humans are almost identical to those found in fish. This should come as no surprise because we inherited them from our fishy ancestors. Similarly the hormones involved in stress responses are very similar across all vertebrates. It would be fair to surmise that pain and stress in all vertebrates is a very similar and highly conserved phenomenon (Brown 2015). While there are those who argue that the psychological aspect of pain may differ between animals, this is also highly unlikely since the physical detection and emotional response to pain evolved side-by-side with the expressed outcome being the long-term avoidance of potentially dangerous stimuli (Brown 2015).

References

Aronson, L.R. (1951). Orientation and jumping behaviour in the Gobiid fish *Bathygobis soporator*. *American Museum Novitates* 1486: 1–22.

Brown, C. (2015). Fish intelligence, sentience and ethics. *Animal Cognition* 18: 1–17.

Brown, C. (2017). A risk assessment and phylogenetic approach. *Animal Sentience: An Interdisciplinary Journal on Animal Feeling* 2 (16): 3.

Brown, G. and Chivers, D. (2006). Learning about danger: chemical alarm cues and the assessment of predation risk by fishes. In: *Fish Cognition and Behaviour*, Fish and Aquatic Resources Series (eds. C. Brown, K. Laland and J. Krause), 49–69. Blackwell Publishing.

Brown, C. and Laland, K. (2011). Social learning in fishes. In: *Fish Cognition and Behaviour* (eds. C. Brown, K. Laland and J. Krause), 240–257. Wiley Blackwell.

Brown, C. and Warburton, K. (1999). Social mechanisms enhance escape responses in shoals of rainbowfish, *Melanotaenia duboulayi*. *Environmental Biology of Fishes* 56 (4): 455–459.

Bshary, R. and Brown, C. (2014). Fish cognition. *Current Biology* 24: R947–R950.

Dittman, A. and Quinn, T. (1996). Homing in Pacific salmon: mechanisms and ecological basis. *Journal of Experimental Biology* 199: 83–91.

Fuss, T., Bleckmann, H., and Schluessel, V. (2014). The brain creates illusions not just for us: sharks (*Chiloscyllium griseum*) can see the magic too. *Frontiers in Neural Circuits* 8: 24.

Guttridge, T. and Brown, C. (2014). Learning and memory in the Port Jackson shark, *Heterodontus portusjacksoni*. *Animal Cognition* 17: 415–425.

Laland, K.N. and Williams, K. (1997). Shoaling generates social learning of foraging information in guppies. *Animal Behaviour* 53: 1161–1169.

Myrberg, A.A., Mohler, M., and Catala, J.D. (1986). Sound production by males in a coral reef fish (*Pomacentrus partitus*): its significance to females. *Animal Behaviour* 34: 913–923.

Reebs, S. (1993). A test of time place learning in a cichlid fish. *Behavioural Processes* 30: 273–281.

Seehausen, O., Terai, Y., Magalhaes, I.S. et al. (2008). Speciation through sensory drive in cichlid fish. *Nature* 455: 620–626.

Warburton, K. (1990). The use of local landmarks by foraging fish. *Animal Behaviour* 40: 500–505.

Ward, A.J. and Hart, P.J. (2003). The effects of kin and familiarity on interactions between fish. *Fish and Fisheries* 4 (4): 348–358.

Part C

More Than A to B: How Zoo Animal Training Programmes Can Impact Zoo Operations and Missions

From Part A we understand the principles by which animal learning can be achieved. From Part B we have learnt how zoos can support and facilitate learning and training. However, providing animals with learning opportunities or implementing training programmes is not always as simple as training an animal to move from A to B. There are many reasons why training programmes are not successful or as commonly implemented as we may think they are or should be. Implementing training programmes can have repercussions beyond the goal behaviours being trained. This section of the book focuses on zoo operations and how they might be impacted by adopting a training programme. Discussions are provided about how zoo animal training programmes impact the needs of visitors, animals, and keepers.

10

Making Training Educational for Zoo Visitors

Katherine Whitehouse-Tedd*, Sarah Spooner*, and Gerard Whitehouse-Tedd

10.1 Introduction

Trained, habituated, or conditioned animals are often integral to zoo education programmes. Even programmes which refrain from using animals directly as part of their lessons or activities delivered, will still utilise animals on exhibit to support and supplement taught theory. The roles of zoo animals in regards to conservation and education programmes have previously been termed 'exhibit', 'breeding', or 'programme' animals (Watters and Powell 2012); the latter are also known as 'encounter', 'outreach', or 'ambassador' animals. Although animals may switch roles, or perform multiple roles at once, it is the ambassadorial role that this chapter will focus on. This role centres on their interaction with humans (either keepers, educators, visitors, or a combination) and the animals involved are typically selected for certain species-specific attributes, which enable them to perform this role effectively. These attributes can be classified as being of educational benefit; for example, displaying unique adaptations to explain biological concepts, rarity to explain conservation threats, biophobic characteristics to overcome fears or dispel myths, or representative of taxonomic differences. Alternatively, animals may be selected for attributes with more practical

benefits (e.g. often related to ease of handling and transport, public safety, aesthetic [biophilic] appeal, or ability to engender empathy in learners). These characteristics can be applied in educational roles for both direct and indirect interactions; whereby the animals are utilised as an educational tool aimed at providing a more intimate, emotional zoo experience. The underpinning philosophy is that these experiences will generate greater concern for the species involved (Routman et al. 2010; Skibins 2015; Skibins and Powell 2013).

However, a challenge is often raised both within and outside of the zoo community when evaluating the ability of zoo education programmes to achieve their goals. This challenge is based on the fact that the primary driver for a zoo visit is more often considered to be recreational, social or entertainment-based, rather than linked to an interest in learning (Ballantyne and Packer 2016; Ballantyne et al. 2007; Reading and Miller 2007; Turley 2001). The recreational role of zoos should not be overlooked given its function in underpinning the economic viability of the majority of zoos. As such, zoos must develop strategies to deliver their educational and conservation messages to an audience whose primary objective for the visit is often not education or conservation. The challenge of meeting educational objectives within the context of a leisure venue is further exacerbated by the apparent dissonance

---

*These authors contributed equally.

Zoo Animal Learning and Training, First Edition. Edited by Vicky A. Melfi, Nicole R. Dorey, and Samantha J. Ward.
© 2020 John Wiley & Sons Ltd. Published 2020 by John Wiley & Sons Ltd.

between visitor expectation of zoos to meet high conservation and animal welfare standards, and their concurrent expectation for the convenient and ease of viewing a range of different exotic species (Kellert 1996). Added to this is the increase in consumer demand for an entertainment-factor in their purchasing decisions, including how they spend their leisure time (Balloffet et al. 2014). This necessitates that zoos find ways to combine entertainment with education (Ballantyne et al. 2007; Reading and Miller 2007), but avoid the risks of this so-called 'edutainment' (Balloffet et al. 2014).

The approach of combining education with entertaining experiences manifests in a range of visitor experiences that involve trained or habituated animals. In this regards, animal presentations offer a unique and viable marker between education and entertainment in zoos, and are themselves viewed as an important component of visitor expectations. In a large survey of European school children, 75% of the respondents cited contact with animals or watching animals in shows as being one of the positive roles that zoos play (Almeida et al. 2017). However, the terminology applied to these presentations can be as diverse as the programmes themselves, and some even have undesired connotations. For example, zoos may prefer to avoid the word 'show' or 'performance' for fear that it portrays an exploitation of the animals, or be considered too similar to circus-type acts that zoos may wish to distance themselves from, which may determine the policies under which zoos operate. The importance of the terminology used to market and promote these educational programmes may, to some, be overestimated; simply reflecting a case of unnecessary semantics. However, to others it is imperative that programme terminology aligns with the ethical standpoint of the zoo, and aims to focus public perception on the animals as willing participants in their ambassadorial role. For the purposes of this chapter, we have attempted to define three types of animal demonstrations (Table 10.1), although it

is acknowledged that these scenarios can overlap, be used interchangeably, or occur along a spectrum, rather than within discrete categories. Wherever possible we have attempted to delineate the areas of overlap or discrepancy, but felt it necessary to establish consistent uses for the terms involved.

With all three types of demonstrations, there is potential for significant overlap whereby a mixture of two or all three types of activities may be used in a single educational programme or event. A classic example of this would be sea lion displays that include trained performances of natural behaviours (e.g. swimming, porpoising) alongside an interpreted educational commentary, with elements of unnatural behaviours (e.g. ball balancing). This may be used for educational purposes (e.g. demonstrating the use of the vibrissae) or more entertainment purposes, followed by a member of the audience being selected for a personal 'meet and greet' at the end of the show (e.g. a kiss, flipper-shake, or stroke). Likewise, in a number of cases, the demonstration may be no more than an animal training session, often for medical husbandry purposes, that is interpreted by a commentator (Anderson et al. 2003; Price et al. 2015; Szokalski et al. 2013; Visscher et al. 2009) and which therefore does not strictly fall into any of the categories. In these scenarios the behaviours being trained may appear unnatural (e.g. presenting an arm for injection, or a foot for nail clipping or hoof trimming), but the welfare-based rationale for their performance would suggest these are better classified as presentations than performances.

10.2 The Species Involved

The species used in educational demonstrations range from small invertebrates such as Madagascan hissing cockroaches (*Gromphadorhina portentosa*), to much larger vertebrates such as tigers (*Panthera tigris* spp.) and sharks (Selachimorpha clade). Although zoos are perhaps best known for their housing

Table 10.1 Definitions, uses and distinctive features of terms used to describe education events involving animals, as hosted by zoos.

Term	Definition	Uses	Distinctive features
Presentation (or show, display)	Demonstrations of animals exhibiting *natural behaviours* (i.e. those which may be performed under natural, free-living conditions), trained to be performed on cue.	The behaviours are typically interpreted through some form of educational commentary, delivered either by the animal trainers themselves or by a separate interpreter/commentator.	These range from almost ad hoc presentations in which the animals have considerable choice and flexibility in the behaviours displayed and the subsequent interpretation provided, to highly scripted story-telling productions. The key distinction between this and other categories of animal demonstrations is that the behaviour(s) demonstrated can be considered part of the animals' species-specific natural behaviour repertoire.
Performance	These represent scenarios which include (at least some) *unnatural behaviours* (i.e. those which would not be performed under natural, free-living conditions), trained to be performed on cue, or natural behaviours performed in an unnatural context. As per the dictionary definition, 'performances' may include comical activities, dancing, or the use of costumes, props or other types of acting.	Typically the behaviours are incorporated into a series of events with the predominant focus on entertainment.	Wildlife theatre is a term being used with increasing frequency, and despite its name, is generally more aligned to our definition of presentation, but the highly scripted and story-telling nature of these theatre productions could also be described as performances. However, the distinction must be made as to whether these theatre-type productions are presentations or performances based on the type of behaviours that the animals are asked to perform. Those including only natural behaviours would be termed 'presentations', whilst those incorporating at least one unnatural behaviour would be termed a 'performance' for the purposes of this chapter. Wildlife theatre may also include alternatives to live animals such as puppetry and actors, as such these non-animal productions would fall into the performance category. Other author definitions may vary.
Encounter	These events involve some form of *contact* (or interaction) between visitor and animal. As defined here, these activities may also be called 'interactions.' The animals involved in these encounters may be termed 'ambassadors', 'programme animals', or 'encounter animals'.	The encounter may be indirect (e.g. interaction through a barrier, as part of a hand-feeding activity, or being in close proximity to the animal without any barrier but whilst also avoiding direct contact), or more direct (e.g. physical contact between the animal ranging from a short touch (pat, kiss, stroke), to a more sustained interaction (e.g. walking with the animal, having the animal climb onto the visitor, or for extended periods of patting or stroking).	Visitor–animal contact (or interactions) are the distinctive feature of these events. Unlike the two previous categories, visitors will be able to interact with the animal(s) at some point during the event. An encounter may be included as part of a performance. Given that interaction with humans could be considered an unnatural behaviour, encounters would not be included as part of a presentation, under this current definition. Our definition aligns with that of Watters and Powell (2012), with one minor amendment. Previously 'programme' animals were defined as those which were involved in guest interactions outside of their enclosure (Watters and Powell 2012), whereas we include interactions occurring inside the animal's enclosure in our definition of an encounter. Other author definitions may vary.

of non-domestic species, many will utilise domestic species (e.g. rabbits, goats, sheep) in their educational programmes. Unfortunately the use of native species appears to be somewhat rarer in educational programmes, although it is certainly not unheard of (it is more common in Australasia). An informal survey of zoo websites from each continent of the world (Table 10.2) indicates the predominant use of mammalian species, in particular pinnipeds, giraffes (*Giraffa* spp.) and big cats, followed by a range of birds, whereby far fewer examples of reptilian, fish, or invertebrate species were highlighted in marketing material for zoo displays or encounter programmes. In a recent review of the use of zoo animals in encounter programmes (Whitehouse-Tedd et al. 2018), a bias towards smaller, less active animals was determined for these encounters. This is likely as a result of increased handling ability and safety (Fuhrman and Ladewig 2008), but is somewhat at odds with the known ability of larger, more active animals to hold visitor attention (Fuhrman and Ladewig 2008; Ward et al. 1998). However, in this respect, it is unsurprising that large, active animals such as marine mammals, giraffe, big cats, and birds of prey were frequently advertised as integral parts of zoo presentations and performances.

10.3 The Behaviours and Training Involved in Public Displays

Behaviours performed by animals in zoo education and entertainment programmes range from simple, natural behaviours such as locomotion, to more complex natural behaviours performed in either natural or unnatural contexts, and up to entirely unnatural behaviours (Table 10.3). The justification for incorporating entertaining and anthropomorphic animal behaviours is often that it draws visitors in; the hope being that if visitors are interested in something they will want to know more about it (Moss and Esson 2010). For this reason, Flamingo Land (Yorkshire, UK) still dedicates a sizable

section of its bird show to parrots talking and performing impressions. The audience find these performances entertaining and return specifically to see the speaking parrots (Spooner 2017). Whilst knowing that a parrot can talk is not the desired outcome of the show, it is hoped that although visitors came to see a talking parrot they may also have gained some biological or ecological understanding about parrots and other species. Likewise, unnatural behaviours can be used to raise awareness about activities such as recycling by getting parrots to model responsible behaviours (picking up litter and putting it into an appropriate bin).

When considering animal performances, one of the most well-known types is that of sea lion (e.g. *Zalophus californianus*) shows. At Flamingo Land the sea lion show is one of the most popular events, attracting just under 25% of the park visitors (Spooner 2017). The sea lion show (like the bird show) runs as a separate unit within the theme park under the management of APAB Ltd (UK), and aims to convey a mixture of behaviours to entertain the visitors whilst imparting animal facts. Sea lions were previously encouraged to balance balls and bowling pins on their noses whilst a trainer explained that they are only able to do this due to their vibrissae (Figure 10.1). All non-natural behaviours such as catching hoops and balls were explained thoroughly by comparing to the animals' natural behaviours in the wild. Whilst these examples have a clear entertainment basis, they are attempting to bring key environmental issues to the general public on their day out.

The most basic animal presentations involve a zoo professional providing commentary on what the animal is doing within an enclosure. This is the least disruptive type of display as the commentator is merely providing a narrative and explanation of the animal's natural behaviours, with the animal being free to choose which behaviour to display (within the context of a captive environment). Animal talks fall within our presentation classification as they do not require

Table 10.2 Examples of species used in zoo presentations or performances across continents based on an informal survey of online marketing material from the largest zoos in the region.

	North America	South America	Europe	Asia	Africa[a]	Australasia
Mammals	Sea lion, cheetah, sloth, giraffe, lemur, rhino, camel, great apes, zebra, monkeys, binturong	Sea lion, giraffe, monkeys, tiger, zebra	Sea lion and seal, dolphins, orcas, elephants, meerkats, rhino, monkeys, otters, koalas, cheetah, hedgehogs, ferrets, wolf, coatis, pandas	Sea lion, elephants, rhino, giraffe, tiger, lion, hyena, monkeys, otters, raccoons, lemurs, anteaters, alpaca, tapir, bear, jaguar	Lion, leopards, cheetah, monkeys, rabbits, honey badger	Tiger, meerkats, elephants, lemurs, koalas, kangaroo, monkeys, great apes, giraffe, baboon, seals, cheetah, red panda, dingo, marsupials (range), otters, camel
Birds	Penguins, parrots, songbirds, flamingo	Penguins, hornbills, owls, ostrich	Penguins, songbirds, parrots, keas, ibis, vultures, eagles, pigeons	Penguins, cormorants, birds of prey, parrots, pelicans	Penguins, pelicans, birds of prey, cormorants, parrots, owls	Keas, parrots, birds of prey
Reptiles	Not highlighted	Crocodiles, tortoises	Komodo dragon, bearded dragon, geckos, tortoises, snakes	Alligator	Bearded dragon, tortoise, snakes	Tortoise, crocodiles, range
Fish	Stingrays, aquarium touch pools (range of species), aquarium dives	Not highlighted	Aquarium touch pools	Not highlighted	Not highlighted	Not highlighted
Invertebrates	Not highlighted	Not highlighted	Insects (range of species), spiders, cockroaches, stick insects	Not highlighted	Not highlighted	Not highlighted

[a] Information only available for South African zoos.
Species are listed in approximate order of frequency of appearance.

Table 10.3 Behaviours trained for educational and entertainment purposes.

	Natural/unnatural	Training method	Demonstration type (presentation, performance, or encounter)	Examples
Remaining stationary	Natural, but position or duration may be unnatural	Stationing	All	Tiger stationed to sit on a log Sealions on stands
Walking, running	Species dependent	Extensions of natural movements Successive approximation Marker and target Use of lures	Presentation, performance	Bear walking (natural when quadrupedal but unnatural if trained to walk bipedal) Macaw walking naturally, riding a bicycle unnatural but same movement Seriema running in front of audiences in order to find food
Climbing, burrowing	Species dependent	Successive approximation Targeting	Presentation, performance	Burrowing owls and Patagonian conures going through tunnels Big cats climbing trees for food recovery Caracara demonstrating tunnelling behaviour as part of hunting
Sitting, standing, lying	Species dependent	Extension of natural behaviour with trained commands led by human Stationing Targeting	All	Big cats standing on trainers' shoulders (trained as though standing against a tree)
Holding, catching, grasping, carrying	Species dependent	Targeting	All	Sealions/cetaceans catching balls with flippers
Swimming, swinging, jumping, flying	Species dependent	Extension of natural behaviours Stationing Targeting	All	Sealions swimming, birds flying, squirrel jumping Bird trained to fly to glove to get food reward
Feeding, hunting	Natural	Extension or control of a natural behaviour Animals feeding controlled and demonstrated at particular times	Presentation, performance	Falcon flying to the lure, hawk chasing a dummy bunny Serval and caracal jumping for lure or food reward Cheetah running to lure
Balancing	Species dependent	Over extension or control of a natural behaviour	Performance	High wire or tight rope walking. Tigers, bears, apes & other primates riding horses, ponies or elephants

Category	Natural/Unnatural	Description	Type	Examples
Manipulating an object	Species dependent	Over extension or control of a natural behaviour Natural behaviours extended for dramatic effect and visibility from a large audience	Presentation, performance	Capuchin opening a nut Sealions/cetacean balancing a ball on their nose Animals riding scooters and all terrain vehicles Animals painting pictures Parrots posting coins into donation boxes Parrots sorting coloured objects (to demonstrate intelligence and ability to see in colour) Seriema bashing plastic lizard to demonstrate hunting displays
Body or body-part contortion	Unnatural	Forced, can result in injury	Performance	Unnatural movements and tricks Sea lions catching hoops
Risk-taking	Typically unnatural	Extension or control of a natural behaviour Behaviour trained for display purposes	Performance	Jumping through hoops of fire Raptors flying through person's legs and arms
Displaying anthropomorphic actions	Unnatural	Extension or control of a natural behaviour Behaviour trained for display purposes Forcing the behaviour by subjecting the bird to loop tapes in a controlled environment	Performance	Sea lions clapping or catching balls, parrots nodding 'yes' or counting Birds, parrot and corvids talking and singing Animals painting
Social interaction with other species that would not naturally occur	Unnatural	Raising /working animals together from an early age Forced interaction	Presentation, performance	Interaction with other animals not normally found together in the wild state. Lions, tigers, bears together in one enclosure Tigers nursing piglets Cheetahs and dogs housed together
Social interaction with humans	Unnatural	Harness, lead, collar training Training to ride in boxes and vehicles Conditioning from an early age Successive approximation Increased exposure to noise and novel experiences to enable handling	All (primarily encounter)	In hand obedience, walking amongst crowds and groups of people, sitting or lying in a certain place or position Trainers riding or balancing on any part of the animal Training to ride in boxes and vehicles Walking on a harness with a member of the public Riding, sitting on big cats, surfing on cetaceans
Display of body parts	Natural	Marker and target Natural behaviours extended and linked to command signals	Presentation	Medical husbandry techniques: open mouth, presentation of torso or limb, injection acceptance or desensitisation Semen/blood collection Birds holding wing open to demonstrate wingspan

Figure 10.1 Sea lion show at Blackpool Zoo (UK) using unnatural objects to demonstrate adaptations such as hunting. *Source:* Sarah Spooner.

the animal to perform in an unnatural way. Often some form of animal feeding accompanies these types of talks, which has the advantage of increasing the likelihood that the visitor will have a clear view of the animals (Figure 10.2). However, since zoos cannot fully mimic wild environments, the feeding behaviours displayed in captivity may not be entirely natural (i.e. may represent modifications of natural feeding behaviours, or incomplete repertoires). For example, live vertebrate prey items are not allowed to be given to captive carnivores on welfare grounds in many (but not all) countries (DEFRA 2012), necessitating less natural food provisioning (e.g. dead and processed meat items), and preventing the full repertoire of feeding behaviours from being exhibited by the animal. A diverse range of feeding enrichment strategies are available and used by most zoos in order to promote more natural feeding behaviours.

Similarly, the regular provision of food at a specific time and place can risk the development of anticipation behaviours (and potentially stereotypies) where the animal associates the presence of a keeper or an increasing crowd to herald the arrival of food (Watters 2014). This anticipation behaviour can be problematic as animals disrupt their behaviour patterns in preparation for the food stimulus and, as a consequence, the behaviour presented to the public may not be one which would be found in the wild (Jensen et al. 2013), thereby becoming a performance as per our previous definition. Some zoos, such as Chester Zoo (UK) aim to reduce anticipatory behaviours by randomising their talk and feeding times (Bazley 2018). This insures that when a combined talk and feed is given, the animals are more likely to present in a natural way. It is possible the animals still learn some stimuli associated with talk times but these are limited. In contrast, the downside is that, for visitors who seek information in the form of a talk on a particular species, these are restricted and variable, depending on the day.

Increasing the complexity of a presentation beyond simply interpreting a feeding event typically requires the use of some trained animals. Free flying bird shows are popular in zoos and bring animals up close to the audience in order to better display specialised adaptations. Examples of this include

Figure 10.2 Giraffe feeding at Cotswold Wildlife Park, UK. *Source:* Sarah Spooner.

Figure 10.3 Barn owls used in free flight demonstrations of silent flight (left) and adaptations (right), Kalba Bird of Prey Centre (UAE). *Source:* K. and G. Whitehouse-Tedd.

the presentation of trained barn owls to audiences; their facial disks and silent flight are explained via the commentary provided (Figure 10.3). These displays may still present animals in a relatively natural way, however, often encouraging natural behaviours to be performed on demand or for extended periods.

In another example, macaws (*Ara* spp.) at Birdworld (Surrey, UK) were trained to open their wings and raise their feet on command (Figure 10.4). This enabled visitors to see the

bird's wingspan as well as features such as the zygodactyl feet. Although the birds would not normally hold these positions for such extended or repeated durations, they were not unnatural movements to perform. Likewise, seriemas (Cariamidae family) were encouraged to demonstrate jumping ability in order to receive a food reward and to sing on command. This was trained by playing a recording of a seriema to which the bird would automatically respond. The seriemas were also trained to demonstrate hunting

Figure 10.4 Macaw displaying wingspan (left) and feet (right) at Birdworld (UK). *Source:* Colin McKenzie.

Figure 10.5 Trained birds on donation boxes during the free flight bird show at Taronga Zoo, Sydney. *Source:* Host 2008.

behaviours by hitting a plastic lizard against a rock in return for a food reward. These behaviours were trained using positive reinforcement whereby if the bird performed the action they were rewarded with food.

At Taronga Zoo, Sydney (Australia), parrots were trained to collect monetary donations from the audience of the Free Flight Bird Show, and to then place the money in specially-designed boxes as part of the show (Figure 10.5). The funds raised were used for an in situ project aimed at breeding and releasing little penguins (*Eudyptula minor*) into the Sydney Harbour. Using positive reinforcement, trainers conditioned native Australian parrot species to collect, carry, and then deposit money (both paper notes and coins) from the viewing audience (Host

2008). First seen being performed by ravens and crows at a bird show in Texas, the Australian bird trainer, Claudia Bianchi, developed a training programme for a galah (*Eolophus roseicapilla*) and white-tailed black cockatoo (*Calyptorhynchus* spp.). Although corvids were considered potentially more suitable for this routine, Taronga Zoo wanted to utilise existing birds from their collection (Host 2008), and continue to involve mainly native Australian birds in their show.

The birds selected had previously received basic conditioning training, but had not experienced any shaping or training of complex behaviours (Host 2008). In order to perform the donation-collecting activity, the birds would need to sit on the donation box and accept money in their beak (Host 2008) without attempting to swallow the money (or bite the donor!). Desensitisation to the show environment (crowds of up to 1000 people, music, and the free flying presence of other trained birds), was also necessary (Host 2008). This latter aspect proved to be the most challenging, especially since the show programme changes over time with new birds being introduced (Host 2008). Another significant issue was the tendency of one of the birds to throw the money away after picking it up (Host 2008), which was obviously contrary to the message that the show was attempting to convey. Although the birds were also slower to collect and cache the money than the corvids witnessed in Texas, they were considered successfully trained over the course of 10 months (Host 2008).

10.4 Animal Welfare Considerations

Research using faecal cortisol levels suggests that birds used in public performances may be less stressed than birds who are simply displayed to the public in exhibits (Robson 2002). This is potentially due to the positive reinforcement by trainers, meaning that the animals experience shorter periods of boredom or inactivity, and the fact that, whilst on view to a large audience, the birds are performing familiar tasks with individuals with whom they know. However, benefits were only seen when birds participated in a moderate number of shows (maximum of four per day) and faecal corticosteroid levels were significantly higher when show numbers peaked.

Findings in a range of species handled for educational purposes (hedgehogs, red-tailed hawks, and armadillos) also revealed a strong correlation between elevated faecal glucocorticoid metabolite concentrations, as well as increased undesirable behaviours, and increased amount of handling (Baird et al. 2016). Welfare concerns were also identified by Taronga Zoo in their trained parrots. Therefore the zoo no longer offers the interactive money collection activity described above, despite positive benefits for conservation donations (Kemp et al., submitted). This highlights the need to fully evaluate both the animal- and visitor-responses achieved by any non-husbandry or health-related training activity. Nonetheless, it is also important to note that faecal cortisol provides a cumulative indicator of adrenal gland activity and therefore does not provide an immediate indication of acute stress levels in the animal. As such, research is currently being conducted using salivary cortisol from sea lions to elucidate hormone levels before and immediately after public performances (Bloom P., pers. comm.).

Housing environment (enclosure size and substrate depth) was also linked to welfare indicators (Baird et al. 2016), whereby the conditions under which 'programme' or 'education' animals are housed may not be as optimal as 'exhibit' or 'breeding' animals. Many UK zoos include collections of animals suitable for public contact. Various zoo guidelines provide advice on which species are best for handling and whilst this is beneficial in terms of welfare, it restricts the types of animals which can be used and does not always consider educational value (DEFRA 2012; EAZA Felid TAG 2017; European Association of Zoos and Aquaria 2014). The exception to this is the World Association of Zoo and Aquarium's guidance on the use of animals in presentations, in which an educational component is listed as mandatory (World Association of Zoos and Aquariums 2003). Nonetheless, many of the handling collections include animals commonly kept as pets such as corn snakes. Although handling an animal, regardless of species, may have benefits such as reducing phobias and increasing concern for the species, the ease of availability of such animals may increase public desire for keeping such animals as pets; missing the overall messages of biodiversity and conservation. Furthermore, individuals who handle animals in these sessions are often praised for their 'bravery' at touching the animals, which reinforces concepts that the animal is dangerous. Some collections such as Chester Zoo (UK) have opted to move away from live animal handling on welfare grounds, as handling animals are often kept in different conditions to the other animals in a collection. However, they do encourage visitors to collect and handle wild animals in their grounds such as invertebrate species (Bazley 2018). This indicates a shared belief that first hand contact with animals is a crucial part of understanding nature and echoes ideas of learning through discovery and touch (Piaget 1973). There needs to be a balance between delivering high impact educational activities with conservation value, and maintaining high animal welfare standards. This may not be easily achieved in practice, especially since animal welfare must be assessed on an individual scale, whilst education and conservation impact is typically measured on much larger, population-level scales. However, by prioritising the needs of the individual animals, and developing training, housing, and display methods in-keeping with these needs, it is possible to deliver more sincere, legitimate conservation-education messages.

10.5 Impact on Visitor Learning, Attitude, and Behaviour

As recently reviewed (Whitehouse-Tedd et al. 2018), the ability of animal presentations to elicit an increased knowledge retention has been reported in a number of studies (Ballantyne et al. 2007; Hacker and Miller 2016; Reading and Miller 2007), and even linked to conservation intent (Hacker and Miller 2016). Presentations or performances using trained or conditioned animals offer experiential learning opportunities. By increasing visitor engagement through the assimilation of education within an entertaining or enjoyable experience, these programmes align with classical learning theories such as Piaget's theory of discovery and play, and Vygotsky's social constructivism theory (Piaget 1973; Vygotsky 1978). This combination of affective and cognitive learning is often key to zoo mission statements. The aim of many zoo education programmes is typically to enable visitors to develop a concern for the natural world, and ultimately to increase their commitments towards its preservation (Kellert 1996). To this end, interactions with animals can create a more personalised education experience in which conservation messages may be better received, potentially achieving increased personal meaning or relevance, which is considered an important dimension to learning in a zoo setting (Falk and Dierking 2000). This emotional response to experiences is often associated with increased learning, empathy, or connectedness. In zoos, the ability to see animals performing active behaviours, as well as having 'up-close' encounters with the animals (Figure 10.6) has been linked to increased positive affective responses and predicted their ability to make meaningful connections to concepts (Luebke et al. 2016). Self-reported emotional arousal was also highest for visitors observing animals performing active behaviours such as during a bird of prey flying presentation or lion feeding event (Smith et al. 2008b). Moreover, this self-reported data was supported by findings for changes in the heart and respiration rates of these visitors, indicating a physiological response to these zoo experiences.

It follows that animal training presentations, as well as the use of trained animals in zoo education presentations, have a role to

Figure 10.6 An animal encounter experience at London Zoo (UK). *Source:* K. Whitehouse-Tedd.

play in visitor learning. Although active participation is known to increase learning (Sim 2014; Sterling et al. 2007), it is possible that the observation of other people participating in an activity can be equally important in a learning experience. Therefore, the viewing of animal–trainer interactions may be an underlying educational component. For example, observing the protected-contact training of big cats was perceived to provide increased educational opportunities for the public, and the relationship between trainers and animals thought to be particularly important in this context (Szokalski et al. 2013). Increased visitor attention during live animal presentations or performances (Figure 10.7), compared to keeper-only presentations or

static exhibit viewing, indicates the potential for enhanced visitor learning opportunities (Alba et al. 2017). Information can be presented in an engaging manner during static or training presentations with live animals, potentially increasing the feeling of connectedness between visitors and animals they are observing (Szokalski et al. 2013) (i.e. using the trainer as a proxy). In another study, more positive experiences (scores for enjoyment, educational experience, and statements of value) were reported by visitors observing a training demonstration compared to those involved in passive exhibit-viewing or interpretation-only presentations (Anderson et al. 2003). Likewise, a greater knowledge gain was reported by visitors viewing an animal

Figure 10.7 Formal education class at Bristol Zoo Gardens (UK) using a live snake for demonstration purposes. *Source:* K. Whitehouse-Tedd.

presentation compared to those that hadn't (Price et al. 2015).

An important point here is that none of these studies tested learning in these contexts and therefore rely on the visitors' perceived learning. Moreover, where presented, it appears that only superficial learning was occurring (i.e. based on the recall of facts without the ability to apply or evaluate concepts) (Bloom et al. 1956; Crowe et al. 2008), such that the true educational value of animal training presentations remains to be determined. Similarly, whilst zoo visitors often desire animal interactions and view them as enabling a more holistic learning experience, the actual encounters are often recalled in terms of the pleasure experienced from petting and feeding animals, with only occasional recall of education, preservation

or conservation (Bulbeck 2004). Research on the 2016–2017 sea lion show at Flamingo Land demonstrated that whilst some facts were recalled from the sea lion shows, visitors left the performance with a weaker understanding of the natural uses for a sea lion's whiskers and an increased recall of unnatural behaviours such as balancing objects (Spooner 2017). This suggested that using artificial examples of natural behaviours may have been misleading to visitors. In response to research findings, the show adapted the objects used for balancing to appear more natural. For example, instead of using balls and bowling pins, the show uses a model fish (APAB Ltd, pers. comm.) which the sea lions balance to demonstrate how they would hunt under water, in order to move away from misconceptions that sea lions balance objects on their nose in the wild. However, this has yet to be evaluated regarding the visitors response and its impact on educational value.

Controversies also exist in regards to the value or justification of some types of training programmes. A survey of the perceptions of Australasian zookeepers involved in free- or protected-contact with big cats revealed a general consensus that any contact with big cats should be restricted to keepers (Szokalski et al. 2013). However, public interactions with these carnivores do occur in zoos around the world (e.g. posing for photos with tigers, walking with lions, stroking cheetahs). Concerns have been raised including sending the 'wrong' message, including misinformed perceptions of animal behaviour, the safety of interactions, or even promoting exotic pet ownership (Ballantyne et al. 2007; Bulbeck 2004; Szokalski et al. 2013). In contrast, others feel that animals would not be 'as interesting' without closer contact (Szokalski et al. 2013).

With that in mind, observing free-contact interactions between an animal and trainer was shown to increase visitor dwell time as well as provide evidence of more deeper learning via voluntarily offered statements related to animal behaviour and conservation status (Povey and Rios 2002). This use of the animal trainer as a proxy for visitors' personal experiences was also demonstrated in a study of visitor attitudes (Hacker and Miller 2016). Where visitors considered themselves to have experienced an 'up-close encounter' with an elephant (without any physical contact), a strong link was demonstrated to an improved attitude towards the importance of wild elephant conservation (Hacker and Miller 2016). Other studies have also determined animal training presentations to be effective in influencing visitor attitude (Miller et al. 2012; Price et al. 2015), as well as conservation intentions achieved from the use of trained animals in visitor presentations (Miller et al. 2012; Swanagan 2000).

When assessing the efficacy of animal presentations, the importance of the inclusion of a personalised experience, and appropriate interpretation, is apparent. The study of the elephant training presentation mentioned above revealed that viewing this activity alone did not influence visitor attitude or conservation intent (Hacker and Miller 2016), suggesting that the experience of an encounter must be an integral component of a successful education experience. The elephant presentation took place in a large amphitheatre, which may have reduced the intimacy felt by some visitors, depending on where they were seated. In another example, observing a rhino training presentation without any associated factual or interpretive commentary resulted in a high failure rate when visitors were quizzed on their knowledge after the demonstration (Visscher et al. 2009). Visitors that saw the same training presentation preceded by a keeper/educator explanation, had nearly 100% pass rates on the post-viewing quiz, illustrating the necessity for interpretation alongside animal training (Visscher et al. 2009). Extrapolation of individual study conclusions regarding the effectiveness or value of animal presentations in general should therefore be performed with caution, with sufficient acknowledgment of case-specific attributes in terms of experience quality and characteristics. Contextualisation is key (Falk and Dierking 2000), and educational impact will vary according to the programme design, the animal handlers, the location, and the safety of the performance (Orams 1997; Szokalski et al. 2013).

Differentiating between knowledge gain and attitude or conservation intention is equally important, as these outcomes are not always integrally or consistently associated. Visitors demonstrated higher knowledge scores (compared to their entry scores) three months after either the viewing of a presentation using trained animals, or an encounter between visitor and trained animals (Miller et al. 2012). However, only the visitors who participated in an animal encounter had maintained an increased attitude and behavioural intention score; the group viewing the presentation had returned to baseline in terms of their attitudes and conservation intentions (Miller et al. 2012). As such, whilst memorability may be a precursor to understanding, developing understanding from the recalled facts is not necessarily automatic.

Evidence of the effectiveness of animal presentations to increase support for conservation was shown in a study of visitors who participated in an interactive elephant presentation, compared to visitors viewing elephants in their enclosures (Swanagan 2000). During this study, the viewing of the animal presentation was associated with the highest rate of petitions for elephant conservation (Swanagan 2000). Similarly, Taronga Zoo's performance using donation-collecting parrots (described in Section 10.3) raised over £20 000 within its first year (Host 2008); the role of the birds in this activity appeared to increase the proportion of the audience donating, compared to when no birds were involved (Kemp et al., submitted). However, the amount donated (per visitor) was significantly less when a bird was used, compared to donations received without visitor–bird contact, and it has been postulated that visitors were seeking an interaction with the bird for a minimal cost, rather than making a true donation for the sake of conservation (Kemp et al., submitted). Another bird show was also determined to motivate an increased participation in conservation behaviours, or a reinforcement or supplementation of the knowledge (Smith et al. 2008a). However,

when participants were contacted six months after their zoo visit, most had failed to start new behaviours (Smith et al. 2008a). Similar short-term increases in conservation intentions had failed to be retained by visitors a few months after a visit to a 'conservation station' exhibit at Disney's Animal Kingdom (in which live animal presentations, and interactions were available) (Dierking et al. 2004). This lack of retention indicates that although short-term benefits may be achieved from trained animal presentations, an isolated zoo visit is challenged in its ability to instil longer-term behavioural changes.

Nonetheless, other examples exist to demonstrate the successful incorporation of conservation messaging in trained animal presentations at zoos, with impacts on visitor conservation awareness and intention to perform positive behaviours. Zoos Victoria, Melbourne (Australia) promoted marine habitat conservation via donations and motivation for recycling as part of their live animal presentation (Mellish et al. 2017). Likewise, Wellington Zoo (New Zealand) implemented a pledge to perform a conservation behaviour as part of a live animal presentation, and determined that by asking visitors to sign a pledge, as opposed to just being informed of the behaviour, significantly greater uptake of the behaviour could be achieved (Macdonald 2015).

Case Studies

There are a variety of techniques used for training and most trainers have their own preferences depending on the species and behaviour being trained. The following are some case studies selected to demonstrate the range of messages delivered, and the training techniques involved.

Biological and Ecological Facts

<u>Free Flown Exotic Bird Show and Birds Used for Film-work; Pleasure Wood Hills (UK)</u>
Biological fact: owl's sense of hearing.
Presentation: free flight showing natural flight and hunting behaviour.
The power of owls hearing can be shown or demonstrated by training the bird to a buzzer. Simply described, the owl is fed and the buzzer activated, the bird then identifies the sound with food. The buzzer is then placed in a location and activated; the bird flies to the point of the sound and is rewarded. During the demonstration multiple buzzers can be used to fly the bird to and from various points around the stage/demonstration area. Visitors can interact and choose the point of landing before the bird arrives, which will show that the bird is not flying to the same point every time and is in fact going to the sound.

Message: the bird is using its sense of hearing to locate its prey using the buzzer as a target.

Animals in Action; London Zoo (UK)

Biological facts: wingspan and scavenging feeding ecology.

Presentation: a turkey vulture is flown from point to point using a food reward.

As with most birds they are weighed daily; each bird will have its own flying weight and is fed accordingly. A bone on the stage table is also laced with a food reward. The routine is simple and easy for the bird to learn. Many birds used in shows are hand reared which can make bird–trainer interactions a lot easier as the bird should have no fear of humans and may even consider the trainer as a sibling or a parent.

Message: showing wingspan and flight of this species, the scavenging ecology is demonstrated with the use of a bone as a prop.

Cheetah Run; San Diego Zoo Safari (USA)

Biological facts: speed and hunting ability.

Presentation: this is a very natural behaviour that clearly displays the speed of the cheetah.

Training is simple as the cat has a natural inclination to chase moving objects. Starting as a cub with a lure on a pole as a toy, and at a later stage transferring the lure to a machine is a very simple way of achieving this behaviour. However, the choice of object/lure should be thought about carefully. For example, using a child's cuddly toy many not be appropriate; should the animal see such a toy in a pram or being held by a child, this could invoke a chasing response.

Message: evolutionary adaptations for speed and use of chase in this species' hunting strategy.

The Savannah Ecosystem; Cincinnati Zoo and Botanical Gardens (USA)

Ecological facts: the interactions between species.

Presentation: in this demonstration a range of savannah species are displayed, including a serval (which demonstrates its jumping ability) and a cheetah (its running ability as described above), as well as a livestock guarding dog (to demonstrate a conservation strategy).

The serval is a hand reared cat and like many animals can easily be taught to follow the trainer by being rewarded at intervals with food. Training the cat to step up onto a platform is simply a matter of putting the reward on the platform and the cat goes up to get it, and the same routine is used on the tree branch. The jump is a natural behaviour and the ball at the end of the pole is the target. The target can be introduced at an early age as a toy to the kittens. Being naturally playful they will get to identify with it quickly. The target is attached to the pole and gradually the height is increased; it may be that the cat receives a food reward after it has made the jump or shortly after to reinforce the target.

Message: displaying to the visitor a range of species found in a savannah ecosystem, their unique adaptations and a conservation strategy.

Ecological Information; Various Examples (Authors' Own)

Animals and birds can be displayed in many different ways in a show. A squirrel entering and exiting from a hole in a tree, or a meerkat coming out of a pipe to look like a burrow will give the visitor an immediate idea of where it lives and some of its nesting behaviours. It is perfectly natural for a meerkat to want to go into a hole at ground level, a food reward on entering will reinforce the behaviour. Training such a behaviour can be as easy as simply feeding the animal in the location, box or prop.

Message: showing the bird or animal with a natural prop (hole in a tree stump) gives the visitor a visual idea about this species' ecology.

Conservation Messages and Behaviours

Ocean Warrior Mascot, Rusty the Penguin; uShaka Marine World, Durban (South Africa)

Conservation behaviour: appropriate waste disposal.

Presentation: something as simple as an animal or bird putting rubbish into a bin can send a strong message to visitors.

Ocean warrior Rusty the penguin puts litter into a dust bin. First he will be taught to pick up the paper and rewarded each time he does, once he is holding the paper for a few seconds the bin is placed close so when he drops the paper it falls directly in; again he receives the reward. As the behaviour progresses the bin can be moved farther away and the bird will go to the bin to deliver the paper and is rewarded for doing so.

Message: the visitor will be reminded to dispose of their rubbish in a suitable manner.

10.6 Conclusion

Whilst further research is needed into the extent of educational impact from visitor–animal interactions and presentations, it is clear that humans desire contact with animals. Extending natural behaviours to facilitate ease of visibility and focus learning appears a valid use of animal training and has benefits in terms or veterinary and husbandry practices. Despite non-natural behaviours having the potential to mislead visitors, these behaviours have been used in some settings to raise public awareness of environmental issues and model positive behaviours. What is still not clear is where to draw the line as to what is in the animal's best interests and represents them as a species, versus the added value of coming into direct or close contact with an animal in terms of conservation concern and increased awareness.

References

Alba, A.C., Leighty, K.A., Pittman Courte, V.L. et al. (2017). A turtle cognition research demonstration enhances visitor engagement and keeper-animal relationships. *Zoo Biology* 36 (4): 243–249. https://doi.org/10.1002/zoo.21373.

Almeida, A., Fernández, B.G., and Strecht-Ribeiro, O. (2017). Children's opinions about zoos: a study of Portuguese and Spanish pupils. *Anthrozoos* 30 (3): 457–472. https://doi.org/10.1080/08927936.2017.1335108.

Anderson, U.S., Kelling, A.S., Pressley-Keough, R. et al. (2003). Enhancing the zoo visitor's experience by public animal training and oral interpretation at an otter exhibit. *Environment and Behavior* 35 (6): 826–841. https://doi.org/10.1177/0013916503254746.

Baird, B.A., Kuhar, C.W., Lukas, K.E. et al. (2016). Program animal welfare: using behavioral and physiological measures to assess the well-being of animals used for education programs in zoos. *Applied Animal Behaviour Science* 176: 150–162. https://doi.org/10.1016/j.applanim.2015.12.004.

Ballantyne, R. and Packer, J. (2016). Visitors' perceptions of the conservation education role of zoos and aquariums: implications for the provision of learning experiences. *Visitor Studies* 19 (2): 193–210. https://doi.org/10.1080/10645578.2016.1220185.

Ballantyne, R., Packer, J., Hughes, K., and Dierking, L. (2007). Conservation learning in wildlife tourism settings: lessons from research in zoos and aquariums. *Environmental Education Research* 13 (3): 367–383. https://doi.org/10.1080/13504620701430604.

Balloffet, P., Courvoisier, F.H., and Lagier, J. (2014). From museum to amusement park: the opportunities and risks of edutainment. *International Journal of Arts Management* 16 (2): 4–16. https://www.gestiondesarts. com/fr/from-museum-to-amusement-park-the-opportunities-and-risks-of-edutainment/#.XOK86G5Kg2x.

Bazley, S. (2018). General discussion on animal handling. BIAZA Regional Educators Meeting, Northern Region. Chester Zoo, Chester, UK: BIAZA.

Bloom, B.S., Engelhart, M.D., Furst, E.J., and Hill, W.H. (1956). *Taxonomy of Educational Objectives, Handbook I: The Cognitive Domain*. New York, USA: David McKay Co Inc.

Bulbeck, C. (2004). *Facing the Wild: Ecotourism, Conservation and Animal Encounters*. Oxon: Earthscan: Taylor and Francis.

Crowe, A., Dirks, C., and Wenderoth, M.P. (2008). Biology in bloom: implementing bloom's taxonomy to enhance student learning in biology. *CBE Life Sciences Education* 7: 368–381. https://doi. org/10.1187/cbe.08.

DEFRA (2012). Secretary of State's Standards of Modern Zoo Practice. Department for Environment, Food and Rural Affairs. https://assets.publishing.service.gov.uk/government/uploads/system/uploads/attachment_data/file/69596/standards-of-zoo-practice.pdf.

Dierking, L., Adelman, L.M., Ogden, J. et al. (2004). Impact of visits to Disney's Animal Kingdom : a study investigating intended conservation action. *Curator* 47 (3): 322–343.

EAZA Felid TAG (2017). Demonstration Guidelines for Felid Species. https://www. eaza.net/assets/Uploads/CCC/2017-Felid-TAG-demonstration-guidelines-final-approved.pdf.

European Association of Zoos and Aquaria (2014). EAZA Guidelines on the Use of Animals in Public Demonstrations. https://www.eaza.net/assets/Uploads/Guidelines/Animal-Demonstrations-2018-update.pdf.

Falk, J.H. and Dierking, L. (2000). *Learning from Museums: Visitor Experience and the Making of Meaning*. American Association for State and Local Book Series. Lanham, USA: Rowman & Littlefield Publishers, Inc.

Fuhrman, N.E. and Ladewig, H. (2008). Characteristics of animals used in zoo interpretation: a synthesis of research. *Journal of Interpretation Research* 13 (2): 31–42.

Hacker, C.E. and Miller, L.J. (2016). Zoo visitor perceptions, attitudes, and conservation intent after viewing African elephants at the San Diego Zoo Safari Park. *Zoo Biology* 35 (4): 355–361. https://doi.org/10.1002/zoo.21303.

Host, B. (2008). Conservation campaigning through animal training. *Journal of the International Zoo Educators Association* 44: 12–13.

Jensen, A.-L.M., Delfour, F., and Carter, T. (2013). Anticipatory behavior in captive Bottlenose Dolphins (*Tursiops Truncatus*): a preliminary study. *Zoo Biology* 32 (4): 436–444. https://doi.org/10.1002/zoo.21077.

Kellert, S. (1996). *The Value of Life: Biological Diversity and Human Society*. Washington D.C.: Island Press.

Luebke, J.F., Watters, J.V., Packer, J. et al. (2016). Zoo visitors' affective responses to observing animal behaviors. *Visitor Studies* 19 (1): 60–76. https://doi.org/10.1080/10645578.2016.1144028.

Macdonald, E. (2015). Quantifying the impact of Wellington Zoo's persuasive communication campaign on post-visit behavior. *Zoo Biology* 34 (2): 163–169. https://doi.org/10.1002/zoo.21197.

Mellish, S., Sanders, B., Litchfield, C.A., and Pearson, E.L. (2017). An investigation of the impact of Melbourne Zoo's 'Seal-the-Loop' donate call-to-action on visitor satisfaction and behavior. *Zoo Biology* 36 (3): 237–242. https://doi.org/10.1002/zoo.21365.

Miller, L.J., Zeigler-Hill, V., Mellen, J. et al. (2012). Dolphin shows and interaction programs: benefits for conservation education? *Zoo Biology* 32 (1): 45–53. https://doi.org/10.1002/zoo.21016.

Moss, A. and Esson, M. (2010). Visitor interest in zoo animals and the implications for collection planning and zoo education programmes. *Zoo Biology* 29 (6): 715–731. https://doi.org/10.1002/zoo.20316.

Orams, M.B. (1997). The effectiveness of environmental education: can we turn tourists into 'Greenies'. *Progress in Tourism and Hospitality Research* 3: 295–306.

Piaget, J. (1973). *To Undertand Is to Invent: The Future of Education*. New York, USA: Grossman.

Povey, K.D. and Rios, J. (2002). Using interpretive animals to deliver affective messages in zoos. *Journal of Interpretation Research* 7 (2): 19–28. https://www. interpnet.com/nai/docs/JIR-v7n2. pdf#page=19.

Price, A., Boeving, E.R., Shender, M.A., and Ross, S.R. (2015). Understanding the effectiveness of demonstration programs. *Journal of Museum Education* 40 (1): 46–54. https://doi.org/10.1179/1059865014Z. 00000000078.

Reading, R.P. and Miller, B.J. (2007). Attitudes and attitude change among zoo visitors. In: *Zoos in the 21st Century: Catalysts for Conservation?* (eds. A. Zimmermann, M. Hatchwell, L. Dickie and C. West), 63–91. Cambridge, UK: Cambridge University Press.

Robson, M. (2002). *A Non-Invasive Technique to Assess Stress in Captive Macaws*. York, UK: Central Sciences Laboratories.

Routman, E., Ogden, J., and Winsten, K. (2010). Visitors, conservation learning, and the design of zoo and aquarium experiences. In: *Wild Mammals in Captivity: Principles & Techniques for Zoo Management* (eds. D.G. Kleiman, K.V. Thompson and C.K. Baer), 137–150. Chicago, USA: The University of Chicago Press.

Sim, G. (2014). Learning about biodiversity: investigating children's learning at a museum environmental centre and a live animal show. Doctoral thesis. University of London. http://discovery.ucl.ac. uk/10021761.

Skibins, J.C. (2015). Ambassadors or attractions: disentangling the role of flagship species in wildlife tourism. In: *Animals and Tourism: Understanding Diverse Relationships* (ed. K. Markewll), 256–273. Bristol, UK: Channel View Publications.

Skibins, J.C. and Powell, R.B. (2013). Conservation caring: measuring the influence of zoo visitors' connection to wildlife on pro-conservation behaviors. *Zoo Biology* 32 (5): 528–540. https://doi. org/10.1002/zoo.21086.

Smith, L., Broad, S., and Weiler, B. (2008a). A closer examination of the impact of zoo visits on visitor behaviour. *Journal of Sustainable Tourism* 16 (5): 544–562. https://doi. org/10.1080/09669580802159628.

Smith, L., Weiler, B., and Ham, S. (2008b). Measuring emotion at the zoo. *Journal of the International Zoo Educators Association* 44: 27–31.

Spooner, S.L. (2017). Evaluating the effectiveness of education in zoos. PhD thesis. University of York.

Sterling, E., Lee, J., and Wood, T. (2007). Conservation education in zoos: an emphasis on behavioral change. In: *Zoos in the 21st Century: Catalysts for Conservation?* (eds. A. Zimmermann, M. Hatchwell, L. Dickie and C. West), 37–91. Cambridge, UK: Cambridge University Press.

Swanagan, J.S. (2000). Factors influencing zoo visitors' conservation attitudes and behavior. *Journal of Environmental Education* 31 (4): 26–31. https://doi.org/10.1080/ 00958960009598648.

Szokalski, M.S., Litchfield, C.A., and Foster, W.K. (2013). What can zookeepers tell us about interacting with big cats in captivity? *Zoo Biology* 32 (2): 142–151. https://doi. org/10.1002/zoo.21040.

Turley, S.K. (2001). Children and the demand for recreational experiences: the case of zoos. *Leisure Studies* 20 (1): 1–18. https:// doi.org/10.1080/02614360122877.

Visscher, N.C., Snider, Ã.R., and Stoep, G.V. (2009). Comparative analysis of knowledge gain between interpretive and fact-only presentations at an animal training session: an exploratory study. *Zoo Biology* 28: 488–495. https://doi.org/10.1002/zoo.

Vygotsky, L.S. (1978). *Mind in Society: Development of Higher Psychological Processes*. Massachusetts, USA: Harvard University Press.

Ward, P.I., Mosberger, N., Kistler, C., and Fischer, O. (1998). The relationship between popularity and body size in zoo animals. *Conservation Biology* 12 (6): 1408–1411.

Watters, J.V. (2014). Searching for behavioral indicators of welfare in zoos: uncovering anticipatory behavior. *Zoo Biology* 33 (4): 251–256. https://doi.org/10.1002/zoo.21144.

Watters, J.V. and Powell, D.M. (2012). Measuring animal personality for use in population management in zoos: suggested methods and rationale. *Zoo Biology* 31: 1–12: https://doi.org/10.1002/zoo.20379.

Whitehouse-Tedd, K., Spooner, S., Scott, L., and Lozano-Martinez, J. (2018). Animal ambassador encounter programmes in zoos: current status and future research needs. In: *Zoo Animals: Husbandry, Welfare and Public Interactions* (eds. M. Berger and S. Corbett), 89–139. Hauppauge, NY, USA: Nova Science Publishers.

World Association of Zoos and Aquariums (2003). *WAZA Code of Ethics and Animal Welfare*. World Association of Zoos and Aquariums.

11

Welfare Implications of Zoo Animal Training
Vicky A. Melfi and Samantha J. Ward

Training can improve and compromise zoo animal welfare; the outcome is dependent on a great many variables and often situations which can only be judged on a case by case basis. This can make the relationship between training and animal welfare complex.

11.1 Setting You up to Succeed

As described by many zoo professionals, training is just one of many different 'tools' in the husbandry 'tool box'. Zoo animal management aims to ensure good animal welfare through a myriad of applications, from ensuring genetic diversity, high quality and appropriate nutrition, a suitable social group (in terms of composition and number), preventative and remedial veterinary care, appropriate housing, and as we've learned in this book, learning opportunities which can be afforded in different ways, e.g. environmental enrichment, zoo environment, or training (see Chapters 3, 5, 6).

Training has been applied to the lives of zoo animals in countless ways for almost as many reasons. To support preventative veterinary care, administer drugs, and remedial veterinary support, to move animals between areas within their enclosure, to separate animals temporally or for prolonged periods of time, to aid the introduction and/or translocation of the target animal or those who will receive a new animal, to maximise enclosure use by animals, to increase animal visibility to the public, to facilitate human–animal interactions, to support education programmes and outreach which might include animals leaving the zoo site, to ensure appropriate nutritional intake, to increase or reduce the expression of different behaviours, to entertain visitors, to facilitate breeding programmes and the use of artificial reproductive techniques... the list could go on and on!

When we consider the many ways that training can be applied to the lives of zoos animals, it appears that there are three overarching goals: improving the individual animals' welfare; facilitating zoo operations; and achieving the zoos' mission, which includes a conservation imperative (Barongi et al. 2015). These three goals are important to the success of the zoo, but unhelpfully they might not be congruent with respect to their impact on animal welfare. In much the same way that zoo professionals consider the ultimate goal of zoos to be species conservation (Fa et al. 2014), we know that tools and techniques which might yield good conservation goals can themselves compromise welfare (Beausoleil et al. 2014; Keulartz 2015).

In this chapter we hope to: clarify how animal welfare science can be used to better understand the impact of training; enable zoo professionals to take an evidence-based approach to whether training is the best tool for a given a situation; explore methods of evaluating the impact of training on welfare;

Zoo Animal Learning and Training, First Edition. Edited by Vicky A. Melfi, Nicole R. Dorey, and Samantha J. Ward.
© 2020 John Wiley & Sons Ltd. Published 2020 by John Wiley & Sons Ltd.

and initiate discussion and hopefully motivate investigation of the relationship between animal welfare and the provision of husbandry training in zoos. Given that we anticipate that readers of this chapter have or will read other chapters in this book, we have not provided and overview of the impact different training techniques may have on the welfare of animals, as this is considered throughout this book, especially in Chapters 4 and 7. So the obvious misconceptions, conveyed by language, that any training which includes the term negative or punishment, is also detrimental to animal welfare is not further discussed.

11.2 Animal Welfare Science vs Ethical Considerations

Ethics are the 'moral principles that control or influence a person's behaviour' (OED 2018). The study of animal ethics relates to how we as humans use animals within society and whether these uses can be justified. The study of ethics is based on theoretical philosophical reasoning of concepts such as, the moral status of animals and the nature and significance of species, as well as considering practical applications, in terms of how animals are used in today's society for example in farming systems or within zoos (reviewed Beauchamp and Frey 2011). By contrast, animal welfare describes the animals' state, along a continuum (from great to awful) at a particular time, which manifests as a result of its ability (or lack thereof) to meet physical and psychological challenges within its daily life. Animal welfare science is a relatively new multidisciplinary subject which aims to empirically evaluate and consider the welfare state of animals; a good review of the history of animal welfare science was undertaken by Broom (2011). There can be some confusion whether a topic is an ethical or welfare issue. It is important to clarify how these perspectives differ, as the study and outcomes of them can lead to different results and thus application. Simply, issues relating to whether we should use animals and how is the realm

of ethical debate. Whereas measuring the impact of what we do to animals is the realm of animal welfare science.

There are various texts which specifically consider the ethics of zoo animals (e.g. Norton et al. 2012; Gray 2017). Other than wanting to draw the readers' attention to the distinction between animal ethics and welfare, this chapter aims to focus on animal welfare as it relates to zoo animal training. There are many ethical considerations which relate to zoo animal training, not least should we train animals? Does training impact the animals' integrity (telos; Figure 11.1)? Does training zoo animals further remove them from their wild conspecifics and does it matter? If zoo animals are trained does it influence our perception, treatment, and consideration of them? As can happen within the field of ethics, the number of questions which arise can become overwhelming and the discourse required to do justice to these important issues lengthy! As pragmatists, and let's be honest zoo animal welfare scientists rather than philosophers, we have decided it would be really valuable to provide an overview of the implications training zoo animals might have relative to their welfare state. This overview is intended to support you, the reader, to make evidence based decisions about training, which might

Figure 11.1 An animal's telos typifies what it is to be that animal: a hummingbird illustrated here hovering typifies what it is to be a hummingbird. *Source:* https://www.pexels.com/photo/animal-avian-beak-bird-349758.

also contribute to ethical discussion; the latter will likely vary between countries, cultures, and organisations.

11.3   Animal Welfare Science

Central to the debate of whether animals have the capacity to suffer is whether they possess consciousness and/or sentience (Dawkins 1980). Historically few have attempted to empirically study animal consciousness and emotion; as much as anything the study of the subjective affective world of humans is difficult to traverse, so this undertaking in animals is also very difficult (Dawkins 2008). To move the study of animal welfare forward, Dawkins, acknowledging that the study of animal consciousness is central to considerations about animal welfare, suggested that we can infer much about animal welfare through two simple questions: (i) Are the animals healthy? and (ii) Do they have what they want? (Dawkins 2004). Following in this direction a number of studies prior to this, focussed on animals' preferences, or how hard they might 'work' for resources within their environment (review Dawkins 1990; Fraser and Matthews 1997); these studies were considered to address the question 'What do animals' want?' More recently studies of cognitive bias, testing whether animals are optimists or pessimists, and how the consequential choices they make are affected (Harding et al. 2004) have been instrumental into shedding light on the emotional capacity of animals. Further work in this area suggests that the majority of vertebrate animals are sentient beings and are able to feel both positive and negative emotions and affective states (Guesgen and Bench 2017; Paul and Mendl 2018).

It is likely that the difficulties inherent in studying animal emotional experience, led to the large body of work and bias in animal welfare science, which has focussed on the biological functioning of animals (Barnett and Hemsworth 1990). This functioning approach is often considered to 'refer[s] to

the state of an individual in relation to its environment, and this can be measured' (Broom 1991). These studies generally use physiological or psychological (usually measured through behavioural expression) metrics to measure an animals' stress response in different situations (Broom 2011). This area of animal welfare science has dominated the methods chosen to study welfare, the questions asked and the prevailing attitude that animal welfare science can be largely determined by the extent to which an animal thrives in captivity; evidenced for the most part on production indices, i.e. mortality, morbidity, fecundity, and growth rates. This approach is underpinned by our understanding of stress biology; the extent to which animals' are able to meet (or cope with) challenges in their environment (Moberg and Mench 2000). In the most extreme situations, captive animals that are not able to adapt their physiological or psychological response appropriately to challenges will die. For a great entertaining overview of stress biology, which relates to humans but has biological relevance to zoo animal welfare too, check out Sapolsky (1998); it has been written explicitly to be palatable and engage all audiences.

A third avenue adopted in the study of animal welfare is to appreciate and respect the animals' telos, 'the set of needs and interests which are genetically based and environmentally expressed, which collectively constitute or define the form of life or way of living by that animal, and whose fulfilment or thwarting matter to the animal' (Rollin 2003). An easier way of considering this view was proposed by environmental philosopher Bernard Rollin, who suggested that 'Birds gotta fly and fish gotta swim' and as animals deviate from these features which typify them, so too would their welfare state (Rollin 1990) (Figure 11.1).

Taken together, these three pillars within the field of animal welfare science have been referred to as the study of the animals' mind, body, and nature (Duncan and Fraser 1997) (Figure 11.2). Others have suggested that an animal's natural history and evolutionary

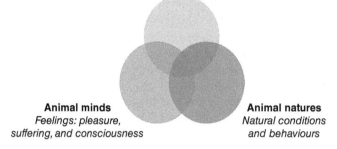

Animal bodies
*Functioning: health,
growth, disease, reproduction*

Animal minds
*Feelings: pleasure,
suffering, and consciousness*

Animal natures
*Natural conditions
and behaviours*

Figure 11.2 Mind, body, and nature are the three most commonly used directions to study animal welfare.

biology be considered when evaluating animal welfare, to provide parameters of what animals might have evolved to tolerate and the mechanisms they use in doing so (e.g. Barnard and Hurst 1996). Alternatively, we can consider welfare by allostasis (Korte et al. 2007); a concept whereby we consider biological and psychological structure and function. Following an allostatic approach, we might venture that given many animals possess similar biological structures as ourselves, that the function of these is also similar. A logical endpoint is that if we have consciousness and other animals possess the same biological apparatus as us, that they too will likely have consciousness too (Griffin 2013). As you can imagine depending on the approach taken, different data need to be collected and evaluated, and importantly different questions about welfare are likely to be addressed. That we all come to the study and issue of animal welfare with 'baggage', due to different cultural views, personal experiences, societal values, and many other influencing factors, needs to be remembered, given the questions we ask about animal welfare will determine how we measure animal welfare and how we interpret an animal's welfare state (Fraser 2008). Furthermore, though objective indices of welfare have, and are, considered to be the most reliable and robust tools for assessing animal welfare, studies using validated subjective ratings are proving to be incredibly sensitive and informative i.e. qualitative

behavioural assessment (Wemelsfelder and Lawrence 2001; Wemelsfelder et al. 2001).

This is an overview, and by no means considered an exhaustive review of animal welfare science. It is provided to help identify how animal welfare can be approached, which is essential when we move forwards in this chapter to explore how training and/or learning might impact on zoo animal welfare.

11.4   Choice and Control

Embedded within zoo animal welfare science is the precept that providing control to animals is essential for them to experience a good welfare state (Hill and Broom 2009) and that control in the environment can be achieved by offering choice in that environment. Consequently, offering choice and control within captive animal management has become a generic goal and studies have demonstrated welfare benefits when animals are afforded some degree of control over their environment (reviewed Whitham and Wielebnowski 2013). Providing zoo animals with opportunities to gain some control within their environment, can be achieved in one of two main ways: through minor changes or additions to the enclosure itself i.e. the provision of enrichment which animals choose to use or not (e.g. Carlstead and Shepherdson 2000), and through changes to husbandry, i.e. providing animals with the choice of where to spend their time (e.g. Ross 2006).

There are many studies where similar types of changes to housing and husbandry have been associated with measurable benefits to zoo animal welfare, and authors have attributed, in part, the success of the intervention with the reasoning that it conveys a degree of 'control' to the animal within its environment (e.g. Carlstead and Shepherdson 2000; Fernandez et al. 2009). Whether the mechanism for the success of these interventions is that they bestow control is arguably difficult to empirically identify. What is clear, is that these studies do offer some choices to animals.

Some years ago, Broom (1991) warned that the choices animals made, might not necessarily equate to good welfare. Broom's (1991) examples where a diet could be chosen and expression of self-injurious behaviour performed, both of which might be choices, but that both would reduce health and also welfare. The provision of choice alone it seems, does not necessarily result to elevated animal welfare. Furthermore, qualities of the items being offered to be chosen between, are fundamental to whether they might convey welfare benefits; a choice of two options, neither of which are appealing or appropriate, does not do justice to the concept of providing choice (Fraser and Matthews 1997). We add this preamble, because in many ways, efforts made to improve the welfare of zoo animals have become ubiquitous in their assertion that they are providing choice and control within the zoo environment but that this detail has almost become perfunctory. And how does all this relate to zoo animal learning and training?

It is often suggested that training programmes enhance zoo animal welfare by offering them choice and control within their environment (Westlund 2014). An often quoted benefit of positive reinforcement training, is that animals 'voluntarily' take part or that they 'cooperate' with the training programme; both of these processes would give us cause to believe, and certainly many trainers do, that the participating animals therefore have a 'choice' to be involved and they exert 'control' in their environment by taking part (e.g. Bloomsmith et al. 1998, 2003). If indeed animals did have the choice to take part in the training programme, we agree it would deliver control to the animal over its environment. For this to be true we would expect that participation in training would not affect other aspects of housing and husbandry. Where this scenario becomes blurred, starts with the definition of training versus learning: 'Learning can be broadly defined as a change in behaviour resulting from practice or experience; when practice or experience is dictated by humans, the process is called training' (Mellen and Ellis 1996). In a situation where humans dictate what is learned, there seems limited opportunity for animal choice. The irrationality of the scenario being voluntary, is expounded when we consider that there is an expectation when a training programme is initiated, that the acquisition of prescribed behaviours is required, rather than hoped for. It seems unreasonable and untenable to believe that zoo professionals will be provided with the resources to train a behaviour, just in case the animal wants to cooperate and join in. To be clear, there are circumstances where 'training' goals are loose and different types of behaviours might be reinforced (e.g. animals being rewarded for 'inventing' behaviours). However, part of our scepticism about the voluntary nature of the animals' participation in training, results from observation of what happens if the animal chooses not to engage. From experience, there might be situations where different animals are chosen for training, so indeed voluntary participation is possible, but in the majority of situations that we are aware of, the training programme would be altered to 'set-up the animal for success'.

We acknowledge that training zoo animals can lead to positive welfare outcomes (see Figure 11.3). There is no malice intended by drawing attention to the lack of choice or control in most training programmes, or that the terminology used belies that humans are dictating and directing what the animal is

Figure 11.3 A trained dolphin having a catheter fitted. *Source: Katharina Herrmann.*

expected to do; versus providing animals with free choice to take part in their own behavioural journey. We would argue that training, as it is often implemented, removes choice and control from the lives of captive animals, but this does not detract from the benefits which it can achieve through promotion of animal welfare, facilitating husbandry, and achieving the zoo's mission (as referred to above). What we hope to highlight, is that though good training programmes are not physically forcing animals to cooperate, animal participation is gained in such a way that despite what the training terminology would suggest, animals' are seldom free to opt out. In fairness, choice and control within our own lives isn't all it seems, in fact less choice can yield greater satisfaction (Iyengar and Lepper 2000). Many of us might feel we have made many choices in our lives which lead us to buying a certain car, living in a certain neighbourhood, indulging in a specific lifestyle; but even those 'alternative' lifestyle choices have been crafted and marketed to us via the emerging field of neuroeconomics (Hansen and Christensen 2007; Hodgson 2003). With people viewing thousands of brand placements and other adverts daily, ubiquitous marketing in our world is akin to trainers in the lives of zoo animals; if we believe we have choice, maybe they do in the zoo too.

Instead of being voluntary we would venture that zoo animal training is a form of

benevolent dictatorship; whereby as zoo professionals we hold absolute power over the actions and opportunities afforded to the animals in our care, but hopefully choose to use this power to support their needs so that they can attain good welfare. For example, shaping a behaviour is a method of training an animal to perform a desirable end-goal behaviour, which would otherwise be difficult, by presenting it as small changes of a previously learned and thus accepted behaviour. As good trainers we offer choices to the animal which will enable us to teach the animals the behaviours we need them to learn. We might undertake preference tests prior to training, to ensure that the reinforcers used are highly prized resources; whether this is a food item, activity, or access to social interaction (e.g. Clay et al. 2009). Highly prized resources might be withheld and only available during training sessions. Social relationships might be used to increase motivation and engagement, whether that represent human–animal interactions or peers, and use the animal's motivation to be in proximity with that key person or conspecific to set them up to succeed. The animal's environment might be modified so that it becomes a less desirable place to be, so the alternative training option is favoured. Taking part in the training programme might be facilitated with physical barriers, which make it hard for the animal to not comply or withdraw once training has begun. How we set animals up to succeed is, to a large extent, limited by our own imagination, knowledge of the species biology and the particular individual animal. Setting an animal up to succeed isn't malicious, far from it; instead the goal is to make participation in the training programme appear favourable and in managing the animals' perception to positively engage with the training programme, we are removing choice at a fundamental level; especially if the training we feel is necessary might lead to pain, injury, or distress.

Learning can be viewed as a series of associations, which when repeated often can cause changes in brain morphology

(Merzenich et al. 2013); the more a behaviour is repeated the easier the brain finds it to repeat that behaviour in the future. Most of us are more than aware of the power of habitual behaviour to occur without conscious thought or to override it (Lally and Gardner 2013). When we establish zoo animal training programmes to 'encourage' animals to perform the behaviours we require, regardless of the circumstances (internal and environmentally) the animal finds itself in, we are in effect aiming to create behavioural habitats, wherein by definition, animals no longer choose whether they take part or not in the requested behaviour. This also means that once a training programme has successfully established a behaviour, that an animal has learned after a specified cue a particular behaviour is required, any and all positive ramifications associated with training due to it providing an learning opportunity can no longer be justified, i.e. that it is enriching (e.g. Melfi 2013).

With all that we have discussed in this section, and to be honest a theme within this chapter more generally, there is huge scope for variation in how training is delivered by us and perceived by the animals we are working with. To suggest that there is one hard and fast rule by which to interpret whether training programmes provide control and choice to the animals involved, would be to ignore the richness which encompasses the art and science of animal training; and the differences in those people practising training and the animals taking part. What we want to highlight, is that animal training programmes need to be considered by the activities performed rather than by the professional terminology often used to describe training whether to peers or those outside of the profession.

11.5 What Behaviours Should Be Trained?

Zoo animals can be trained to perform a wide variety of different behaviours. Throughout this book other authors have considered how these behaviours are trained, whereas in this chapter we're concerned with the impact that the expression of these behaviours might have on the animals' welfare. There are obvious safety implications which need to be considered when planning and implementing zoo animal training programmes, for example whilst performing the behaviour the animal should had adequate physical space to manoeuvre and actions should be taken to ensure falls or slips are unlikely (see Chapter 13). There are also a number of other potentially less obvious considerations, which might determine what behaviours are likely to be trained.

Despite the large number and variety of animals which are trained in zoos, to achieve a large number of different goals, the types of behaviours most commonly trained are quite small. Frequently animals are taught to move (sometimes referred to A–B), or stay still (station), to present a part of their body, or perform a naturally occurring behaviour on cue. Successfully achieving this small number of behaviours can support the promotion of animal welfare (i.e. veterinary checks), facilitate operational demands (i.e. moving animals within their enclosure), and achieve wider zoo missions (i.e. educational activities with conservation messaging). How the expression of behaviours affects the animals' welfare, can be affected by context, i.e. there might be social implications, but it is not influenced by how we as people perceive the behaviour. We feel it important to consider this latter point early in this discussion, as we can sometimes confuse how we feel about an animal performing a behaviour, with the likely impact it has on the animal. For example, if we consider the simple A–B behaviour. We might watch an animal being moved between areas of its enclosure as part of husbandry, being achieved using a cue and reward, and consider training to have positively progressed animal welfare. Especially when historical alternatives might have included brooms, loud noises or water jets, different forms of negative reinforcement, or positive punishment to move the animal out of the area it was in. However, seeing that same animal, perform the same behaviour in a show which doesn't contain any conservation messaging, might make some reflect that the 'show' is little more than a four letter word (Martin 2000) and the animal's welfare has been compromised. To be clear, the welfare of the animal in these two different scenarios is equal with respect to the behaviour it has been trained to perform; there might be welfare ramifications resulting from visitor pressure in the show, but for the most part, the animal has been similarly psychologically and physically challenged in both situations. What has changed is how we have perceived the behaviour. It is worth

noting that animal welfare is not affected by the narration, what is said to human visitors when animals are performing trained behaviours does not impact the animals welfare, only our perception of whether it is acceptable (see Figure 11.4).

11.5.1 Natural vs Non-natural Behaviours

In much the same way as we can confuse our interpretation of a behaviour, with how it might affect animal welfare (see above), there is also a tendency to consider that 'natural' behaviours are inherently good for animal welfare, whereas 'artificial' behaviours will

Figure 11.4 Whether we call it a show, presentation, or demonstration, and whether there be a conservation message or not, the impact on the animal's welfare is the same. *Source: Katharina Herrmann.*

compromise welfare. As with the example above about narration, whether trained behaviours are within the animals' natural repertoire might affect how we interpret them, and potentially might impact on education messaging (see Chapter 10), but doesn't necessarily impact the animals' welfare. It is important to remember that not all trained behaviours support an animal's behavioural ecology, survival skills, or enable an animal to thrive in the wild. Instead, trained behaviour might still have merit if they improve the animal's life whilst in captivity, facilitate operations or achieve zoo missions. Three main clauses need to be satisfied when we consider the impact of trained behaviours on animal welfare: the behaviours need to be within the animals physical ability, so that it doesn't cause strain or injury; likewise the behaviour needs to be within the psychological aptitude of the animal, so that attempts to achieve the behaviour don't lead to frustration (McGreevy and Boakes 2011); and the behaviour should not lead to pain, injury, and distress. There might be exceptional circumstances to these clauses, but when a trained behaviour contravenes any one of these, a discussion should take place to consider why compromising the animal's welfare by expression of a trained behaviour is necessary.

11.5.2 Pain, Injury, and Distress

It seems unlikely that in a profession where we work tirelessly to improve welfare, we would consider training an animal to perform a behaviour which might lead to pain, injury, or distress. Some training programmes, the prerogative of which was to improve welfare, do so by reducing welfare to some degree as animals may experience pain, injury, and distress performing behaviours that we hope will lead to a positive outcome. Akin to a spoonful of sugar, in some instances it is hoped that the performance of these behaviours, will support veterinary care, which will have a long-term benefit over the short-term welfare insult caused. For example, training

programmes which aim to support veterinary injections or other invasive tests, will lead to some pain (Otaki et al. 2015; Reamer et al. 2014). It is successfully argued that this short lived painful event which occurs by performing a trained behaviour, constitutes a better welfare outcome than the alternative, which might require removing the animal from its social group, sedation, handling, and moving. However it is important to always remember, just because we can train an animal to perform a behaviour, it might not always be in the animal's best interest. As outlined in Section 11.4, participation in a training programme doesn't suggest that the animal is cooperative, doesn't experience pain, doesn't 'care' that it is injured, or can't experience distress. Instead, when training programmes include behaviours, which are likely to cause pain, injury, and distress, the situation has likely been set up appropriately to ensure animal participates. Or as we have seen in practice, the training programme parameters are frequently changed if an animal becomes non-compliant, so that engagement is achieved. We would venture, that if a training programme has to be modified frequently to out compete an animal which 'chooses' not to take part, the behaviours being sought are likely causing pain, injury, or distress.

11.5.3 Using Training to Modify Daily Behavioural Expression

In some instances, we choose to train animals to perform behaviours which we feel benefit the animal's welfare, mostly because we prefer the look of the trained behaviour compared to what the animal might be inclined to do by choice. We might choose to train an animal to move around its enclosure and thus be more physically active, because we consider physically fit animals will have better welfare (Bloomsmith et al. 2003; Veeder et al. 2009). Operant conditioning was included within husbandry in this fashion by Markowitz and his colleagues and termed environmental engineering (e.g. Forthman-Quick 1984; Markowitz and

Spinelli 1986). The aim of which was to train animals to perform behaviours which would improve their welfare; this developed into the phenomenon of environmental enrichment as practised widely now (see Chapter 6). We might train animals to show fewer aggressive behaviours towards us (Minier et al. 2011), as we don't like to think that the animals don't want to be in proximity to us. Another argument for training animals to perform behaviours we like the look of, and/or are conducive to good animal welfare, is that by performing these behaviours, the animal changes and becomes more like the behaviour it has been trained to display. Data do exist within the laboratory primate literature, where primates have been trained to be more social; this was executed by using positive reinforcement training of 'less socially engaged' primates when they sat next to or made physical contact with another primate (e.g. Schapiro et al. 2003).

Coleman and Maier (2010) found the inclusion of a training programme reduced the performed behaviour and stereotypies in a group of laboratory housed rhesus macaques. The macaques were not explicitly trained not to show stereotypies, but instead were trained for venepuncture. The authors suggested that the training might have functioned by alleviating boredom and stress. Shepherdson et al. (2013) also noted that

polar bears were less likely to show stereotypic behaviour if a training programme were included in their husbandry routine. Similar findings were reported by Shyne and Block (2010) in African wild dogs which were trained for veterinary care, but during periods when training was provided, the dogs spend less time performing stereotypic behaviours. In a study where both training and play (via human–animal interactions) were provided to a group of zoo housed gorillas, a variety of behavioural indicators associated with poor welfare were seen to reduce (stereotypies, behaviour directed at visitors, aggression, and inactivity), whilst behaviours associated with good welfare were seen to rise (affiliative and social play-related behaviours; Carrasco et al. 2009). Importantly this final study found that the positive ramifications associated with the training programme were seen in all animals in the group, even though training was not provided to all animals in the group. Interestingly, in another group of zoo housed gorillas, behavioural change (reduced ear covering and keeper-directed aggression) was observed not during a period when training was provided but afterwards (Figure 11.5). The authors suggest that a combination of training and increased human–animal interactions were likely the cause the behavioural change.

Figure 11.5 Studies have shown that training and human–animal interactions can increase play behaviour in gorillas. *Source: Ryan Summers* https://www.flickr.com/photos/oddernod/35106444790.

11.6 The Impact of Training

Thus far, this chapter has outlined how we might apply some learning principles to better understand how it affects animal welfare; and importantly how to recognise when our feelings prejudice our view of what the animal might be. Considering the impact of training on animal welfare is an epic challenge, not least because there are different animal learning principles, which underpin many different training techniques, which are interpreted differently and thus actioned differently; and that is just the variation in how we provide training! There are of course, as many differences in the animals we choose to train, whether they exist at the species or individual level. To provide a meaningful and coherent consideration of how animal training has been empirically demonstrated to enhance zoo animal welfare, we too are restricted by the published literature; few studies have detailed training provision or the impact of training on either physiological or psychological parameters. Instead, much of the literature focusing on zoo animal training evidences success of a training programme, by the expression of the desired behaviour.

11.6.1 Overview of Positive Impacts Associated with Training

There are a number of ways that training zoo animals can improve their welfare. The performance of the behaviour might itself have beneficial welfare repercussions. For example, we read in Chapter 3 that learning can result in all types of physiological and psychological benefits. So we might be safe to assume that whilst animals are learning in a training programme, the welfare of the animals might be improved through positive ramifications of learning, such as brain development (reviewed in Chapter 3). The animal might also directly benefit from the behaviours it performs during a training programme, if they themselves serve to promote welfare in and of themselves, or if the animal performs behaviours associated with good welfare, outside of the training session, as a consequence of the training programme.

There are studies which suggest that taking part in a training programme can confer behavioural changes outside of the training programme itself and thus welfare can be promoted. For example, Pomerantz and Terkel (2009) observed that laboratory housed chimpanzees displayed a higher rate of behaviours associated with positive welfare and few behaviours associated with poor welfare, after the implementation of a training programme. The authors suggested that training led to positive changes in welfare which were sustained and occurred outside of the training programme. Similar findings have been reported in zoo situations, for different species. For example, zoo housed ring-tailed lemurs, during periods where training was provided (for cognitive research), were observed to display higher rates of affiliative behaviour and lower rates of aggressive behaviours (Spiezio et al. 2017). The inclusion of a training programme into the husbandry of fur seals was thought to be the reason why no stereotypies were displayed (Wierucka et al. 2016); though with no data relating to the animal prior to the implementation of the training programme is it difficult to appreciate whether the authors are observing an association or causal relationship between training and stereotypic behaviour.

Another way training can improve animal welfare, is indirectly by virtue of a successful training programme facilitating the role of the zoo professional, and facilitating husbandry. When animals are trained to facilitate husbandry requirements, the animals can benefit from better veterinary/health care, nutrition, access to resources and much more. The ways in which training can facilitate husbandry are many, but some examples are listed here: proactive and preventative veterinary care i.e. training for venepuncture, urine sample collection, or a nebuliser (Aldabra tortoise, Weiss and Wilson 2003; marine mammals, Ramirez 2012; zoo chimpanzee, Gresswell and Goodman 2011; primates, Savastano et al. 2003); facilitating enclosure management, i.e. moving animals between enclosure areas, crate training (laboratory chimpanzees, Bloomsmith et al. 1998; giant panda, Bloomsmith et al. 2003; laboratory sooty managbeys, Veeder et al. 2009; marabou stork, Miller and King 2013). Not only can training make achieving husbandry goals easier for the zoo professional, but there are some data from the laboratory community which suggests that trained compliance can be much less stressful for the animal than the alternative. For example Lambeth et al. (2006) measured stress reaction, via parameters available in a blood sample (e.g. total white blood count, glucose), in laboratory housed chimpanzees trained for venepuncture. The trained chimpanzees displayed significantly lower levels of stress compared to conspecifics which had blood taken 'traditionally' – by physical restraint. I appreciate, that as mentioned previously, there might exist some differences in zoo and laboratory housing and husbandry, but it seems likely that where trained compliance appears less stressful for the animal and the zoo professional, it is likely physiologically less stressful too.

11.6.2 Considering Short- and Long-term Welfare

Within animal welfare science there is an appreciation of short-term versus long-term welfare, whereby we recognise the welfare state of an animal in the present (short-term) against what it might be at some point in the future (Dawkins 2006). There are many examples, where a short-term welfare impairment is considered justified for a long-term welfare goal. For example, we might consider the pain experienced during venepuncture training sessions is justified against the alternative negative welfare impacts in the long-term if the animal went undiagnosed with a disease they are susceptible to. The factors which need to be weighed against each during this type of cost–benefit analysis are: what is the frequency, duration, and intensity of welfare impairment over the short term? How likely will the long term welfare be impaired if no action is taken? What is the nature and intensity of the long term welfare impairment if it arises? For example, we might consider weekly venepuncture, which appears to cause intense but short lived pain, justified versus the debilitating impact of disease. A cost–benefit analysis is a systematic evaluation of what can be gained or compromised through an action; widely applied in the field of animal use.

Again there are few empirical studies which can guide us, as to the pain, injury, or distress which animals can or should withstand during training sessions. However, ensuring that an assessment has been carried out to justly compare the accumulated short-term insults against the potential long-term insult is important; and that the risk of the long-term insult occurring is appropriately calculated.

11.6.3 The Social Side of Training

Training programmes can have social consequences which impact animal welfare, including human–animal interactions (reviewed Chapter 9), as well as conspecific social interactions. With respect to interactions with people, training programmes provide opportunities for an increased frequency of human–animal interactions, which have been suggested to likely improve animal welfare (Ward and Melfi 2015). It has been suggested that high frequencies of positive human–animal interactions aid in the development of positive human–animal relationships, which are associated with good animal welfare (reviewed Ward and Sherwin 2019). Interestingly, training may result in fewer human directed behaviours displayed by animals (Melfi and Thomas 2005), which might result from animals more accurately predicting when to engage with people, i.e. only during training sessions when positive reinforcements follow. Training can also be used to habituate animals to the presence of humans (Carrasco et al. 2009), which can have the positive welfare consequences that the animal's social interactions are directed within the animal group rather than at visitors. This might be especially beneficial in animals which display begging behaviour, which would likely have been reinforced by visitors feeding the animals, but in a situation when animals beg they then can cue and encourage further provisioning by visitors.

Ironically, there is the potential for negative welfare repercussions if positive human-animal relationships and especially if bonds are formed during training. When animals form bonds with people, we understand that those interactions take on special significance (Hosey and Melfi 2018). Whether the trained animal perceives there to be positive relationships which are shared with all the people that train them, or if there is a one person they especially 'like', interrupts in the delivery of these interactions that may be viewed negatively. This might be especially true for animals which have had formalised training as part of their husbandry for many years, or since they were young individuals, and therefore come to favourably predict or anticipate the training sessions. Predictable

routines for zoos animals can promote the development of positive anticipatory behaviours, which have been suggested to indicate positive animal welfare (Watters 2014). Clegg et al. (2018) observed that dolphins (*Tursiops truncates*) displayed a higher frequency of positive anticipatory behaviours in association with human–animal interactions with familiar trainers, compared to when they received access to toys in a pool. The authors suggested the observed anticipatory behaviours indicated that the dolphins perceived the training event with the familiar person to be positive; and in this context more positive than access to toys (Figure 11.6). It could be argued that if training sessions, which are positively anticipated are discontinued the animals involved may suffer. It is speculation how this might manifest, but it is plausible to consider that animals might need to find alternative ways to spend their time, and/or become frustrated, aggressive and/or bored (Chamove 1989).

Training programmes can also affect conspecific social interactions, either directly when social interactions with conspecifics

Figure 11.6 Ripley and China (adult Atlantic bottlenose dolphins) demonstrating the squawk vocalisation during a chat at the Indianapolis Zoo's Marsh Dolphin Adventure Theatre. *Source: Durin* https://nv.m.wikipedia.org/wiki/E%CA%BCelyaa%C3%ADg%C3%AD%C3%AD:IndyZoo-DolphinsBlowBubbles.jpg.

are the goal of training, or indirectly during training for other behaviours. A common tool when training socially housed animals is to 'station' them to one spot; reinforce them for staying still in a specific location in the enclosure. The reasons for instigating this type of training could include: ensuring that all animals could be trained effectively despite being socially housed; ensuring that all animals were able to eat separate portions of food without competition; to reduce the risk of aggression during times considered to be problematic; to separate animals when they are engaged in aggressive behaviour towards each other; maximising animal visibility within an enclosure. Bloomsmith et al. (1998) trained an adult male chimpanzee during meal times to sit and not perform aggressive behaviours towards cage-mates. They observed that this training did not generalise to conditions outside of the training situation (feeding time) and yet enabled all members of the chimpanzee group to feed without aggression. This study demonstrates welfare benefits can be conferred by training social restraint; which no doubt go further in yield gains for husbandry and mission goals. It would be interesting to see further empirical data, about the impact of such trained social restraint in other species. It would be especially interesting to explore how training might impact the integrity of the social hierarchy, which would ordinarily be set by the dominant animal(s) in the group, and instead removes that animal(s) control of the group and translates effective governance of the group to the trainer. Within the paradigm of the human–animal interactions, it is unclear, whether this exchange of 'dominant' would mean that the trainer is perceived to be part of the social hierarchy, which would have implications of its own, or if the dominant animal's thwarted attempts to gain priority access to all resources might convey negative ramifications. Either way, it seems reasonable that altering the social hierarchy will have impacts on behaviour and welfare outside of the training event.

11.7 Monitoring Animal Welfare as Part of Training Programmes

To enable zoo professionals to assess and evaluate whether a training regime is beneficial to an individual or not, the ability to measure zoo animal welfare is key. Currently there are very few species-specific tools to measure welfare. The welfare tool for elephants was developed as a response to concerns over elephant welfare in UK zoos (Asher et al. 2015) and was designed for use by elephant keepers to provide quick and reliable daily monitoring to be evaluated over time. Other species-specific tools such as the one developed for dolphins (*Tursiops truncatus*) (Clegg et al. 2015) or Dorcas Gazelles (*Gazella dorcus*) (Salas et al. 2018) were based on farm animal welfare protocols such as Welfare Quality® and have a mixture of animal-based (parameters of an animals' behaviour/physiology) and resource-based (parameters including the environmental conditions or provisions) measures. With only these few specific toolkits available, reliably assessing animal welfare on multiple grounds is difficult, let alone with a specific focus on training.

Take Away Message

Training provides zoo professionals with a tool by which they can effectively manipulate the behaviour of animals in their care. This has led to many welfare benefits being associated with the use of training. The welfare impacts of training however, should not be determined by whether its application was intended to benefit the participating animals, but result from empirical monitoring of its application. Training is an extremely strong tool, which enables you to exert control over the actions of an animal, by determining which behaviours you feel it should perform. It is important therefore, to fully understand the impact you're having when you train; and what you might add, or remove, from the life of that animal as a consequence.

References

Asher, L., Williams, E., and Yon, L. (2015). *Developing Behavioural Indicators, as part of a Wider Set of Indicators, to Assess the Welfare of Elephants in UK Zoos.* Department for Environment, Food and Rural Affairs.

Barnard, C.J. and Hurst, J.L. (1996). Welfare by design: the natural selection of welfare criteria. *Animal Welfare* 5: 405–434.

Barnett, J.L. and Hemsworth, P.H. (1990). The validity of physiological and behavioural measures of animal welfare. *Applied Animal Behaviour Science* 25: 177–187.

Barongi, R., Fisken, F.A., Parker, M., and Gusset, M. (2015). *Committing to Conservation: The World Zoo and Aquarium Conservation Strategy.* Gland: WAZA Executive Office.

Beauchamp, T.L. and Frey, R.G. (2011). *The Oxford Handbook of Animal Ethics.* Oxford University Press.

Beausoleil, N.J., Appleby, M.C., Weary, D.M., and Sandøe, P. (2014). Balancing the need for conservation and the welfare of individual animals. In: *Dilemmas in Animal Welfare* (eds. M.C. Appleby, P. Sandøe and D.M. Weary), 124–147. CABI Publishing.

Bloomsmith, M.A., Jones, M.L., Snyder, R.J. et al. (2003). Positive reinforcement training to elicit voluntary movement of two giant pandas throughout their enclosure. *Zoo Biology* 22: 323–334.

Bloomsmith, M.A., Stone, A.M., and Laule, G.E. (1998). Positive reinforcement training to enhance the voluntary movement of group-housed chimpanzees within their enclosures. *Zoo Biology*: Published in affiliation with the American Zoo and Aquarium Association 17 (4): 333–341.

Broom, D.M. (1991). Animal welfare: concepts and measurement. *Journal of Animal Science* 69 (10): 4167–4175.

Broom, D.M. (2011). A history of animal welfare science. *Acta Biotheoretica* 59 (2): 121–137.

Carlstead, K. and Shepherdson, D. (2000). Alleviating stress in zoo animals with environmental enrichment. *The biology of animal stress: Basic principles and implications for animal welfare*: 337–354.

Carrasco, L., Colell, M., Calvo, M. et al. (2009). Benefits of training/playing therapy in a group of captive lowland gorillas (*Gorilla gorilla gorilla*). *Animal Welfare* 18 (1): 9–19.

Chamove, A.S. (1989). Environmental enrichment: a review. *Animal Technology* 60: 155–178.

Clay, A.W., Bloomsmith, M.A., Marr, M.J., and Maple, T.L. (2009). Systematic investigation of the stability of food preferences in captive orangutans: implications for positive reinforcement training. *Journal of Applied Animal Welfare Science* 12 (4): 306–313.

Clegg, I., Borger-Turner, J., and Eskelinen, H. (2015). C-well: the development of a welfare assessment index for captive bottlenose dolphins (*Tursiops truncatus*). *Animal Welfare* 24: 267–282. https://doi.org/10.7120/09627286.24.3.267.

Clegg, I.L.K., Rödel, H.G., Boivin, X., and Delfour, F. (2018). Looking forward to interacting with their caretakers: dolphins' anticipatory behaviour indicates motivation to participate in specific events. *Applied Animal Behaviour Science* 202: 85–93.

Coleman, K. and Maier, A. (2010). The use of positive reinforcement training to reduce stereotypic behavior in rhesus macaques. *Applied Animal Behaviour Science* 124 (3–4): 142–148.

Dawkins, M.S. (1980). *Animal Suffering: The Science of Animal Welfare.* London: Chapman & Hall.

Dawkins, M.S. (1990). From an animal's point of view: motivation, fitness, and animal welfare. *Behavioral and Brain Sciences* 13 (1): 1–9.

Dawkins, M.S. (2004). Using behaviour to assess animal welfare. *Animal Welfare* 13 (1): 3–7.

Dawkins, M.S. (2006). A user's guide to animal welfare science. *Trends in Ecology & Evolution* 21 (2): 77–82.

Dawkins, M.S. (2008). The science of animal suffering. *Ethology* 114 (10): 937–945.

Duncan, I.J.H. and Fraser, D. (1997). Understanding animal welfare. In: *Animal Welfare* (eds. M. Appleby and B. Hughes), 19–31. Wallingford, UK: CABI Publishing.

Fa, J.E., Gusset, M., Flesness, N., and Conde, D.A. (2014). Zoos have yet to unveil their full conservation potential. *Animal Conservation* 17 (2): 97–100.

Fernandez, E.J., Tamborski, M.A., Pickens, S.R., and Timberlake, W. (2009). Animal–visitor interactions in the modern zoo: conflicts and interventions. *Applied Animal Behaviour Science* 120 (1–2): 1–8.

Forthman-Quick, D.L. (1984). An integrative approach to environmental engineering in zoos. *Zoo Biology* 3 (1): 65–77.

Fraser, D. (2008). Understanding animal welfare the science in its cultural context. In: *UFAW Animal Welfare Serices*. UK: Wiley.

Fraser, D. and Matthews, L.R. (1997). Preference and motivation testing. In: *Animal Welfare* (eds. M.C. Appleby and B.O. Hughes), 159–173. New York: CAB International.

Gray, J. (2017). *Zoo Ethics. The Challanges of Compassionate Conservation*. Ithaca: Cornell University Press.

Gresswell, C. and Goodman, G. (2011). Case study: training a chimpanzee (*Pan troglodytes*) to use a nebulizer to aid the treatment of airsacculitis. *Zoo biology* 30 (5): 570–578.

Griffin, D.R. (2013). *Animal Minds: Beyond Cognition to Consciousness*. University of Chicago Press.

Guesgen, M. and Bench, C. (2017). What can kinematics tell us about the affective states of animals? *Animal Welfare* 26: 383–397. https://doi.org/10.7120/09627286.26.4.383.

Hansen, F. and Christensen, L.B. (2007). Dimensions in consumer evaluation of corporate brands and the role of emotional response strength (NERS). *Innovative Marketing* 3 (3): 19–27.

Harding, E.J., Paul, E.S., and Mendl, M. (2004). Animal behaviour: cognitive bias and affective state. *Nature* 427 (6972): 312.

Hill, S.P. and Broom, D.M. (2009). Measuring zoo animal welfare: theory and practice. *Zoo Biology* 28 (6): 531–544.

Hodgson, G.M. (2003). The hidden persuaders: institutions and individuals in economic theory. *Cambridge Journal of Economics* 27 (2): 159–175.

Hosey, G. and Melfi, V. (2018). *Anthrozoology: Human-animal Interactions in Domesticated and Wild Animals*. Oxford University Press.

Iyengar, S.S. and Lepper, M.R. (2000). When choice is demotivating: can one desire too much of a good thing? *Journal of Personality and Social Psychology* 79 (6): 995.

Keulartz, J. (2015). Captivity for conservation? Zoos at a crossroads. *Journal of Agricultural and Environmental Ethics* 28 (2): 335–351.

Korte, S.M., Olivier, B., and Koolhaas, J.M. (2007). A new animal welfare concept based on allostasis. *Physiology & Behavior* 92 (3): 422–428.

Lally, P. and Gardner, B. (2013). Promoting habit formation. *Health Psychology Review* 7 (sup1): S137–S158.

Lambeth, S.P., Hau, J., Perlman, J.E. et al. (2006). Positive reinforcement training affects hematologic and serum chemistry values in captive chimpanzees (*Pan troglodytes*). *American Journal of Primatology* 68 (3): 245–256.

Markowitz, H. and Spinelli, J.S. (1986). Environmental engineering for primates. In: *Primates* (ed. K. Benirschke), 489–498. New York, NY: Springer.

Martin, S. (2000). The value of shows. http://naturalencounters.com/site/wp-content/uploads/2015/11/The_Value_Of_Shows-Steve_Martin.pdf (accessed 9 December 2018).

McGreevy, P. and Boakes, R.A. (2011). *Carrots and Sticks: Principles of Animal Training*. Australia: Darlington Press, Sydney University Press.

Melfi, V. (2013). Is training zoo animals enriching? *Applied Animal Behaviour Science* 147: 299–305.

Melfi, V.A. and Thomas, S. (2005). Can training zoo-housed primates compromise their conservation? A case study using Abyssinian colobus monkeys (*Colobus guereza*). *Anthrozoos* 18: 304–317.

Mellen, J. and Ellis, S. (1996). Animal learning and husbandary training. In: *Wild Animals in Captivity: Principles and Techniques* (eds. D. Kleiman, M. Allen, K. Thompson and S. Lumpkin), 88–99. Chicago: University of Chicago Press.

Merzenich, M., Nahum, M., and van Vleet, T. (2013). *Changing Brains: Applying Brain Plasticity to Advance and Recover Human Ability*, vol. 207. Elsevier.

Miller, R. and King, C.E. (2013). Husbandry training, using positive reinforcement techniques, for Marabou stork *Leptoptilos crumeniferus* at Edinburgh Zoo. *International Zoo Yearbook* 47 (1): 171–180.

Minier, D.E., Tatum, L., Gottlieb, D.H. et al. (2011). Human-directed contra-aggression training using positive reinforcement with single and multiple trainers for indoor-housed rhesus macaques. *Applied Animal Behaviour Science* 132: 178–186.

Moberg, G.P. and Mench, J.A. (2000). *The Biology of Animal Stress: Basic Principles and Implications for Animal Welfare*. CABI.

Norton, B.G., Hutchins, M., Maple, T., and Stevens, E. (eds.) (2012). *Ethics on the Ark: Zoos, Animal Welfare, and Wildlife Conservation*. Smithsonian Institution.

OED (2018). *Oxford Learner's Dictionaries*. Oxford Univesrity Press https://www.oxfordlearnersdictionaries.com/definition/english/ethic.

Otaki, Y., Kido, N., Omiya, T. et al. (2015). A new voluntary blood collection method for the Andean bear (*Tremarctos ornatus)* and Asiatic black bear (*Ursus thibetanus*). *Zoo Biology* 34: 497–500.

Paul, E.S. and Mendl, M.T. (2018). Animal emotion: descriptive and prescriptive definitions and their implications for a comparative perspective. *Applied Animal Behaviour Science* 205: 202–209.

Pomerantz, O. and Terkel, J. (2009). Effects of positive reinforcement training techniques on the psychological welfare of zoo-housed chimpanzees (*Pan troglodytes*). *American Journal of Primatology* 71 (8): 687–695.

Ramirez, K. (2012). Marine Mammal Training: the history of training animals for medical behaviors and keys to their success. *Veterinary Clinics of North America: Exotic Animal Practice* 15 (3): 413–423.

Reamer, L.A., Haller, R.L., Thiele, E.J. et al. (2014). Factors affecting initial training success of blood glucose testing in captive chimpanzees (*Pan troglodytes*). *Zoo Biology* 33: 212–220.

Rollin, B.E. (1990). Animal welfare, animal rights and agriculture. *Journal of Animal Science* 68 (10): 3456–3461.

Rollin, B.E. (2003). Oncology and ethics. *Reproduction in Domestic Animals* 38 (1): 50–53.

Ross, S.R. (2006). Issues of choice and control in the behaviour of a pair of captive polar bears (*Ursus maritimus*). *Behavioural Processes* 73 (1): 117–120.

Salas, M., Manteca, X., Abáigar, T. et al. (2018). Using farm animal welfare protocols as a base to assess the welfare of wild animals in captivity—case study: Dorcas Gazelles (*Gazella dorcas*). *Animals* 8: 111.

Sapolsky, R.M. (1998). *Why Zebras Don't Get Ulcers: an Updated Guide to Stress, Stress Related Diseases, and Coping*, 2e. W. H. Freeman.

Savastano, G., Hanson, A., and McCann, C. (2003). The development of an operant conditioning training program for New World primates at the Bronx Zoo. *Journal of Applied Animal Welfare Science* 6 (3): 247–261.

Schapiro, S.J., Bloomsmith, M.A., and Laule, G.E. (2003). Positive reinforcement training as a technique to alter nonhuman primate behavior: quantitative assessments of effectiveness. *Journal of Applied Animal Welfare Science* 6 (3): 175–187.

Shepherdson, D., Lewis, K.D., Carlstead, K. et al. (2013). Individual and environmental factors associated with stereotypic behavior and fecal glucocorticoid metabolite levels in zoo housed polar bears. *Applied Animal Behaviour Science* 147 (3–4): 268–277.

Shyne, A. and Block, M. (2010). The effects of husbandry training on stereotypic pacing in captive African wild dogs (Lycaon pictus). *Journal of Applied Animal Welfare Science* 13 (1): 56–65.

Spiezio, C., Vaglio, S., Scala, C., and Regaiolli, B. (2017). Does positive reinforcement training affect the behaviour and welfare of zoo animals? The case of the ring-tailed lemur (Lemur catta). *Applied Animal Behaviour Science* 196: 91–99.

Veeder, C., Bloomsmith, M., McMillan, J. et al. (2009). Positive reinforcement training to enhance the voluntary movement of group-housed sooty Mangabeys (*Cercocebus atys atys*). *Journal of the American Association for Laboratory Animal Science* 48: 192–195.

Ward, S.J. and Melfi, V. (2015). Keeper-animal interactions: differences between the behaviour of zoo animals affect stockmanship. *PLoS One* https://doi.org/10.1371/journal.pone.0140237.

Ward, S.J. and Sherwin, S. (2019). Human–Animal Interactions in the Zoo. In: *Anthrozoology* (eds. G. Hosey and V. Melfi), Oxford, UK: Oxford University Press.

Watters, J.V. (2014). Searching for behavioral indicators of welfare in zoos: uncovering anticipatory behavior. *Zoo Biology* 33: 251–256.

Weiss, E. and Wilson, S. (2003). The use of classical and operant conditioning in training Aldabra tortoises (*Geochelone gigantea*) for venipuncture and other husbandry issues. *Journal of Applied Animal Welfare Science* 6 (1): 33–38.

Wemelsfelder, F., Hunter, T.E., Mendl, M.T., and Lawrence, A.B. (2001). Assessing the 'whole animal': a free choice profiling approach. *Animal Behaviour* 62 (2): 209–220.

Wemelsfelder, F. and Lawrence, A.B. (2001). Qualitative assessment of animal behaviour as an on-farm welfare-monitoring tool. *Acta Agriculturae Scandinavica Section A Animal Science* 51 (S30): 21–25.

Westlund, K. (2014). Training is enrichment— and beyond. *Applied Animal Behaviour Science* 152: 1–6.

Whitham, J.C. and Wielebnowski, N. (2013). New directions for zoo animal welfare science. *Applied Animal Behaviour Science* 147 (3–4): 247–260.

Wierucka, K., Siemianowska, S., Woźniak, M. et al. (2016). Activity budgets of captive Cape fur seals (*Arctocephalus pusillus*) under a training regime. *Journal of Applied Animal Welfare Science* 19 (1): 62–72.

12

Training Animals in Captivity or the Wild, so They Can Return to the Wild

Jonathan Webb

12.1 Introduction

The earth is currently experiencing an unprecedented extinction crisis. To conserve threatened species, many zoos and wildlife conservation organisations maintain captive breeding programmes, or maintain animals in the wild in predator free refuges. The goal of these programmes is to provide an insurance population against extinction, and ultimately, to provide a source of animals for reintroduction to the wild once the threatening processes have been identified and removed (Kleiman 1989). However, captive reared animals often lack appropriate skills that they need to survive in the wild. Such skills include locomotion (climbing, crawling, or flying), interacting with conspecifics, finding suitable shelter sites, finding and processing food, and identifying and responding appropriately to predators (Shepherdson 1994; Reading et al. 2013). Consequently, the survival of captive-born animals is often significantly lower than wild-born animals following release to the wild (Griffith et al. 1989; Beck et al. 1994; Fischer and Lindenmayer 2000; McCleery et al. 2013). Clearly, the significant investment of money, time, and energy by zoos and wildlife institutions that carry out captive breeding programmes is wasted if the released animals fail to survive in the wild.

One solution to this problem is to train animals prior to reintroducing them to the wild. Training can be done in captivity, in semi-natural arenas at the release site, and can also be done post-reintroduction (Beck et al. 1994). Environmental enrichment can play a key role in preparing captive reared animals for release to the wild (Reading et al. 2013). Ideally, enrichment should be provided to animals early in life, when learning is most effective, and under appropriate conditions (i.e. in the presence of older relatives or social group members, parents, with appropriate food, etc.). If enrichment occurs in captivity, the conditions in captivity should mimic the conditions experienced at the release sites. Alternatively, animals can be maintained in caged areas in the wild prior to release, to allow them to develop important social, sensory, food gathering, and locomotor skills (Reading et al. 2013). The importance of environmental enrichment for learning life skills is discussed in Chapter 6, and will therefore not be discussed further here.

12.2 Teaching Animals to Develop Hunting Skills

There is increasing evidence that juveniles of many carnivores progressively learn hunting skills during growth and development

Zoo Animal Learning and Training, First Edition. Edited by Vicky A. Melfi, Nicole R. Dorey, and Samantha J. Ward.
© 2020 John Wiley & Sons Ltd. Published 2020 by John Wiley & Sons Ltd.

(Caro and Hauser 1992). For example, in African meerkats (*Suricata suricatta)* the young pups rely on their parents and older group members to provide them with prey. The young meerkats are initially given dead or disabled prey, but as they acquire the necessary hunting skills, they are provided with more active prey (Thornton and McAuliffe 2006; Thornton 2007). In domestic cats (*Felis catus*) mothers also alter their hunting behaviour as their offspring develop. Typically, mothers bring back dead prey to very young, non-mobile kittens. However, once the kittens are mobile, mothers present live prey to their offspring, and they actively recapture any fleeing animals to ensure that their kittens have ample opportunity to develop their hunting skills. When kittens are older, mothers observe prey catching behaviours in their offspring, but they take little part in the pursuit, capture, and killing of prey (Leyhausen 1979). A detailed study on wild cheetahs in the Serengeti revealed remarkably similar maternal provisioning behaviours (Caro 1994). When cheetah cubs are young (up to two months old), mothers usually kill the prey before presenting it to their cubs. However, as the cubs grow older (2.5–3.5 months), the mothers begin to release live animals in the presence of cubs, providing the young cheetahs with opportunities to learn to overpower and kill the prey (Caro 1994).

These examples illustrate the complexities involved with training carnivores to hunt prey. Ideally, intervention training of captive bred animals should be done inside an enclosure at the release site, to provide animals with opportunities to learn to identify and locate prey, and to acquire hunting and killing skills (Biggins et al. 1999). If possible, young animals should be trained to hunt in the presence of experienced mothers, siblings or conspecifics. Finally, training should involve the introduction of live prey during ontogeny. The use of live prey raises ethical and animal welfare issues, but it is crucial to the success of training and subsequent reintroduction to the wild (Jule et al. 2008).

A recent study on orphaned cheetah cubs demonstrates how large felids can be trained to hunt live prey (Houser et al. 2011). Three orphan cheetah cubs were raised in captivity at the Cheetah Conservation Botswana facility in southern Botswana, Africa, where they had minimal human contact. Initially, the cubs consumed a diet of meat and bone, but as they grew older, they were offered live chickens; three months later, they were offered live rabbits; and seven months later, they were offered dead impala (*Aepyceros melampus,* a medium-sized antelope). One month after the cheetahs had learnt to consume the dead impala, they encountered injured impala that they successfully captured, killed by suffocation, and consumed. Even in the absence of their mothers, the cheetahs hunting skills improved over time.

Given that the cheetahs had learnt to hunt large prey in captivity, they were fitted with radio collars and released into a 100 ha fenced enclosure on the Kwalata Game Farm. The farm contained suitable habitats for cheetah, and contained free ranging wild herbivores including impala and tsessebes (*Damaliscus lunatus*). The cheetahs successfully hunted game in this enclosure (mostly, impala), and seven months later they were fitted with global positioning satellite (GPS) collars and released onto the 9000 ha Kwalata Game Farm. The free ranging cheetahs continued to hunt successfully on the game farm, and their behaviour was very similar to their wild counterparts, suggesting that the pre-release training was very successful (Houser et al. 2011). However, longer-term monitoring revealed that the cheetahs migrated from the game reserve to pastoral areas, where landholders subsequently killed them. Clearly, we need to remove the threatening processes, in this case, persecution by humans, prior to reintroducing large carnivores to the wild. This problem is not unique to Africa, and indeed, human persecution has plagued many attempts to reintroduce large carnivores across the globe.

12.3 Training Northern Quolls (*Dasyurus hallucatus*) to Avoid Eating Toxic Cane Toads

The introduction of invasive, feral species can pose serious problems for native predators, particularly if the invasive species is highly toxic. If the invader possesses novel toxins to which the predator species is evolutionary naive, then predators that attack or consume the invader can die from poisoning. A classic example of this problem is the introduction of cane toads (*Rhinella marina*) to Australia. Cane toads are native to South America, and were introduced to north-eastern Australia in the 1930s to control the grubs of the cane beetle that were devastating sugar cane crops (Lever 2001). After multiple introductions, the cane toads established viable populations, and they then began spreading across the continent, at an accelerating rate (Phillips et al. 2006). Cane toads contain powerful toxins (bufodienolides) that are pharmacologically very different to the typical toxins found in Australian native frogs (Daly and Witkop 1971). Australia has no native frogs or toads in the genus *Rhinella* (toads native to Central and South America), and consequently most native Australian predators lack physiological mechanisms for detoxifying toad toxins. Both adult and juvenile cane toads look very similar to palatable native frogs, and thus, many toad-naive predators treat cane toads as prey, with disastrous and fatal results. Large goannas, freshwater crocodiles, elapid snakes, and marsupial predators can die after mouthing, attacking, or ingesting large toads (Covacevich and Archer 1975).

The spread of cane toads across the top end of northern Australia has caused massive population declines of large varanid lizards (Doody et al. 2009), freshwater crocodiles (Letnic et al. 2008), and a marsupial predator, the northern quoll (*Dasyurus hallucatus*) (Burnett 1997; Woinarski et al. 2010). Quolls are spotted, carnivorous marsupials (Figure 12.1) that are roughly the size of a small domestic cat and they were once widespread across northern

Figure 12.1 A female northern quoll, *Dasyurus hallucatus*, with one of her juvenile offspring that were raised in captivity in a purpose built quoll facility at the Territory Wildlife Park, near Darwin, Northern Territory Australia. *Source:* Jonathan Webb.

Australia. Quolls are particularly sensitive to cane toad toxins, and die after mouthing large toads (Covacevich and Archer 1975; O'Donnell et al. 2010). Northern quolls are short lived, and both sexes attain maturity at around 10 months of age. Males typically die after mating, and they seldom live longer than a year in savannah woodlands. Female survival is also low, and many females do not live past the age of two years (Oakwood 2000). These unusual life history traits (short lived, with male die off after mating) make this species especially vulnerable to extinction. As cane toads spread across Australia's Northern Territory, quoll populations crashed, local extinctions occurred, and the species was listed as endangered (Rankmore et al. 2008). To protect quolls from extinction, researchers collected quolls from the Darwin and Kakadu region, housed them at purpose built enclosures at the Territory Wildlife Park, and introduced them to two toad-free islands. Quolls flourished on these toad and predator-free islands, and the populations rapidly increased (Rankmore et al. 2008). Unfortunately, cane toads are exceptionally good at rafting on floodwater debris, or hitch hiking in boots and camping gear, and they have colonised several islands, causing quoll populations to crash (Woinarski et al. 2011b). Thus, island quoll populations are not necessarily secure from the threat of cane toads.

What can we do to prevent northern quoll extinctions? One promising approach is to use conditioned taste aversion (CTA) to train quolls not to eat cane toads. CTA is a powerful form of learning that occurs when animals ingest a novel, toxic food, experience illness, and subsequently associate the smell or taste of that food with illness, and avoid ingesting the toxic food for long periods (Garcia et al. 1974). Unlike classical conditioning, whereby animals require multiple trials to learn the appropriate response (Chapter 1), CTA usually involves one-trial learning; that is, after a single bout of illness, animals learn to avoid the food that induced illness. After its initial discovery, conservation biologists realised that they could change the feeding behaviour of predators by adding nausea inducing chemicals to meat baits. Initial pen trials on coyotes (*Canis latrans*) were encouraging; coyotes that consumed sheep baits paired with lithium chloride became ill, and these coyotes subsequently refused to attack live lambs (Gustavson et al. 1974). Likewise, crows that ate green-coloured chicken eggs injected with a nausea-inducing chemical subsequently refused to eat untreated green chicken eggs, but continued to eat white eggs (Nicolaus et al. 1983) Despite these encouraging results, the use of CTA to alter predator behaviour was embroiled in controversy (Gustavson and Nicolaus 1987). Most of this controversy surrounded the use of CTA baits to train wild coyotes to avoid attacking lambs; field trials produced equivocal results (Bourne and Dorrance 1982) and the use of CTA was abandoned in favour of alternative (and often, lethal) methods to control coyote numbers (Gustavson and Nicolaus 1987; Conover and Kessler 1994). Thus, until recently CTA has largely been overlooked as a tool for training animals.

To see if we could train quolls not to eat cane toads, in 2009 my colleagues and I fed toad-naive, captive reared quolls a small, non-lethal sized (<2 g) dead toad infused with the odourless, tasteless, nausea-inducing chemical thiabendazole (a chemical used to deworm livestock which has a nausea

inducing chemical component). The quolls that consumed the toads became mildly ill, and thankfully, none showed any signs of serious toad poisoning. Next, we offered these trained quolls a small, live cane toad in an open topped jar, in front of a video camera, to see whether they would attack the toads. Remarkably, the trained quolls sniffed the toad, but refused to attack it, whereas our control toad-naive quolls quickly consumed the live toad. The aversion to live cane toads lasted a week in captivity, suggesting that the toad-trained quolls might have the skills necessary to survive in toad-infested areas.

Did toad-aversion training confer survival benefits to quolls in toad-infested habitats? To answer this question, 30 trained 'toad-smart' and 30 untrained 'toad-naive' quolls were fitted with radio collars and reintroduced to suitable rocky habitats in toad-infested areas near Darwin, in northern Australia (Figure 12.2). A large team of volunteers was enlisted to help follow each quoll during the first few hours after release, and locate them daily thereafter, so that we could determine their fate. The radio-tracking work revealed that untrained quolls often encountered cane toads within hours of release, and several of these toad-naive quolls attacked large toads, and died from toad poisoning. By contrast, the trained 'toad-smart' quolls encountered toads, sniffed them, and

Figure 12.2 Photograph of a trained northern quoll fitted with a radio-collar. Both untrained and trained northern quolls were released in the wet season at sites around Darwin, where toads were abundant. *Source:* Jonathan Webb.

rejected them as prey, and consequently, the trained quolls had higher survival than the control group during the 10 day monitoring period (O'Donnell et al. 2010). Encouragingly, several trained quolls that we reintroduced to Mary River Park, a tourist accommodation site surrounded by monsoonal vine thicket and savannah woodland, survived long term, and one female survived to breeding age and raised a family of quolls in the caravan park owner's machinery shed.

These encouraging results suggested that it might be possible to reintroduce 'toad-smart' quolls to toad-infested sites where northern quoll populations have crashed or gone locally extinct. To test this idea, 50 captive-born quolls (28 males, 22 females) from the Territory Wildlife Park were trained not to eat toads by feeding them a small (<2 g) dead toad coated with thiabendazole at a dose rate of $300 \, \text{mg} \, \text{kg}^{-1}$ predator mass. The quolls were released to suitable habitats (rock outcrops) near East Alligator Ranger Station in Kakadu National Park. Previous studies showed that quolls were abundant at East Alligator prior to the arrival of cane toads, but the population crashed after toads invaded and was on a path to extinction (Oakwood and Foster 2008; Woinarski et al. 2010). The longer-term survival of trained 'toad-smart' quolls was monitored by trapping three times each year (usually, in March, May, and November) over four years. Tissue samples were taken from all recaptured quolls, and DNA parentage analyses were run to determine the identity of the mothers and fathers of juveniles in the population.

Monitoring of quolls revealed that toad aversion training conferred long-term benefits for some quolls. Most males disappeared from the study sites shortly after release, but four males that were recaptured survived for an average five months (range 1–10 months). Males typically only live for a year, so these males survived long enough to breed with females. Females survived longer than males, and seven females that were recaptured on the study site survived for an average of nine months (range 2–22 months). Three of these

captive reared toad-trained female quolls successfully raised litters of young, and importantly, DNA analyses of parentage revealed that the offspring of one of these females also survived and reproduced (Cremona et al. 2017b). Thus, some juvenile quolls learnt to avoid eating cane toads, and social learning may be important in this respect. Quolls have a high degree of maternal care, and the young often travel on their mother's backs at night. By radio-tracking female quolls and their offspring, and placing remotely triggered infrared cameras near the den sites of mothers when their offspring were denning independently, we discovered close associations between juveniles and mothers. Notably, young quolls that were denning independently of their mother were photographed with their mother foraging at night (Figure 12.3), suggesting that juveniles might have opportunities to learn what to eat, and what not to eat, by observing their 'toad-smart' mothers hunting prey and sniffing and rejecting toads as prey. Whether such information is transmitted to offspring remains unknown, but seems likely (Galef and Laland 2005). Indeed, the extended duration of maternal care in quolls provides an opportunity for social learning in this species. Alternatively, juvenile quolls might naturally learn to avoid eating toads by ingesting small non-lethal sized toads that induce nausea and long-lasting aversions in other small carnivorous marsupials (Webb et al. 2008, 2011).

Although for the second generation of quolls bred in the wild, the reintroduction was not successful. Even though cane toads were no longer a major source of mortality for toad-trained quolls, predation by free-ranging domesticated dogs and dingoes were major sources of mortality for quolls on the study site. Indeed, predation by canids is probably preventing the reintroduced population of quolls from recovering (Cremona et al. 2017). Unfortunately, the loss of traditional burning practices in Kakadu has led to an increase in hot late dry season fires (Russell-Smith 2016); these fires incinerate

Figure 12.3 Photograph of an adult female northern quoll foraging with her offspring in Kakadu National Park. *Source:* Graeme Gillespie, Flora and Fauna Division, NT Department of Environment and Natural Resources.

shelter sites (hollow logs) and create large open expanses of bare ground between rock outcrops (Figure 12.4), further exacerbating predation risk for quolls. Lack of vegetation cover at ground level increases the risk of predation by dingoes and dogs, particularly during the late dry season when juvenile quolls begin leaving their mother's dens in search of food. Indeed, a previous study on quolls at Kapalga in Kakadu National Park found that most predation by dingoes occurred in burnt habitats (Oakwood 2000). Given these problems, careful management of fire and predators will be necessary to facilitate the recovery of northern quoll populations in Kakadu National Park.

12.4 Training Wild Animals to Avoid Eating Novel Foods or Crops

CTA may provide a potential non-lethal method for helping to solve a wide range of human–wildlife conflicts. In many national parks, large carnivores often raid campsite rubbish bins or unsecured foodstuffs, and these 'nuisance' animals can pose a substantial risk to human safety. Black bears (*Ursus americanus*) are well known for entering

camping sites and stealing food, and they also raid military bases to pilfer military food rations, also known as 'meals-ready-to-eat' (MRE). Studies of problem bears at Camp Ripley Military Reservation revealed that just three female bears were responsible for >80% of the nuisance activity (Ternent and Garshelis 1999). To train the bears not to eat MRE, Ternent and Garshelis (1999) laced the entrée and beverage portions of the MRE with thiabendazole and fed the treated MRE to the problem bears. During subsequent tests, the problem bears either ignored the MRE, or tasted and rejected the MRE as food, and this aversion to MRE by the problem bears lasted a year. However, the bears were not trained at the food depot and they failed to develop conditioned place avoidance; thus, they continued to visit the food depot to seek alternative foods, scaring numerous army personnel in the process (Ternent and Garshelis 1999). Nonetheless, CTA, if done in conjunction with other techniques (reducing access to food, education, repellents), might help to reduce the problem of nuisance carnivores where humans are also present.

Wild herbivores are a global problem for farmers because they cause massive damage to crops. Some particularly destructive herbivores that consume crops are elephants,

Figure 12.4 Photograph of the habitat in Kakadu National Park, Australia, where 'toad-smart' quolls were reintroduced. Note that the photograph was taken in December, when young quolls begin foraging in woodland. Inappropriate burning of habitat at the study site has created large open tracts of woodland between the rock outcrops. The lack of a suitable understorey of shrubs and grasses means that young quolls that forage in woodland have little protection from predators such as dingoes and feral cats. *Source:* Jonathan Webb.

wild boar, kangaroos, and badgers. Lethal control methods for problem herbivores are often ineffective and their use has raised ethical and conservation concerns in recent years. Thus, there has been an increasing interest in developing alternative methods for discouraging herbivores from damaging crops. Studies on European badgers (*Meles meles*) suggest that it ought to be possible to train wild animals not to eat certain crops (Baker et al. 2005, 2008). In England and Wales, badgers create substantial financial losses for cereal farmers by flattening and eating the crops (Moore et al. 1999). To train badgers not to eat maize, researchers offered free-ranging badgers maize treated with ziram (a potent emetic) paired with a novel odour (clove oil). The badgers that ate the maize treated with the ziram/clove oil became ill, and these animals subsequently avoided the maize treated with clove oil (Baker et al. 2008). Thus, it may be possible to protect cereal crops from badger damage by deploying cereal baits paired with a tasteless, odourless nausea-inducing chemical and a non-toxic novel odour just before the grain ripens. When the crop ripens, farmers could spray the non-toxic novel odour onto the crop to deter badgers from eating the grain (Baker et al. 2008).

12.5 Training Animals to Recognise Predators

Animals living in isolation from predators, either on predator-free offshore islands or in captivity, often lack the ability to detect or respond appropriately to predators (Griffin et al. 2000). Consequently, predator-naive animals often suffer high mortality from predation following their reintroduction to their native geographic range (Griffin et al. 2000), which has contributed to the poor success rate of reintroductions involving such animals (Fischer and Lindenmayer 2000; Jule et al. 2008). To overcome this problem, there has been a renewed interest in training captive reared or wild animals to recognise predators prior to reintroducing them to the wild (McLean et al. 2000; Blumstein et al. 2002; Crane and Mathis 2011; Gaudioso et al. 2011; Teixeira and Young 2014).

Numerous studies have demonstrated that predator-naive fish, birds, and mammals can be trained to recognise predators as dangerous by pairing the sight or smell of the predator (the conditioned stimulus) with a frightening stimulus or conspecific alarm call (the unconditioned stimulus). For example, in a pioneering study, Ian McLean and colleagues trained wild New Zealand robins (*Petroica australis*) to respond fearfully to invasive ferrets. The researchers located wild female robins with chicks, and simulated an attack from a stuffed ferret by moving it on a string in the presence of a stuffed robin in an aggressive posture paired with robin alarm and distress calls. After the training, robins reacted fearfully to the predator (McLean et al. 1999). In another study, Andrea Griffin and colleagues trained predator-naive tammar wallabies (*Macropus eugenii*) to associate the sight of a model fox (a taxidermic mount) on a trolley with an aversive stimulus (a hooded human chasing the wallaby with a net). After one or two trials, the wallabies learnt to associate the model fox with danger (Griffin et al. 2001). It is also possible to train fish to avoid predators. Many fish possess chemicals in the epidermis (alarm chemicals) that are released after attacks by predators, and these chemicals elicit dramatic antipredator responses by prey when they are detected (Chivers and Smith 1998). Thus, predator-naive fish can learn to identify novel predators as dangerous when the sight or smell of the predator is paired with the odour of injured conspecifics (Wisenden 2003).

These examples show that it is possible to train predator-naive animals to identify single predators. However, most animals encounter multiple predators in the wild, which raises an important question for conservation biologists: how can we train predator-naive animals to avoid or respond appropriately to multiple predators? The phenomenon known as 'generalisation of learned predator recognition' provides a potential serendipitous solution to this problem (Ferrari et al. 2007, 2008). Several studies have demonstrated that when predator-naive prey encounter a dangerous predator paired with an aversive stimulus, the prey subsequently generalise their antipredator response not only to the dangerous predator, but also, to ecologically similar predators (Ferrari et al. 2007, 2008). For example, the tammar wallabies that Andrea Griffin and colleagues trained to avoid the model fox subsequently generalised their antipredator behaviours in the presence of a model cat, but not in the presence of a non-threatening herbivore, a stuffed goat (Griffin et al. 2001). Likewise, fathead minnows that were trained to recognise the odour of lake trout as a predator (by pairing the odour of lake trout with minnow skin extracts) subsequently generalised their antipredator responses to odours of brook trout and rainbow trout, but not to odours of the more distantly related pike (Ferrari et al. 2007). The key concept here is that prey often perceive ecologically similar predator species (i.e. species that share similar visual or chemical cues) as dangerous (Blumstein 2006). For example, prey may show similar antipredator responses to raptors that share the same silhouettes, or venomous snakes from the same family that share similar chemical cues (Webb et al. 2009). As long as we use suitable predator cues during the training trials, then predator-naive prey are likely to extend their predator recognition to multiple predators.

Before embarking on predator training, it is important to identify the key predators that are responsible for causing mortality of the prey species. Once these predators are known, suitable models, or live predators (or substitutes), can be used as the conditioned stimulus for training. During training, models should be paired with a suitable aversive stimulus, and for example, the odour of the predator so the predator does not need to be present. Odour is important because many animals use both visual and olfactory cues to detect the presence of predators, and olfactory cues can enable prey to take elusive action (e.g. hiding, entering a burrow) in the presence of a hidden predator. Models are advantageous because they provide a standard conditioned stimulus and there are few ethical problems with their usage. Aversive stimuli can include firing rubber bands at the test subject, or simulating a frightening predation attempt by scaring or capturing the animal (Griffin et al. 2000). Using models avoids the potential problems associated with the use of live predators, such as the spread of disease, injuries to prey, and variation in the unconditioned stimulus.

Although models are widely used to train animals to avoid predators, they may lack crucial chemical cues that animals use to identify or locate predators; for example, rodents show strong fear responses to the odour emanating from worn cat collars, yet taxidermic mounts lack such chemicals (McGregor et al. 2002). Hence, it may be necessary to add appropriate chemicals to taxidermic mounts to train prey to associate the smell of predators with danger. Predator urine and faeces may not be particularly useful in this respect, since these odours may fail to elicit antipredator responses in wild animals (Apfelbach et al. 2005). It would be far better to identify the predator chemicals that the prey respond to, and pair these chemicals with suitable models during predator training. Finally, in some circumstances, it may be more appropriate to use a live predator, such as a well-trained domestic dog trained to chase, but not catch animals (McLean et al. 2000), particularly if that predator closely resembles a dangerous predator that prey are likely to encounter in the wild (e.g. dingoes in Australia, wolves in North America).

12.6 Does Predator Training Facilitate Survival Following Reintroduction to the Wild?

Many studies have successfully trained predator-naive animals to recognise and respond to predators, but few studies have demonstrated that pre-release predator training enhances the survival of animals following reintroduction to the wild (Ellis et al. 1977; Beck et al. 1988). To test whether predator training enhances survival, predator-naive (the control group) and predator-trained individuals (experimental group) should be monitored following release to the wild using radio-telemetry, or intensive mark-and-recapture trapping techniques, to provide robust estimates of survival (Lebreton et al. 1992). Ideally, researchers should locate animals daily to track their movements, and to identify the causes of mortality. Previous studies have shown that most predation on animals occurs within weeks of release to the wild (Parish and Sotherton 2007), so it is crucial to monitor animals daily during this period. Because many reintroduced animals often disperse away from reintroduction sites (Armstrong and Seddon 2007), large sample sizes may be necessary to provide robust estimates of survival. Such studies are expensive, require good planning and organisation, and require a dedicated team on the ground, and in the air, to track daily movements of animals. Given that such studies are expensive, and time consuming, it is not terribly surprising that few studies have rigorously tested whether predator training provides survival benefits following reintroduction.

Nonetheless, several research groups have not only trained animals to respond to predators, but have also monitored the fate of the animals following release to the wild. Below, I describe some of these projects in more detail, to illustrate the logistical difficulties involved with this sort of research.

12.7 Training Houbara Bustards to Recognise Foxes

The houbara bustard *(Chlamydotis undulata)* is a medium-sized ground-dwelling bird that occurs in semi-desert and shrub-covered arid plains of West Asia and North Africa. The bustard is highly prized by falconers, and has been a traditional food of indigenous people for centuries. Bustard populations have declined substantially in recent decades due to overhunting, overgrazing, and urbanisation (Tourenq et al. 2005; Riou et al. 2011). In Saudi Arabia, populations had declined to such an extent that the National Commission for Wildlife Conservation and Development (NCWCD) was established in 1986 to conserve bustards in that country. The aim of the programme is to establish protected areas and viable populations of houbara bustards to allow sustainable hunting of the species. Part of the programme has involved reintroducing captive reared animals to the wild (Seddon et al. 1995; Combreau and Smith 1998). The main factor that has affected the success of these reintroductions is predation by the red fox (*Vulpes vulpes*) (Combreau and Smith 1998).

In an effort to increase post release survival of captive reared bustards, van Heezik and colleagues attempted to train bustards to recognise foxes as dangerous predators (Van Heezik et al. 1999). Initial trials in 1995 involved training groups of bustards with a taxidermic model of a red fox on a trolley, which rushed into the pen and lunged at the birds repeatedly for one minute. At the same time, the researchers played back alarm calls of wild adult houbara bustards. Birds were trained on three consecutive days, but they rapidly habituated to the model, and so the researchers switched to one training session. Note that no aversive stimulus (such as firing rubber bands at the birds) was paired with the model fox during the training sessions, which may explain why the birds did not treat the taxidermic fox as dangerous. Following release to the wild, there was no difference in the survival of control birds and birds trained with the stuffed fox (Van Heezik et al. 1999).

Given these poor results, in 1996 the researchers used a live hand-reared red fox to train the bustards. Training consisted of introducing the fox, which was wearing a muzzle and a lead, into the bustard cage at dawn or dusk. The fox handler then attempted to control the fox's movements with the lead. Trials lasted 40 seconds to 15 minutes depending on how quickly the fox began to stalk and chase the birds, and all trials were paired with the alarm calls of a wild bustard. Groups of five birds per cage were trained with three trials over three consecutive days. Initially, training was carried out in circular pens (5 m diameter), but larger rectangular pens (15 m × 40 m) were used in later trials to minimise injuries to birds. Unfortunately, the fox dislodged its muzzle during several trials, and it bit two birds which received minor puncture wounds, and it broke the leg of a third bird. Two other birds broke their wings whilst attempting to flee from the fox, but this problem was rectified by carrying out subsequent training sessions inside the larger training pen. Despite the difficulties associated with using the live fox, the bustards trained with the fox showed stronger antipredator responses than did untrained control birds (Van Heezik et al. 1999).

Following their release to the wild, the bustards trained with the live fox had much higher long-term survival than did untrained bustards. Of 22 predator-trained birds, predators killed eight and nine survived long-term (up to 196 days). By contrast, of 22 untrained birds, predators killed 15 birds, and only two birds survived long-term. Predation occurred rapidly following reintroduction; all birds killed by predators were dead within 19 days following release to the wild. These results provide compelling evidence that training provides tangible benefits to captive reared animals, and highlights why it is necessary to train animals to recognise and avoid predators prior to reintroduction. The study also highlights some of the issues involved with using a fox. In this case, it could have been more appropriate to use a trained dog of similar size to the fox; previous studies have demonstrated that naive prey trained with a dangerous predator often generalise their antipredator responses to anatomically similar predators (Griffin et al. 2001). Thus, training with a live dog may have evoked similar responses to live foxes.

12.8 Training Captive Reared Little Owls to Avoid Predators

The little owl *Athene noctua* is a small (to 210 g) owl that occurs in western Europe, north Africa and central Asia. Little owls are not a threatened species, but populations in Europe have declined over the past few decades. To date, nearly all introductions of little owls to Europe have had poor success due to high rates of predation on newly released owls (Van Nieuwenhuyse et al. 2008).

In an attempt to redress this problem, researchers trained captive reared fledgling chicks to avoid two predators: rats and goshawks (Alonso et al. 2011). Shortly after hatching, the chicks were transferred to outdoor cages where they were raised by adult foster parents. This study used both live rats and a stuffed goshawk in the flight position as the conditioned stimuli, and a digital recording of the alarm call of the little owl as the unconditioned stimulus. In the rat trials, a trained rat rapidly crossed a mesh-covered corridor on the floor of the owl's cage, whilst in the goshawk trials, a stuffed goshawk rapidly moved along a cable suspended above the owl's cage. Goshawk trials occurred during the day, and rat trials occurred at night. In both trials, the predator presentation lasted a few seconds and was paired with the recording of the little owl alarm call. In this study, the researchers carried out two to four trials each week, from fledgling age until release. Importantly, the outdoor cages contained natural vegetation, high perches, and nest boxes so that the owls could display appropriate antipredator responses, such as

freezing, hiding, or flying during the trials (Alonso et al. 2011).

After antipredator training was completed, all the owls were carefully monitored for two weeks to ensure that they had the necessary hunting skills for obtaining food. The trained owls were fitted with radio-transmitters, and released to suitable habitats in Madrid where little owls naturally occur. In this study, the researchers released two groups of little owls (nine trained and seven untrained controls) to the wild. The team monitored the owls four times per week during the first six weeks after release; previous studies have shown that most predation on birds reared in captivity occurs during the first few weeks following release to the wild (Parish and Sotherton 2007). The only problem with this study is that due to logistical difficulties, the researchers were unable to train or track the fate of control and trained animals during the same years. Instead, they monitored the fate of control owls in 2007, and the fate of trained owls in 2010 (Alonso et al. 2011). Because predation rates often vary through time, this design means that it is difficult to determine whether training conferred survival benefits to the birds.

Nonetheless, the results of this study were encouraging. In the first few weeks following release, predators killed four of six control birds in 2007, whereas predators only killed two of seven trained birds in 2010. The main predators responsible for little owl deaths included a goshawk (*Accipiter gentilis*), a sparrowhawk (*Accipiter nisus*), two tawny owls (*Strix aluco*), a least weasel (*Mustela nivalis*), and a genet (*Genetta genetta*). The longer-term fate of the little owls used in this study was not reported; clearly, longer-term monitoring will be required to demonstrate that training provides long-term benefits to this species.

12.9 Training Wild Bilbies to Avoid Predators

Australia has a particularly poor record of mammalian extinctions, and has lost at least 29 species of mammal since Europeans colonised the continent. Most of these extinctions have coincided with the spread of invasive foxes across the continent (Short and Smith 1994), but feral cats and changed fire regimes are implicated in the recent collapse of mammals across northern Australia (Woinarski et al. 2011a; Fisher et al. 2014). One species that has suffered major declines is the greater bilby (*Macrotis lagotis)*, a medium-sized (to 2.5 kg) omnivorous marsupial that was once widespread in Australia's sandy desert regions. Bilbies are solitary, nocturnal animals that live inside spiral shaped burrows up to 2 m deep by day (Moseby and O'Donnell 2003), and emerge at night to search for invertebrates, seeds, and plant material (Johnson 2008). The bilby once occupied nearly two thirds of arid Australia, but it is now restricted to just a small fraction of its former geographic range, and is listed as vulnerable under the Australian Environmental Protection and Biodiversity Conservation Act. The major threats to the greater bilby are changes to traditional Aboriginal burning practices and predation by introduced cats and foxes (Burbidge et al. 1988; Burrows et al. 2006). To reduce predation on bilbies, in April 2000, Moseby and colleagues reintroduced captive-bred bilbies to a predator free fenced area of the Arid Recovery Reserve in South Australia (Moseby and O'Donnell 2003). This population increased steadily, and in 2004, a trial reintroduction of wild bilbies from the predator-free zone to the predator-present zone (aptly named the 'Wild West') was undertaken. However, this reintroduction failed due to predation by feral cats (Moseby et al. 2011).

To determine whether they could train predator-naive free ranging wild bilbies living inside the predator exclusion zone to recognise and respond to predators, Moseby and colleagues carried out a series of elegant training trials. To train bilbies to associate the smell of cats with danger, scientists used a declawed freshly thawed feral cat carcass, paired with a 'cat spray' consisting of cat urine and faecal material. Prior to the commencement of training, the researchers infused the nets, calico bags, and cat carcasses with cat spray. At night, two people armed with torches and nylon fishing nets searched for bilbies on roads, and captured them with nets. A third person then placed the cat carcass atop of the captured bilby to simulate an attack from a feral cat. Each trained bilby was weighed, checked for reproductive status and health, fitted with a microchip and a radio-transmitter, and placed in a calico bag. A scented cat carcass was used to encourage the trained bilby to exit the bag, and two squirts of cat spray were directed towards the bilby as it fled from the bag. The researchers captured the control bilbies in the same manner, or inside cage traps, and processed them as described above, but without the cat spray or cat carcass treatment. In total, the study involved seven trained and seven control bilbies with four males and three females in each group. Following release, all bilbies were monitored daily to determine their burrow use and movements (Moseby et al. 2012).

Six days after the initial capture, training, and release, the trained bilbies experienced a second aversion event paired with the cat spray. This treatment was designed to mimic a predator attempting to capture the bilby by digging up its burrow. In this part of Australia, foxes, dingoes, and large varanid lizards are the main predators that dig up bilby burrows. Trained bilbies were located, and a researcher administered three squirts of cat spray prior to digging the burrow entrance with a gardening hoe for three minutes, levelling off the sand, and applying more cat spray. Control bilbies received the same treatment, but without the addition of the cat spray.

Predator training of the wild bilbies living inside the predator-proof enclosure was highly effective. Following training, the predator-trained bilbies used more burrows, and changed burrows more often, than did control bilbies. After the reinforcement burrow dig, five of seven trained bilbies moved from their burrows, whereas all the control bilbies remained inside their burrows. Again, this provides strong evidence that the trained bilbies associated the disturbance with danger.

Armed with these encouraging results, wild bilbies were also trained to avoid predators, and both wild trained and untrained bilbies were reintroduced to the unfenced 'Wild West' where predatory feral cats and foxes were present. Over three consecutive nights in August 2007, researchers captured 20 bilbies within the predator-proof reserve. Trained animals (n = 10) were captured in nets and trained as described above, but without the reinforcement test. Control animals (n = 10) were captured in nets and burrow traps. All the animals were fitted with radio-transmitters and located every day by a team on quad bikes, and occasionally, from light aircraft. Keeping track of the locations of 20 free ranging marsupials is a huge effort, and requires a dedicated team. After four months, the researchers located the burrows occupied by the control and trained bilbies. At dusk, they simulated a predator attack by digging around the burrow with a hoe for three minutes in the presence of the cat spray. The idea here was to see whether the bilbies would move away from the disturbed burrow.

The bilby predator training results were encouraging, but they highlight the difficulties associated with conserving marsupials in Australia. Six months after the bilbies were introduced to the Wild West zone, only one control bilby had been killed by a feral cat, and only one trained bilby had been killed (or possibly scavenged) by a wedge-tailed eagle (Moseby et al. 2012). By contrast, when untrained bilbies were released to the Wild West zone in 2004, six of seven bilbies were killed by cats within 25 days of release (Moseby et al. 2011). Interestingly, there was no difference in burrow use of trained and control bilbies in the 2007 study, suggesting that the control animals may have acquired predator avoidance behaviour by observing the behaviour of conspecifics after exposure to odour cues of cats (Moseby et al. 2012). This 'cultural acquisition' of antipredator behaviour has been demonstrated in birds and mammals (Griffin 2004).

How successful was the training long term? Visual counts of the tracks made by the bilbies in the sand suggested that the population remained stable for 12 months. Encouragingly, females reproduced, and the abundance of juvenile bilbies increased up until May 2008, when the population suddenly declined, coincident with a major drought. Unfortunately, cat tracks were found around juvenile bilby burrows, suggesting that cats had contributed to their demise, potentially because the cats' other major food source (rabbits) had crashed with the onset of the drought. Sadly, the last bilby tracks were observed 19 months after the initial reintroduction in January 2009 (Moseby et al. 2011). The study showed that whilst it is possible to train adult female bilbies to avoid novel predators, this behaviour was not transmitted to their offspring, apparently because the young bilbies do not forage with their mothers after leaving the safety of the pouch (Moseby et al. 2012). Alternatively, the drought probably reduced plant cover to such an extent that the bilbies probably had nowhere to hide whilst foraging. Feral cats are ambush hunters that often specialise on one prey type, so without cover, the bilbies were likely outgunned by a superior hunter.

12.10 Training Black-tailed Prairie Dogs to Avoid Predators

In many animals, social learning is important for the development of foraging skills, social behaviours, and antipredator behaviours. Young animals often learn about predators whilst in the company of parents, family groups, or conspecifics (Curio 1993). Animals can learn by observing a conspecific responding to a stimulus (observational learning) or an experienced conspecific can indirectly focus the attention of the young animal to the stimulus (local enhancement). The ability of offspring to learn antipredator behaviours from parents, siblings or conspecifics will depend on the period of parental care, and the size of the group in which they live. Marsupials that become independent of parents shortly after leaving the pouch (e.g. bilbies) may have little opportunity to obtain social information about predators. By contrast, in primates, extension of the period of parental care increases the opportunity for juveniles to socially acquire antipredator behaviours from their parents. Likewise, juveniles of highly social animals that live in extended family groups will have numerous opportunities to interact with siblings, parents, and conspecifics. In many group living species, the production of offspring is synchronised, and juveniles may therefore have the opportunity to not only learn skills from their parents, but also from other parents and vigilant group members (Thornton and McAuliffe 2006).

For animals that live in groups, it is crucial to incorporate social learning into predator training regimes. Shier and Owings research on black-tailed prairie dogs (*Cynomys ludovicianus*) provides a good example of how this can be done (Shier and Owings 2006, 2007). The black-tailed prairie dog is a colonial species from North America that lives in social groups called coteries. Typically, coteries consist of several adult females, one adult male, yearling males and females, and juveniles. Prairie dog populations have declined dramatically across North America (Kotliar et al. 2006) and efforts to conserve this species have involved reintroducing animals to areas where the species has gone locally extinct (Truett et al. 2001; Long et al. 2006). As with other reintroductions, the success or failure of such programmes will depend on whether or not reintroduced animals can respond appropriately to multiple predators.

Numerous predators, including raptors, snakes, weasels, coyotes (*Canis latrans*), and bobcats (*Lynx rufus*) all prey on prairie dogs. When adult prairie dogs detect raptors or mammalian predators, they bark repeatedly to warn offspring and other group members about the approaching predator. Bark alarm calls elicit scanning behaviours by other group members, and if the predator is detected, individuals typically run to a burrow mound, and either hide inside the burrow, or call whilst facing the predator (Hoogland 1995). The prairie dogs responses to snake predators are quite different. When a rattlesnake is encountered, the adults approach the snake, and head-bob or jump away whilst making 'jump-yip' calls, which can be accompanied by foot drumming (Owings and Owings 1979). After juveniles emerge from their burrows, they remain close to the burrow entrances, and exhibit jump-yip calls but they rarely bark or foot drum until they attain several months of age. Thus, juveniles likely learn to identify and respond appropriately to predators via social interactions with mothers and experienced group members.

To see whether training juvenile prairie dogs with experienced adults enhanced the juveniles' ability to learn to identify predators, Shier and Owings (2007) trapped 36 wild prairie dogs from 8 females and brought them into captivity shortly after the juveniles emerged from their burrows. They housed each female with her litter in a separate wire meshed field enclosure $(2 \times 2 \times 3\,\text{m})$ with a mesh roof. Each focal juvenile was given a pre-training test (in the absence of littermates or mothers) to assess their antipredator responses to the test stimuli (below). Juveniles were randomly allocated to three groups: (i) trained with experienced adult (either their mother or close relative); (ii) trained with inexperienced sibling; or (iii) trained alone. The researchers then trained the juveniles to avoid predators over five weeks, with two presentations of predators per week. Trials involved exposing animals in each treatment group to the following stimuli for ten minutes: (i) a live black-footed ferret (*Mustela nigripes*); (ii) a moving, stuffed red tailed hawk (*Buteo jamaicensis*); (iii) a live prairie rattlesnake (*Crotalus viridis*); and (iv) a live desert cotton tail (*Sylvilagus auduboni*), which served as a predator control. The live predator or cotton tail was placed in a mesh box in the prairie dog enclosure, whereas the hawk was attached to a wire and was released such that it flew down over the cage five minutes into the test. To ascertain whether juveniles displayed antipredator behaviours, the researchers recorded the following behaviours: total time allocated to vigilance, the frequency of antipredator vocalisations, the time spent in or near a shelter, the time spent active, and whether or not the prairie dog ran away from the stimulus.

Perhaps not surprisingly, juvenile prairie dogs that were trained in the presence of an experienced adult were more wary of predators than were juveniles trained alone or in the presence of an inexperienced sibling (Shier and Owings 2007). Once training was completed, all the prairie dogs were introduced to a newly established prairie dog colony. Survival of prairie dogs was estimated by trapping all the animals one year later during 2002. This design, with only two capture periods does not allow one to distinguish between mortality versus emigration away from the study site (Lebreton et al. 1992); nonetheless, long-distance dispersal by prairie dogs occurs rarely, so the researchers considered that any animals not captured were likely to have perished. In this study, juveniles that were trained to avoid predators in the presence of an experienced mother or close relative were more likely to survive one year after release than juveniles that were trained alone or with an inexperienced sibling (Shier and Owings 2007). This finding highlights the importance of including appropriate social interactions into predator training regimes.

12.11   Final Thoughts

In this Chapter, I have provided some examples of some of the methods that we can use to train captive reared or wild animals. Incorporating an understanding of animal behaviour into conservation biology will increase our ability to conserve threatened taxa, and may also help to solve some of the wildlife problems that humans have created. Given that many reintroduction projects fail

due to predation, we need more research to improve predator-training protocols. Two questions that deserve more attention are: (i) What cues do prey use to identify and/or locate predators from a distance? and (ii) If we incorporate such cues into predator training, will they help animals to survive following reintroduction to the wild? Chemical cues are clearly important in this respect, but often researchers use urine and/or faeces in predator training protocols, yet the biological relevance of these odour cues remains questionable (Apfelbach et al. 2005). In Australia, foxes and cats are the major predators responsible for the poor success rate of mammal reintroductions (Moseby et al. 2011). Identifying the chemicals present in the fur of live foxes and cats that elicit strong fear responses in Australian mammals, and incorporating such chemicals into training protocols, might help to increase the success of future mammal reintroductions. Whilst we can train animals to avoid predators, there is still uncertainty about whether predator training enhances the survival of animals following reintroduction to the wild. We need more rigorous, well-designed studies to evaluate the usefulness of predator training. In this respect, long-term monitoring of reintroduced animals is essential for evaluating the success of such programmes. Finally, although it is possible to train animals to hunt, avoid certain foods, or avoid predators, there is little point releasing trained animals to the wild unless the threatening processes have been identified and eliminated. Ultimately, we need to educate people, and manage habitats at release sites carefully if we are to facilitate the long-term survival of endangered species.

Acknowledgements

I thank Vicky Melfi, Nicole Dorey, and Samantha Ward for inviting me to write this book chapter, and Myfanwy Webb for providing critical comments on an earlier draught of the manuscript.

References

Alonso, R., Orejas, P., Lopes, F., and Sanz, C. (2011). Pre-release training of juvenile little owls *Athene noctua* to avoid predation. *Animal Biodiversity and Conservation* 34: 389–393.

Apfelbach, R., Blanchard, C.D., Blanchard, R.J. et al. (2005). The effects of predator odors in mammalian prey species: a review of field and laboratory studies. *Neuroscience and Biobehavioral Reviews* 29: 1123–1144.

Armstrong, D.P. and Seddon, P.J. (2007). Directions in reintroduction biology. *Trends in Ecology & Evolution* 23: 20–25.

Baker, S.E., Ellwood, S.A., Slater, D. et al. (2008). Food aversion plus odor cue protects crop from wild mammals. *Journal of Wildlife Management* 72: 785–791.

Baker, S.E., Ellwood, S.A., Watkins, R., and MacDonald, D.W. (2005). Non-lethal control of wildlife: using chemical repellents as feeding deterrents for the European badger *Meles meles*. *Journal of Applied Ecology* 42: 921–931.

Beck, B., Castro, I., Kleiman, D. et al. (1988). Preparing captive-born primates for reintroduction. *International Journal of Primatology* 8: 426.

Beck, B.B., Rapaport, L.G., Stanley Price, M.R., and Wilson, A.C. (1994). Reintroduction of captive-born animals. In: *Creative Conservation: Interactive Management of Wild and Captive Animals* (eds. P.J.S. Olney, G.M. Mace and A.T.C. Feistner), 265–286. London: Chapman and Hall.

Biggins, D.E., Vargas, A., Godbey, J.L., and Anderson, S.H. (1999). Influence of prerelease experience on reintroduced black-footed ferrets (*Mustela nigripes*). *Biological Conservation* 89: 121–129.

Blumstein, D.T. (2006). The multipredator hypothesis and the evolutionary persistence of antipredator behavior. *Ethology* 112: 209–217.

Blumstein, D.T., Mari, M., Daniel, J.C. et al. (2002). Olfactory predator recognition: wallabies may have to learn to be wary. *Animal Conservation* 5: 87–93.

Bourne, J. and Dorrance, M.J. (1982). A field test of lithium chloride to reduce coyote predation on domestic sheep. *Journal of Wildlife Management* 46: 235–239.

Burbidge, A.A., Johnson, K.A., Fuller, P.J., and Southgate, R.I. (1988). Aboriginal knowledge of the mammals of the central deserts of Australia. *Australian Wildlife Research* 15: 9–39.

Burnett, S. (1997). Colonizing cane toads cause population declines in native predators: reliable anecdotal information and management implications. *Pacific Conservation Biology* 3: 65–72.

Burrows, N.D., Burbidge, A.A., Fuller, P.J., and Behn, G. (2006). Evidence of altered fire regimes in the Western Desert region of Australia. *Conservation Science of Western Australia* 5: 272–284.

Caro, T.M. (1994). *Cheetahs of the Serengeti Plains: Group Living in an Asocial Species*. Chicago: University of Chicago Press.

Caro, T.M. and Hauser, M.D. (1992). Is there teaching in nonhuman animals? *Quarterly Review of Biology* 67: 151–174.

Chivers, D.P. and Smith, R.J.F. (1998). Chemical alarm signalling in aquatic predator-prey systems: a review and prospectus. *Ecoscience* 5: 338–352.

Combreau, O. and Smith, T.R. (1998). Release techniques and predation in the introduction of houbara bustards in Saudi Arabia. *Biological Conservation* 84: 147–155.

Conover, M.R. and Kessler, K.K. (1994). Diminished producer participation in an aversive conditioning program to reduce coyote predation on sheep. *Wildlife Society Bulletin* 22: 229–233.

Covacevich, J. and Archer, M. (1975). The distribution of the cane toad, *Bufo marinus*, in Australia and its effects on indigenous vertebrates. *Memoirs of the Queensland Museum* 17: 305–310.

Crane, A.L. and Mathis, A. (2011). Predator-recognition training: a conservation strategy to increase postrelease survival of hellbenders in head-starting programs. *Zoo Biology* 30: 611–622.

Cremona, T., Crowther, M.S., and Webb, J.K. (2017a). High mortality and small population size prevent population recovery of a reintroduced mesopredator. *Animal Conservation* 20 (6): 555–563.

Cremona, T., Spencer, P., Shine, R., and Webb, J.K. (2017b). Avoiding the last supper: parentage analysis shows multi-generational survival of a re-introduced 'toad-smart' lineage. *Conservation Genetics* 18 (6): 1475–1480.

Curio, E. (1993). Proximate and developmental aspects of antipredator behavior. *Advances in the Study of Behaviour* 22: 135–238.

Daly, J.W. and Witkop, B. (1971). Chemistry and pharmacology of frog venoms. In: *Venomous Animals and their Venoms* (eds. W. Bucherl and E.E. Buckley), 497–519. New York: Academic Press.

Doody, J.S., Green, B., Rhind, D. et al. (2009). Population-level declines in Australian predators caused by an invasive species. *Animal Conservation* 12: 46–53.

Ellis, D.H., Dobrott, S.J., and Goodwin, J.G. (1977). Reintroduction techniques for masked bobwhites. In: *Endangered Birds: Management Techniques for Preserving Threatened Species* (ed. S.A. Temple), 345–354. London: Croom Helm.

Ferrari, M.C.O., Gonzalo, A., Messier, F., and Chivers, D.P. (2007). Generalization of learned predator recognition: an experimental test and framework for future studies. *Proceedings of the Royal Society B: Biological Sciences* 274: 1853–1859.

Ferrari, M.C.O., Messier, F., and Chivers, D.P. (2008). Can prey exhibit threat-sensitive generalization of predator recognition? Extending the predator recognition continuum hypothesis. *Proceedings of the Royal Society B: Biological Sciences* 275: 1811–1816.

Fischer, J. and Lindenmayer, D.B. (2000). An assessment of the published results of animal relocations. *Biological Conservation* 96: 1–11.

Fisher, D.O., Johnson, C.N., Lawes, M.J. et al. (2014). The current decline of tropical marsupials in Australia: is history repeating? *Global Ecology and Biogeography* 23: 181–190.

Galef, B.G. and Laland, K.N. (2005). Social learning in animals: empirical studies and theoretical models. *BioScience* 55: 489–499.

Garcia, J., Hankins, W.G., and Rusiniak, K.W. (1974). Behavioural regulation of the milieu interne in man and rat. *Science* 185: 824–831.

Gaudioso, V.R., Sanchez-Garcia, C., Perez, J.A. et al. (2011). Does early antipredator training increase the suitability of captive red-legged partridges (*Alectoris rufa*) for releasing? *Poultry Science* 90: 1900–1908.

Griffin, A.S. (2004). Social learning about predators: a review and prospectus. *Learning and Behaviour* 32: 131–140.

Griffin, A.S., Blumstein, D.T., and Evans, C. (2000). Training captive-bred or translocated animals to avoid predators. *Conservation Biology* 14: 1317–1326.

Griffin, A.S., Evans, C., and Blumstein, D.T. (2001). Learning specificity in acquired predator recognition. *Animal Behaviour* 62: 577–589.

Griffith, B., Scott, J.M., Carpenter, J.W., and Reed, C. (1989). Translocation as a species conservation tool: status and strategy. *Science* 245: 477–480.

Gustavson, C.R., Garcia, J., Hankins, W.G., and Rusiniak, K.W. (1974). Coyote predation control by aversive conditioning. *Science* 184: 581–583.

Gustavson, C.R. and Nicolaus, L.K. (1987). Taste aversion conditioning in wolves, coyotes, and other canids: retrospect and prospect. In: *Man and Wolf: Advances, Issues, and Problems in Captive Wolf Research* (ed. H. Frank), 169–203. Boston: Junk.

Hoogland, J.L. (1995). *The Black-tailed Prairie Dog: Social Life of a Burrowing Mammal*. Chicago: University of Chicago Press.

Houser, A., Gusset, M., Bragg, C.J. et al. (2011). Pre-release hunting training and post-release monitoring are key components in the rehabilitation of orphaned large felids. *South African Journal of Wildlife Research* 41: 11–20.

Johnson, K.A. (2008). Bilby. In: *Mammals of Australia* (eds. D. Van Dyck and R. Strahan), 191–193. Sydney: Reed New Hollan.

Jule, K.R., Leaver, L.A., and Lea, S.E.G. (2008). The effects of captive experience on reintroduction survival in carnivores: a review and analysis. *Biological Conservation* 141: 355–363.

Kleiman, D.G. (1989). Reintroduction of captive mammals for conservation. *Bioscience* 39: 152–161.

Kotliar, N.B., Miller, B.J., Reading, R.P., and Clark, T.W. (2006). The prairie dog as a keystone species. In: *Conservation of the Black-tailed Prairie Dog* (ed. J.L. Hoogland), 53–64. Washington: Island Press.

Lebreton, J.D., Burnham, K.P., Clobert, J., and Anderson, D.R. (1992). Modeling survival and testing biological hypotheses using marked animals - a unified approach with case-studies. *Ecological Monographs* 62: 67–118.

Letnic, M., Webb, J.K., and Shine, R. (2008). Invasive cane toads (*Bufo marinus*) cause mass mortality of freshwater crocodiles (*Crocodylus johnstoni*) in tropical Australia. *Biological Conservation* 141: 1773–1782.

Lever, C. (2001). *The Cane Toad. The History and Ecology of a Successful Colonist.* Yorkshire: Westbury Academic and Scientific Publishing.

Leyhausen, P. (1979). *Cat Behavior: The Predatory and Social Behavior of Domestic Wild Cats.* New York: Garland STPM Press.

Long, D., Bly-Honness, K., Truett, J.C., and Seery, D.B. (2006). Establishment of new prairie dog colonies by translocation. In: *Conservation of the Black-tailed Prairie Dog* (ed. J.L. Hoogland), 188–209. Washington: Island Press.

McCleery, R., Oli, M.K., Hostetler, J.A. et al. (2013). Are declines of an endangered mammal predation-driven, and can a captive-breeding and release program aid their recovery? *Journal of Zoology* 291: 59–68.

McGregor, I.S., Schrama, L., and Ambermoon, P. (2002). Not all 'predator odours' are equal: cat odour but not 2, 4, 5 trimethylthiazoline (TMT; fox odour) elicits specific defensive behaviours in rats. *Behavioural Brain Research* 129: 1–16.

McLean, I.G., Holzer, C., and Studholme, B.J.S. (1999). Teaching predator-recognition to a naive bird: implications for management. *Biological Conservation* 87: 123–130.

McLean, I.G., Schmitt, N.T., Jarman, P.J. et al. (2000). Learning for life: training marsupials to recognise introduced predators. *Behaviour* 137: 1361–1376.

Moore, N., Whiterow, A., Kelly, P. et al. (1999). Survey of badger, *Meles meles,* damage to agriculture in England and Wales. *Journal of Applied Ecology* 36: 974–988.

Moseby, K.E., Cameron, A., and Crisp, H.A. (2012). Can predator avoidance training improve reintroduction outcomes for the greater bilby in arid Australia? *Animal Behaviour* 83: 1011–1021.

Moseby, K.E. and O'Donnell, E.O. (2003). Reintroduction of the greater bilby, *Macrotis lagotis* (Reid) (Marsupialia: Thylacomyidae), to northern South Australia: survival, ecology and notes on reintroduction protocols. *Wildlife Research* 30: 15–27.

Moseby, K.E., Read, J.L., Paton, D.C. et al. (2011). Predation determines the outcome of 10 reintroduction attempts in arid South Australia. *Biological Conservation* 144: 2863–2872.

Nicolaus, L.K., Cassel, J.F., Carlson, R.B., and Gustavson, C.R. (1983). Taste-aversion conditioning of crows to control predation on eggs. *Science* 220: 212–214.

Oakwood, M. (2000). Reproduction and demography of the northern quoll, *Dasyurus hallucatus,* in the lowland savanna of northern Australia. *Australian Journal of Zoology* 48: 519–539.

Oakwood, M. and Foster, P. (2008). Monitoring extinction of the northern quoll. *Australian Academy of Science Newsletter* (no. 71, March), p. 6.

O'Donnell, S., Webb, J.K., and Shine, R. (2010). Conditioned taste aversion enhances the survival of an endangered predator imperilled by a toxic invader. *Journal of Applied Ecology* 47: 558–565.

Owings, D.H. and Owings, S.C. (1979). Snake-directed behavior by black-tailed prairie dogs (*Cynomys ludovicianus*). *Zeitschrift fuer Tierpsychologie* 49: 35–54.

Parish, D.M.B. and Sotherton, N.W. (2007). The fate of released captive-reared grey partridges *Perdix perdix*: implications for reintroduction programmes. *Wildlife Biology* 13: 140–149.

Phillips, B.L., Brown, G.P., Webb, J.K., and Shine, R. (2006). Invasion and the evolution of speed in toads. *Nature* 439: 803.

Rankmore, B.R., Griffiths, A.D., Woinarski, J.C.Z., et al. (2008). Island translocation of the northern quoll *Dasyurus hallucatus* as a conservation response to the spread of the cane toad *Chaunus [Bufo] marinus* in the Northern Territory, Australia. Report to The Australian Government's Natural Heritage Trust (February).

Reading, R.P., Miller, B., and Shepherdson, D. (2013). The value of enrichment to reintroduction success. *Zoo Biology* 32: 332–341.

Riou, S., Judas, J., Lawrence, M. et al. (2011). A 10-year assessment of Asian Houbara Bustard populations: trends in Kazakhstan reveal important regional differences. *Bird Conservation International* 21: 134–141.

Russell-Smith, J. (2016). Fire management business in Australia's tropical savannas: lighting the way for a new ecosystem services model for the north? *Ecological Managment and Restoration* 17: 4–7.

Seddon, P.J., Saint-Jalme, M., van Heezik, Y. et al. (1995). Restoration of houbara bustard population in Saudi Arabia: developments and future directions. *Oryx* 29: 136–142.

Shepherdson, D.J. (1994). The role of environmental enrichment in the captive breeding and reintroduction of endangered species. In: *Creative Conservation* (eds. G.M. Mace, P.J.S. Onley and A.T.C. Feistner), 167–177. London: Chapman and Hall.

Shier, D.M. and Owings, D.H. (2006). Effects of predator training on behavior and post-release survival of captive prairie dogs (*Cynomys ludovicianus*). *Biological Conservation* 132: 126–135.

Shier, D.M. and Owings, D.H. (2007). Effects of social learning on predator training and postrelease survival in juvenile black-tailed prairie dogs, *Cynomys ludovicianus*. *Animal Behaviour* 73: 567–577.

Short, J. and Smith, A. (1994). Mammal decline and recovery in Australia. *Journal of Mammalogy* 75: 288–297.

Teixeira, B. and Young, R.J. (2014). Can captive-bred American bullfrogs learn to avoid a model avian predator? *Acta Ethologica* 17: 15–22.

Ternent, M.A. and Garshelis, D.L. (1999). Taste-aversion conditioning to reduce nuisance activity by black bears in a Minnesota military reservation. *Wildlife Society Bulletin* 27: 720–728.

Thornton, A. (2007). Early body condition, time budgets and the acquisition of foraging skills in meerkats. *Animal Behaviour* 75: 951–962.

Thornton, A. and McAuliffe, K. (2006). Teaching in wild meerkats. *Science* 313: 227–229.

Tourenq, C., Combreau, O., Lawrence, M. et al. (2005). Alarming houbara bustard population trends in Asia. *Biological Conservation* 121: 1–8.

Truett, J.C., Dullam, L.D., Matchell, M.R. et al. (2001). Translocating prairie dogs: a review. *Wildlife Society Bulletin* 29: 863–872.

Van Heezik, Y., Seddon, P.J., and Maloney, R.F. (1999). Helping reintroduced houbara bustards avoid predation: effective anti-predator training and the predictive value of pre-release behaviour. *Animal Conservation* 2: 155–163.

Van Nieuwenhuyse, D., Génot, J.C., and Johnson, D.H. (2008). *The Little Owl. Conservation, Ecology and Behavior of Athene noctua*. Cambridge: Cambridge University Press.

Webb, J.K., Brown, G.P., Child, T. et al. (2008). A native dasyurid predator (common planigale, *Planigale maculata*) rapidly learns to avoid a toxic invader. *Austral Ecology* 33: 821–829.

Webb, J.K., Du, W.G., Pike, D.A., and Shine, R. (2009). Chemical cues from both dangerous and non-dangerous snakes elicit antipredator behaviours from a nocturnal lizard. *Animal Behaviour* 77: 1471–1478.

Webb, J.K., Pearson, D., and Shine, R. (2011). A small dasyurid predator (*Sminthopsis virginiae*) rapidly learns to avoid a toxic invader. *Wildlife Research* 38: 726–731.

Wisenden, B.D. (2003). Chemically-mediated strategies to counter predation. In: *Sensory Processing in the Aquatic Environment* (eds. S.P. Collin and N.J. Marshall), 236–251. New York: Springer.

Woinarski, J.C.Z., Armstrong, M., Brennan, K. et al. (2010). Monitoring indicates rapid and severe decline of native small mammals in Kakadu National Park, northern Australia. *Wildlife Research* 37: 116–126.

Woinarski, J.C.Z., Legge, S., Fitzsimons, J.A. et al. (2011). The disappearing mammal fauna of northern Australia: context, cause, and response. *Conservation Letters* 4: 192–201.

Woinarski, J.C.Z., Ward, S., Mahney, T. et al. (2011). The mammal fauna of the sir Edward Pellew island group, Northern Territory, Australia: refuge and death-trap. *Wildlife Research* 38: 307–322.

13

Last but in Fact Most Importantly … Health and Safety

Tim Sullivan

13.1　Introduction

It has been many years since a phone call forever changed my perspective on animal training. Until that call, I had spent my entire career honing my knowledge and skills to train zoo animals. Several years earlier, my job had changed from training animals to training the keepers at my zoo to do the same. I enjoyed this challenge because I knew that my efforts would now have a broader effect in enhancing the care of our animals.

I answered the phone and the caller identified himself as an attorney whose firm was representing a major zoo. A keeper at this zoo had suffered a traumatic injury during a training session with a large carnivore. A civil lawsuit had been filed against the zoo and I was being asked to provide testimony for the zoo as an expert witness. I found myself immediately conflicted, I felt terrible for the keeper that had been injured but I had learned enough about the incident to know that the zoo was not necessarily at fault for what happened. I asked the attorney if I could have a few days to consider this request and he kindly agreed.

I had seen my fair share of TV court dramas where the expert witness is subjected to brutal cross-examination. I was not thrilled about the possibility of being on the receiving end of an interrogation thoughtfully designed to undermine my expertise and character.

I thought 'Why should I subject myself to this when I don't even work for this zoo?' The answer came to me when I learned something very important; information that made this event relevant to me and everyone else in this field. As a result of this accident and the subsequent lawsuit, the zoo in question had temporarily prohibited all animal husbandry training. The lawsuit and an Occupational Health and Safety Administration (OSHA) investigation forced the zoo to examine the risks that animal training brought to their keepers and their business in general.

It was clear to me that a negative finding in either the lawsuit or OSHA investigation could have a terrible ripple effect across the entire zoo community. In today's litigious society, strict government regulations and pressures of rising insurance rates, a single employee or employer operating in an unsafe manner can jeopardise everyone's ability to use animal training to enhance animal welfare. We must all develop and maintain a culture of safety in the workplace, especially during animal training when we are brought in close proximity to potentially dangerous animals.

I did become an expert witness in the above case and spent the coming year sifting through the facts of this case. I studied animal training records, staff training documents and safety policies and procedures. I viewed deposition videos from all the witnesses involved in the case. It was an enlightening

Zoo Animal Learning and Training, First Edition. Edited by Vicky A. Melfi, Nicole R. Dorey, and Samantha J. Ward.
© 2020 John Wiley & Sons Ltd. Published 2020 by John Wiley & Sons Ltd.

experience that taught me a great deal about how to limit or hopefully avoid such accidents. In the end, there was a summary judgement in favour of the defendant (the zoo). It was found that the employee had inadvertently broken an established safety protocol that caused the injury. There were no 'winners' in this case. The only good that can come of it is if the profession learns from this cautionary tale and acts responsibly to protect current and future zoo professionals.

13.1.1   A Note on Incident Severity

When we talk about accidents whilst working with zoo animals, most people immediately think about serious incidents that affect life or limb. Whilst these types of tragedies grab the headlines and must be avoided at all cost, we should also consider the effects of more minor incidents like scratches and close-calls (Hosey and Melfi 2015). These 'lesser' incidents provide important information and must not be overlooked. Remember that the circumstances that led to a domestic cat to scratching a zoo professional's arm can just as easily happen when working with a tiger. Every accident, no matter how small or inconsequential, must be reported, investigated and adjustments made to avoid future incidents. It is also important to note that this foreshadowing effect is not lost on insurance carriers. Workplace incident and injury rates are often used to calculate a company's insurance premiums. So a spate of minor injuries can have a significant negative effect on an employer's finances to the tune of tens if not hundreds of thousands of dollars. Regardless of the severity of a safety incidence, any injury to an employee or animal must be avoided. Beyond the actual injury and negative financial impact, the reputation of the employer can also be put in jeopardy. This can have a chilling effect on animal training programmes or put them at risk altogether. This could result in a reduction of these welfare enhancing programmes across all species or have negative impacts on animal husbandry protocols and ultimately the animals' health and welfare.

13.2   Roles and Responsibilities

Safety is everyone's responsibility from the CEO to the individual front-line employee. Each level in the organisation has different responsibilities but the combined effort reduces the potential for accidents. Each employee should understand their role and the importance of staying safe. The goal should be to develop a culture of safety throughout the organisation. Comprehensive safety programmes can create additional work and make some tasks more cumbersome to complete. These facts can reduce compliance if not supported through staff training that clearly outlines the many benefits of operating safely. The aim should be to create buy-in throughout the staff. Each employee strives to be safe and looks out for the safety of their colleagues. In a culture of safety, a junior employee feels comfortable reminding their boss to stay behind a line or to put on personal protective equipment (PPE) (Occupational Safety and Health Administration Act 1970).

13.2.1   Employers' Responsibilities

- Provide and maintain safe facilities and protective equipment.
- Create comprehensive animal training safety policies and procedures.
- Provide staff training in these policies and procedures.
- Provide adequate supervision to ensure that employees operate in a safe manner.
- Promote and ensure a culture of safety throughout the workplace.
- Investigate all errors and/or accidents to ascertain their cause and make adjustments to avoid future incidents.

13.2.2   Employees' Responsibilities

- Act in a safe manner at all times following prescribed policies and procedures.
- Understand and respect the natural history and physical capabilities of the animals they are training.

- Use all required safety equipment to protect them when training animals.
- Keep focused whilst they are training or acting as a backup; understand and maintain 'situational awareness'.

13.3 Understanding and Maintaining Situational Awareness

Zoo professionals operate in complex and dynamic environments. Training settings can vary depending on the species, but most are different than the ones humans live in. With few exceptions, most species behave and make decisions at a higher tempo than the average zoo professional. Add in multiple animals and numerous environmental variables and the zoo professional's mental resources can quickly be challenged. Effective training is about being prepared and making good, timely decisions in the moment. The ability to do this consistently and in a variety of training situations will lead to good training. These same qualities can also keep you safe. Situational awareness is a concept that grew out of the need to make good, safe decisions in complex and dynamic working environment, like those that airline pilots encounter. A widely accepted definition of situational awareness is, 'knowing what is going on so you can figure out what to do' (Adam 1993). Essentially, situational awareness is having awareness about what is happening around you, in order to make decisions based on that information, now and in the future. In more detail, situational awareness clarifies what is needed for reaching the goals of a specific job by understanding what important information is to be used in the decision-making process. In fact, this means 'only those pieces of information that are relevant to the task at hand are important for Situational Awareness' (Endsley 1993). Formally, situational awareness has been defined by Endsley (1988, 1995, Endsley and Garland 2000) as 'the perception of the elements in the environment within a volume of time and space, the comprehension of their meaning and the projection of their status in the near future'. The formal definition of situational awareness is categorised into three hierarchical phases: perception of elements in current situation; comprehension of current situation; and projection of future status. The relationships between these phases and individual factors including the dynamic state of the working environment, decisions of the individual, and their resulting actions are illustrated in Figure 13.1.

Endsley et al. (1998) have expanded these hierarchical phases as follows:

13.3.1 Level 1 Situational Awareness: Perception of the Elements in the Environment

'The first step in achieving situational awareness involves perceiving the status, attributes, and dynamics of relevant elements in the environment. For example, a pilot needs to accurately perceive information about his/her aircraft and its systems

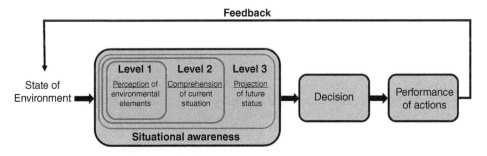

Figure 13.1 Endsley's model of situational awareness. *Source:* adapted from Endsley (1995).

(airspeed, position, altitude, route, direction of flight, etc.), as well as weather, air traffic control clearances, emergency information, and other pertinent elements' (Endsley et al. 1998). Within the animal training domain, zoo professionals should perceive information such as the animal's demeanour and motivation as you arrive for training. Is the animal paying attention to you or is it distracted by other things in the environment, especially other animals or the presence of unfamiliar people? Is the animal responding to cues quickly or is it sluggish in doing so? Is the animal taking reinforcement eagerly but calmly or is it lunging at the food?

13.3.2 Level 2 Situational Awareness: Comprehension of the Current Situation

'Comprehension of the situation is based on a synthesis of disjointed level 1 elements. Level 2 situational awareness goes beyond simply being aware of the elements that are present to include an understanding of the significance of those elements in light of the zoo professional's goals. Based upon knowledge of level 1 elements, particularly when put together to form patterns with other elements, a holistic picture of the environment will be formed, including a comprehension of the significance of information and events' (Endsley et al. 1998). For example, within the animal training domain, to determine whether an animal is becoming frustrated and prone to aggression, zoo professionals should understand conditions and indications like: (i) the amount of success/failure occurring in the session; (ii) the resulting body position of the animal, calm/relaxed or tense/threatening; and (iii) how this particular animal has responded to these same circumstances in the past. This knowledge provides the zoo professional with a mental picture of what they should do to get the animal in a better psychological state to avoid aggression and achieve their training goals.

13.3.3 Level 3 Situational Awareness: Projection of Future Status

'It is the ability to project the future actions of the elements in the environment, at least in the near term, that forms the third and highest level of situational awareness. This is achieved through knowledge of the status and dynamics of the elements and a comprehension of the situation (both level 1 and level 2 situational awareness)' (Endsley et al. 1998). The zoo professional with good situational awareness can synthesise all the relevant information from the animal, the environment, and past experience to make predictions about what will happen next and act accordingly. Although it is stated that improved situational awareness can result in better decision-making, this may not be true in all situations. There are other factors such as strategy, experience, training, personality, and organisational and technical constraints that can also affect the decision-making process (Endsley and Garland 2000). There are cases where situational awareness is lost and individuals can be slower in identifying problems in the situations they find themselves in, resulting in the need for additional time to diagnose the problem and perform corrective actions (Endsley and Kiris 1995). Even small lapses in situational awareness may cause serious problems, zoo professionals must learn what factors can lead to a loss of situational awareness and how to avoid them. Several reasons associated with a loss of situational awareness are listed below.

1) Low and high stress level
   Under situations of low or high stress, zoo professionals are more likely to miss important information about the situation. When the amount of information the zoo professional receives is significantly lower than usual, their attitude may become careless. This usually can occur when training sessions or activities are too routine. A general lack of alertness is associated with missing warning signals and a reduced ability to react quickly and correctly in an emergency. Similarly, when

the amount of information received by the zoo professional is significantly above their capacity to process, zoo professionals will again operate under lower levels of situational awareness, potentially missing critical information. Often labelled 'information overload', these situations can happen when a zoo professional is new to a complicated training regime or there are multiple animals in the environment.

Zoo professionals are often required to operate in new, changing, and sometimes challenging environments. Some of these settings are likely to lead to loss of situational awareness. The most troubling training situations are those that are outside of your normal routine and unusual in nature. Some examples include: i) when guests and VIPs are present at training sessions. Additional people and pressure can distract not only the zoo professional but the animals too. The zoo professional might feel compelled to do more for the guests and push themselves and the animals out of their respective comfort zones. Likewise, the zoo professional's supervisor or even the company CEO might ask for something special that has not been tried nor tested under the best of conditions. It is in these scenarios where the company's culture of safety can be tested. The zoo professional should be empowered to say "no" to any request where safe outcomes are not certain. ii) Another classic example of where situational awareness can be lost is found in the nature of caring for living creatures. Zookeepers regularly find themselves rushing due to high workloads and unexpected tasks that come with the job. It is at this time where a zoo professional must have the wherewithal to know that this may not be the safest time to train. Postponing or cancelling a training session should always be an option in these circumstances. If rushing is an everyday occurrence then this is a more systemic issue that should be addressed by management.

2) Ambiguity
When the information provided and received is ambiguous, decisions and actions of the zoo professional are likely to be based on inaccuracy and could lead to uncertain and potentially dangerous results. For example if a zoo professional sets up training apparatus in the tiger holding area, they might ask a colleague if the tiger shift door is closed or open. If the colleague's response is 'Yes' and the zoo professional assumes that the shift door is shut an accident could occur.

3) Fixation or preoccupation
In certain training situations, the ability to detect important information is lost when the zoo professional is preoccupied, fixated or otherwise distracted. For example, if an elephant keeper is training an elephant to raise a front leg, by cueing the elephant to touch the top of its foot to a hand-held target. The elephant keeper might then be completely focused on the small point in space where the elephant's foot will make contact with the target. During this time, the elephant keeper is less able to perceive important information like the location and movement of the elephant's trunk nor can they see body cues that might help predict aggressive intent.

4) Departure from policies and protocols
Using improper procedures, places zoo professionals in a grey area, where safe outcomes cannot be predicted with certainty. Zoo professionals may not have the experience and, therefore, the proper judgement to understand the risks they might be placing themselves in when deviating from agreed policies and protocols; which have in fact been established to reduce the risks of incidents. Something as innocuous as breaking protected-contact to pet 'Fluffy' the lion could end in tragedy. Consistent and/or blatant violations of rules often reveal other systemic problems within an organisation. A disregard for following established procedures

could indicate a lack of proper supervision or accountability regarding compliance issues and certainly puts everyone at risk.

13.4 'Gut' Feeling

Increasingly research is demonstrating that over 80% of decisions are determined by the subconscious, which was previously attributed to gut feeling (e.g. Kahneman 2011); when we feel or detect subtle stimuli long before we have consciously considered them. The activation of the sympathetic nervous system can occur without conscious knowledge of what triggered it. Learn to recognise your own signs of reflexive discomfort, such as stomach butterflies, increased heart rate, muscle tension, and mood swings. Become sensitive to these internal cues and do not ignore them. Teach yourself to take a step back for a moment to ascertain potential threats in the training environment. Maintaining situational awareness can help you make better decisions and keep you safer, so it is important to know ways to maintain situational awareness; listed below are a few examples.

A) Experience
   Life and work experience create mental files that the mind can draw upon and combine with new information in the working memory, i.e. the system used for temporarily storing and managing the information required to carry out highly detailed cognitive tasks. In complex environments, there are far too many stimuli bombarding the senses, which can lead to difficulty trying to synthesise and interpret the best course of action in an instant. To overcome this situation, the brain stores composite stimuli patterns that are related with certain situations and their eventual outcomes. When these patterns (exact or similar) are recognised again, the mind draws upon this reference to expedite the decision-making process.

B) Positional awareness
   When working with potentially dangerous animals, zoo professionals must always maintain an awareness of their position in relation to the animal and any barrier meant to separate and protect them from each other. Additionally, the zoo professional must understand the capabilities of the animals they are working with. By being keenly aware of an animal's reach and speed, proper positioning can be maintained to avoid being within the animal's reach. When working in the same space as the animal, a zoo professional must know where their egress to safety is and must always protect it. Be vigilant in maintaining your position between the animal and your exit.

C) Physical/mechanical training skills
   Much animal training is based on knowledge of animal behaviour modification techniques and principles, that when applied correctly result in behaviour change. The physical aspects of training are often mechanical in nature and require practice to employ them skilfully. Training tools (Figure 13.2) include: event markers, like clickers and whistles, where the timing of their use is critical; targets of various size and material that must be wielded with precision; containers to hold food that must be conveniently located for quick access; and apparatus like tongs and 'meat sticks' to deliver food safely to the most dangerous of animals.

   With the use of these tools, some training programmes can be quite cumbersome, distracting, and possibly compromise safety. At first, zoo professionals should practice the use of their training tools and techniques away from the animal to develop the mechanical skill and fluency necessary to be effective and safe when training. This can be done using a co-worker to play the part of the animal and in a similar environment to the actual training setting. If you cannot master the use of all the tools you plan to use in practice, consider simpler

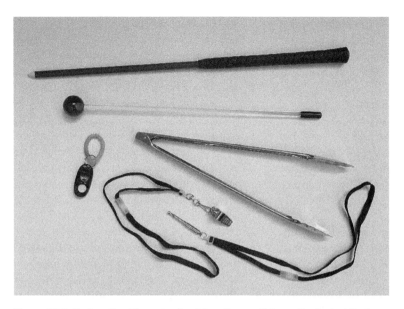

Figure 13.2 Tools utilised for animal training. *Source:* Chicago Zoological Society.

alternatives. Ultimately, the right system is the one that can be mastered and is effective at facilitating the behaviour change you desire whilst remaining safe and not compromising the welfare of the animals.

D) Emotional and physical condition
Your emotional and physical condition can affect your perception of the environment. Heightened emotional states, mental illness, and medical and physical ailments that cause discomfort and/or require medication can negatively affect a zoo professional's ability to maintain situational awareness. Although it might not be apparent, safety can be compromised if these conditions cloud or distort the zoo professional's ability to accurately perceive events or conditions in the training setting. Good judgement and restraint must be used to know when you are unfit to train. Know when to say 'no' to training and err on the side of caution, as missing a session will unlikely negatively affect the animal's learning.

E) Personal attitude
The days of showing off one's scars from animals and sharing these harrowing stories over a beer should be a thing of the past. These incidents should not be considered badges of courage but evidence of mistakes to be learned from and avoided. Professionalism is a matter of attitude, and safety should be paramount above all else. Safety is no accident and is the result of preparation and conscious effort. To be safe, one must always think and act safely.

Understanding and maintaining situational awareness provides insight on how our minds perceive and process information, that can help zoo professionals make safe decisions. Having this knowledge can also help inform the development and implementation of more fundamental and practical aspects required to train zoo animals safely.

13.5 Establishing Safety Policies, Protocols, and Work Rules

The establishment of safety policies, protocols, and work rules is a critical process in any business. Developing and maintaining a culture of safety at your zoo must be a priority. The expectations must be clear, both verbally and in writing. As important as it is for the zoo to train and orient its employees, unless the information provided is written into policies and rules, the zoo could be exposed to legal battles, misunderstandings, and loopholes. These policies may be based on an understanding that has evolved after working together for a period of time, but they are much easier to enforce if they are in writing. Such policies and rules, if written properly, protect the zoo and its employees and provide the company with the flexibility to manage its business needs.

Creating a safe and healthy work environment is not only a requirement for most employers around the world; it is also a 'best practice' amongst top performing businesses. Great businesses know how to get their work done effectively, efficiently, and safely. Formalising a zoo's expectations, including step-by-step instructions for job tasks, is a very effective way to emphasise to employees that it is as serious about their health and safety as it is about animal care, habitat quality, financial stability, and customer service. The following sections can provide you with some practical information and examples to help you assess, update, or create your zoo's safety policies, procedures, and rules for animal training and in general.

13.5.1 Safety Policies

Safety policies cover broad topics and can communicate a zoo's philosophy and more generally, best practice to their staff. Safety policies should be written as guidelines for reference and discussed with employees upon appointment and during orientation/induction. They provide both the organisation's vision and should inspire the employee by highlighting what is believed, valued, and expected within the organisation. Issues and topics that may be included are safety policy statement, employee responsibility and accountability, hazard assessment and correction, and the discipline policy.

13.5.1.1 Safety Policy Statement

An effective way to communicate your zoo's overall safety policy is to develop a Safety Policy Statement that includes the elements that are core to your zoo's focus on staff health and safety. The Safety Policy Statement is typically the introductory statement in a written safety policy, and should reflect the importance of injury prevention. Reviewing your Safety Policy Statement with potential job applicants will set expectations and help them determine if they would be a good fit for the organisation. An example might be as follows:

At Marwood Zoo, we care about the safety, health, and well-being of our employees. We value each and every contribution that our employees make towards our success. Our mission is to promote the conservation of wildlife, nature, and we value honesty, integrity, and teamwork.

We Value Our Employees
Our zoo operates with a goal of zero harm to people, animals, and property. It is our policy to provide safe working conditions. At Marwood Zoo, everyone shares equally in the responsibility of identifying hazards, following safety rules, and operating practices. All jobs and tasks must be performed in a safe manner, as safety is crucial to the quality of our business aims.

Safety Policy
At Marwood Zoo, no phase of the operation is considered more important than accident prevention. It is our policy to provide and maintain safe working conditions and to follow operating practices that will safeguard all employees. No job will be considered properly completed unless it is performed in a safe manner. Marwood Zoo is concerned about the health and good work habits of its employees. In the event you are injured or unable to perform your job, we want to help you obtain the best treatment, so you can return to your regular job as soon as possible.

13.5.2 Safety Procedures

Safety procedures cover specific activities where there might be risk to staff, animals or visitors, with the intention of ensuring, if followed, a safe working environment. These procedures outline the steps to be taken to implement a policy. Some procedures are specific to a job or task whilst others are general descriptions of how to implement a policy. Procedures exist for most aspects of a job whether they are formalised and written down or not. To be effective it is ideal that safety procedures are communicated in writing and verbalised, as well as providing a demonstration. All of this should emphasise common safety procedures, related to training and may include, how to report a hazard or safety concern; basic safety rules; required PPE; emergency plans; and step-by-step examples of how to safely complete specific work tasks. For example, procedures when shifting/securing and training dangerous animals which should cover issues of animal contact/non-contact rules, delivery of food, e.g. use of tongs, feed chutes, no hand-feeding, safety backup staff /'two-person' requirements, chemical defence spray use and emergency communications, e.g. possess walkie-talkies, mobile phones.

13.5.3 Safety Rules

Safety rules list the specific activities of what to do, or avoid in order to complete the job effectively and safely. Essentially, no single list of safety rules is adequate for all zoos or activities and it is important to develop your own list of safety rules based upon standard industry practices and your own accident experience, not based solely on generic lists or examples from other zoo employers. It is key that all rules are communicated clearly to management and staff alike and rules should be strictly and consistently enforced. If written safety rules are not consistently and equitably enforced, the zoo's actual practices outside the written rules may create a legal liability if challenged in a legal or regulatory dispute.

An example of what might be included in safety rules are outlined below:

1) Report to work alert, rested, and in good physical condition.
2) PPE (such as safety glasses, face masks, protective clothing, and footwear) must be worn when required for specific job tasks or work areas.
3) All accidents, incidents, and injuries, regardless of how minor, shall be reported immediately to the supervisor in charge.
4) All work is to be performed in a safe manner according to our written policies and procedures. If you have a concern about the safety of a task, bring it to the attention of your immediate supervisor.
5) Understand your work assignments and perform only the job functions in which you are fully trained. Discuss any unfamiliar work assignments with your supervisor prior to beginning the task.
6) Horseplay or practical jokes are prohibited.
7) Use or being under the influence of, intoxicants or drugs whilst on the job is prohibited and shall be considered cause for dismissal.
8) Always use the proper tool, equipment, or process for the job.
9) All employees shall correct an unsafe condition or practice to the extent of their authority and/or report the hazard to their supervisor.
10) Ignoring safe work practices, policies, procedures, rules, or other safety instruction is cause for disciplinary action up to and including termination of employment.

Developing and implementing safety policies, procedures and rules set the foundation for keeping employees safe. These documents and their communication underscore the philosophy, actions, and detail necessary to protect both the employee and the organisation. They must be considered living documents that should be revisited regularly and revised when necessary. Training potentially dangerous animals presents unique hazards to zoo staff. As zoo animal training continues to grow, best practices emerge to produce better and safer outcomes. In Section 13.6 there are a series of practical training safety tips that should be considered.

13.5.4 Legislation

Animal training is becoming popular and as discussed in this chapter, has strict safety protocols and standard operating procedures that are created and adhered to for obvious reasons. However, legislation that covers the specifics of animal training across the world is sparse. Table 13.1 outlines specific legislation and/or accreditation guidelines that discuss the use of training in zoos.

13.6 Practical Safety Tips

13.6.1 Safety Backup

When training dangerous animals including large carnivores, marine mammals, great apes, and pachyderms (see Defra 2012 for a list of UK classified dangerous animals), having dedicated safety personnel can greatly reduce the chance for accidents. As discussed earlier in this chapter, zoo professionals are often required to focus their attention on a very small aspect of the animal and the working space. This narrow state of perception can leave the zoo professional temporarily exposed to harm. The role of the zoo professional safety backup is to provide for the zoo professional's well-being throughout the session by watching the behaviour of the whole animal, the training environment, and the zoo professional's own actions. This should be their only role in the training session so that they can remain vigilant. The zoo professional safety backup must have adequate experience with both the animal being trained and the training system in general. The zoo professional safety backup must be aware of the species-specific and often subtle behavioural precursors that may indicate that an animal is frustrated which could lead to zoo professional-directed aggression. They must

also be familiar with the training system and the particular training goal to know where and when the potential risks are greatest. The zoo professional safety backup must be well-versed and practiced in emergency procedures if the zoo professional were to come under attack. They should carry any required safety equipment, such as defensive chemical pepper spray, radio communications equipment, etc. and be well-trained in their use. If additional training staff are in attendance at a session, they must clearly communicate as to which person will take on the zoo professional safety backup role.

13.6.2 Respect the Potential of Every Animal

Positive reinforcement training can create strong, positive relationships between zoo professionals and animals (Ward and Melfi 2013). Whilst these relationships can facilitate and improve animal welfare, they can also lead to a dangerous level of complacency in zoo professionals. It must be stressed that regardless of whether your human–animal relationship has been long and positive, the potential for zoo professional-directed aggression always exists (Hosey and Melfi 2015). Age, painful injuries, declining health, and countless environmental factors can cause a sudden and unexpected change in an animal's attitude. The sometimes challenging nature of training itself can trigger a reflexive aggressive response from an otherwise mild-mannered animal. Zoo professionals must continually remind themselves to think about the potential of the *species* first instead of the individual you have grown to love and trust. Even the smallest and cutest of species can inflict harm on a zoo professional that can jeopardise their health and training programmes in general.

13.6.3 Safe Delivery of Reinforcement

Positive reinforcement training relies heavily on the use of food as a consequence of desired behaviour. Getting that food to the animal in a timely fashion is important but can be a

Table 13.1 International legislation directly affecting zoo animal training.

Region	Specific training legislation/guidelines	Details
Australia	Proposed Animal Welfare Standards and Guidelines 2014: Training	Specific to training animals: • The operator must ensure written procedures regarding the health, safety, and behavioural needs of the animal during training are developed, maintained and implemented, and are readily available to staff who train animals. • The operator must ensure training is undertaken by an appropriately experienced trainer or under the immediate supervision of an appropriately experienced trainer. • The operator must ensure training does not compromise the animal's normal physical development, health or welfare. • The operator must ensure training programmes do not exceed the physical capabilities of the animal. • The operator must ensure exhibitions of trained behaviours of animals demonstrate behaviours that are reflective of those expressed in the wild. • Trainers should use operant conditioning. • Punishment should be avoided as a training method. • Animals should be conditioned to accept routine husbandry procedures.
	Proposed Animal Welfare Standards and Guidelines 2014: Interactive Programmes	Specific to visitor–animal interactions: • The operator must ensure interactive programmes are designed to enhance people's appreciation of and respect for animals. • The operator must ensure a proficient keeper is responsible for overseeing, coordinating, and supervising all interactive programmes. • The operator must ensure a risk assessment examining the risks to the animals is undertaken for each interactive programme and is reviewed on a regular basis. • The operator must ensure written procedures for interactive programmes are developed, regularly reviewed and implemented, and are readily available to staff. • The operator must ensure interactive programmes do not have adverse impacts on animal welfare. • The operator must ensure animals that display signs of distress or illness are removed from the interactive programme until such time as they are reassessed by a veterinarian or proficient keeper as being suitable to re-enter the interactive programme.
Canada	Canada's Accredited Zoos and Aquariums Guidelines	CAZA Policy on the Use of Animals in Educational Programming states: • An overall programme animal training protocol providing for frequency of training, the process for qualifying and assessing handlers, including who is authorised to train handlers. • Training content (e.g. taxonomically specific protocols, natural history, relevant conservation and educational messages, presentation techniques, interpretive techniques, etc.).

(continued)

Table 13.1 (Continued)

Region	Specific training legislation/guidelines	Details
Colombia	No specific legislation related to training animals or zoos	However: • Law 1333 (2009) refers to environmental procedures – includes protocols regarding the destiny of animals according to their conditions (quarantine, medical, rehabilitation, release, or relocation to zoos). • Law 1774 (2016) Law of Animal Protection and Welfare is used to support cases of neglect, etc.
Europe	EAZA Guidelines on the use of animals in public demonstrations 2018	Guidelines developed to provide EAZA (European Association of Zoos and Aquaria) members with information on how to ensure best practice and includes aspects such as behaviour demonstrations, human–animal interactions, animal health, housing, and animal selection.
	EU Zoo Directive	Good practice guidelines outline: • Training methods are based on positive operant conditioning, but other forms of training can also be adopted. • Competent trainers have a good understanding of anatomical, behavioural, and cognitive abilities of animals, never using objects, restraining or training methods (e.g. negative reinforcement, punishment) that compromise their welfare. • Food used during training sessions should be part of the daily allowance. • A good training plan will include measures to avoid over-stimulating animals, promoting unnatural behaviour or making them working beyond their capacity.
India	The Prevention of Cruelty to Animals Act 1960	• Section 22 refers to restrictions on training of exhibition animals. • Section 23 refers to the procedure to register desired intent to train exhibition animals. • Section 24 refers to the need to ensure animals undergoing training must only do so without unnecessary pain or suffering. • Section 25 refers to authorised persons being able to enter and inspect the premise where performing animals are being trained or exhibited or kept for training or exhibition.
Indonesia	No specific legislation related to training animals	However, there are directions within certain government regulations as below: • Utilisation of Wildlife resources including breeding, demonstration, and keeping for pleasure (pet) is in National Government Regulation 'Peraturan Pemerintah No. 8 Th. 1990 article 3'. • Demonstration of protected wildlife species is in Minister of Forestry Regulation 'Permenhut No. 52 Th. 2006'. • The Dolphin Demonstration Guidelines are in Minister of Forestry Regulation 'Permenhut No. 16 Th. 2014'. • Animal Ethics and Welfare Guidelines are in Minister of Forestry Regulation 'Permenhut No. 9 Th. 2011'.

Table 13.1 (Continued)

Region	Specific training legislation/guidelines	Details
Japan	No specific legislation related to training animals or zoos	However, there are standards relating to the keeping and custody of animals for exhibition (1976) which is based on 'The Law Concerning the Protection and Control of Animals, 1973'. However these standards specifically refer to animal welfare or animal training.
New Zealand	Zoos code of welfare 2018. Issued under the Animal Welfare Act 1999	Includes: If animals are trained or perform, i) the techniques used must be appropriate for the species and the individual animal's physical and mental capabilities; and ii) sessions must be of a length of time determined by the animal's reaction and condition but without over-working the animal; and iii) food deprivation and/or electric prods must not be used; and iv) methods must be based on immediate positive reinforcement; and v) training and command implements must be used in such a manner that does not cause unreasonable or unnecessary pain, injury or distress to an animal.
Philippines	None	Although there is the Animal Welfare Act (RA 8581, RA 10631 – amendments), these do not specifically relate to animal training. There is no current legislation specific to zoos.
	Southeast Asian Zoos Association	Animal Welfare Standards are currently being written but these do not affect the training of animals.
South Africa	Performing Animal Welfare Act 1935	To regulate the exhibition and training of performing animals for safeguarding. Interested parties are to apply for a licence to enable them to train animals, however, this does not apply to zoological gardens.
United Kingdom	Zoo Licencing Act 1981 (as amended in 2002)	Guidelines to accompany the legislation: Secretary of State's Standards for Modern Zoo Practice (Defra 2012) outlines guidance for training of animals. Zoos are inspected regularly (informal and formal inspections) by government appointed inspectors. Elephant specific standard – in 2017 new elephant specific guidelines were published, as part of the Secretary of State's Modern Zoo Practice. These state 'Each institution must have an elephant training programme … and individual tailored goals for each animal.'
	Performing Animals (Regulation) Act 1975	People or organisations who train animals to perform must register with the local authorities and be subject to inspection. Although if a UK zoo has a licence, they would only be assessed under the Zoo Licencing Act (above).
	The Animal Welfare (Licensing of Activities Involving Animals) (England) Regulations 2018	Part six discusses the keeping or training of animals for exhibition. This legislation covers institutions that house and use animals specifically for this purpose that do not hold a zoo licence (as above).

(*continued*)

Table 13.1 (Continued)

Region	Specific training legislation/guidelines	Details
United States	Association of Zoos & Aquariums (AZA) Accreditation Guidelines	The institution should follow a formal written animal training programme that facilitates husbandry, science, and veterinary procedures and enhances the overall health and well-being of the animals. Explanation: An animal training programme should be based on current animal training best practices in the zoological field and should include the following elements: i) goal setting (what behaviours to be trained, what species/individuals of priority), ii) planning (process for developing and approving training plans), and iii) documentation (record of success). Elephant specific standard – all institutions must have an elephant training programme in place which allows elephant care providers and veterinarians the ability to accomplish all necessary elephant care and management procedures. Each institution will adopt and implement an institutional training methodology that promotes the safest environment for elephant care professionals and visitors and ensures high quality care and management of the elephants for routine husbandry, medical management, physical well-being and overall elephant welfare. Institutions must train their elephant care professionals to manage and care for elephants with barriers and/or restraints in place that provide employee safety.

Information compiled by Samantha Ward and provided by the following: Samantha Ward (UK, Europe, New Zealand, Canada, India, Japan, and South Africa); Tim Sullivan (USA); Willem Manansang (Indonesia); Lester Lopez (Philippines); Nick Boyle (Australia); and Catalina Gomez (Colombia).

risky endeavour for some animals. Placing food reinforcement directly in the mouth or hand of a waiting animal creates the potential for injury of the zoo professional. For instance, many zoos now have policies prohibiting hand-feeding of large carnivores. Protective barriers between the animal and zoo professional can create the illusion of safety but past incidences have proven that this method can still lead to great risk for injury. Even the experienced and well-intentioned zoo professional can make mistakes and break the plane of containment with their fingers. The sheer number of reinforcements that must be delivered through the barrier increases the probability that a mistake might occur. The mouthparts of carnivores are well adapted and extremely adept at snagging and holding on to body parts such as small as a fingertip or the glove that covers it. Whether the animal does so on accident or on purpose, zoo professionals can find themselves in a terrifying situation in an instant. There are several safer alternatives to hand

feeding that are widely used such as the use of commercially available, long-handled tongs or custom-made 'meat-sticks', which provide a safe extension to the zoo professional's hand to safely deliver food. The installation of feeding tubes or chutes on the front of the protective barrier deposit food safely away from the zoo professionals' hand (Figure 13.3). These methods change the timing and position of food delivery and can seem cumbersome at first. The added safety these methods provide more than makes up for the initial inconvenience. The proper conditioning and use of a bridging stimulus may ensure that correct responses are reinforced in a timely way and mitigate any delay in the actual food delivery.

13.6.4 Don't Get Grabbed

Many of the training tools we use can inadvertently provide an animal with a place to grab and pull the zoo professional into harm's way. Containers for holding food reinforcement for training should be considered carefully. The type of container should allow for fumble-free extraction of the food and be within easy reach of the zoo professional but not the animal. If the food container is on the zoo professional's person, the attachment should be of a type that allows easy separation if grabbed by the animal. Whistles used as bridging stimuli should be placed on break-away lanyards if worn around the neck. The same is true for event markers such as clickers if on some form of tether meant to keep it on the wrist or belt. Loose-fitting clothing should be avoided when working around animals and leave dangling jewellery at home or in a locker. Following these suggestions is a good start but zoo professionals must always be aware of their position in relation to the animal (Figure 13.4). It is critical to know the danger zones that are within the animal's reach. A safety line denoting these areas is only effective if zoo professionals stay mindful of them at all times.

Figure 13.3 Food reinforcement provided via a safe, feeding pole for a polar bear. *Source:* Chicago Zoological Society.

13.7 Free-contact or Protected-contact

One of the most hotly debated topics in the training field of recent years is about where and how zoo professionals work with their animals. The confusion starts when people confuse the terms *free-contact* or *protected-contact* as being training systems; they are not. Animal learning and, therefore, animal training does not rely on where the zoo professional is relative to the animal. Additionally, a pervasive

Figure 13.4 Zoo professional in reach of an elephant during husbandry (noted: picture was staged). *Source:* Tim Sullivan.

and unhelpful stigma exists due to the common misconception that free-contact management equates to punishment-based training and that protected-contact management is synonymous with positive reinforcement-based training. In fact, there are no functional restrictions on what type of consequence that can be used in either management system.

Free-contact and protected-contact are animal management practices that define, in principle, whether zoo professionals share the same space with an animal. This simple definition is also misleading because this animal management issue is *not* a binary concept. In practice, a spectrum of choices exist which determine the degree to which zoo professionals share space with animals. Management choices can go from complete unrestricted access to absolutely no contact with animals and everything in between. Protected-contact as its name implies is designed to provide an additional degree of safety for animal care staff and zoo professionals through some form of restrictive barrier. Contact is still allowed but limited by the design of the barrier and policies and protocols written to guide the actions of the staff. Whilst protected-contact can be safer for zoo professionals it is no guarantee. Zoo professionals have gotten injured and even killed in protected-contact. Danger is still present and some may argue greater due to the fact that the barriers in protected-contact can produce a false sense of security. Complacency when working around dangerous animals can cause harm no matter what type of management system is in place. It is up to each individual zoo professional to maintain a heightened awareness and follow safe practices when working around all dangerous animals.

13.8   Summary

Animal training has had a significant, positive effect on the care and welfare of zoo animals under professional care. The process of modifying the behaviour of zoo animals places zoo professionals in close proximity of potentially dangerous animals. It is incumbent on zoo professionals and their employers to safeguard the health and well-being of each individual and every animal involved in the training process. Several health and safety principles and practices can be implemented to provide safe interactions during the training process.

- Establishing a culture of safety: everyone in the organisation must care about and be responsible for a safe work environment. This includes the development of, and adherence to health and safety policies and procedures that keep employees and animals safe.
- The process of training animals must be grounded in safety: the work areas and training tools must provide for the safe interactions with animals during training. Direct physical interaction with animals must go through a thoughtful risk assessment to ensure the continued safety of zoo professionals. These individuals must always remember that they are working with potentially dangerous animals and this fact does not change because they live under human care.
- Whilst training animals, zoo professionals must maintain situational awareness: keeping one's senses keen allows for an accurate perception of the state of the animal and the training environment. This focus allows for the comprehension of the current situation that leads to good decision-making and actions that can keep the zoo professional and the animal safe.
- As important, the zoo professional must guard against conditions that can reduce situational awareness. These include one's own physical and mental condition, distractions and stressors in the training environment, and fixation on the training task at hand. The zoo professional must quickly recognise these threats and immediately act to be safe, which could include not starting or aborting a training session that is in progress.

By promoting a culture of safety, zoos and aquariums can realise the full benefit of animal training programmes thereby ensuring their continued success into the future.

References

Adam, E.C. (1993). Fighter cockpits of the future. *Proceedings of 12th IEEE/AIAA Digital Avionics Systems Conference (DASC)*. IEEE, pp. 318–323.

Defra (2012). Secretary of State's Standards of Modern Zoo Practice. Department for Environment, Food and Rural Affairs. Appendix 12, p. 62. https://assets.publishing.service.gov.uk/government/uploads/system/uploads/attachment_data/file/69596/standards-of-zoo-practice.pdf.

Endsley, M.R. (1988). Design and evaluation for situation awareness enhancement. *Proceedings of the Human Factors Society 32nd Annual Meeting*. Santa Monica, CA: Human Factors Society, pp. 97–101.

Endsley, M.R. (1993). A survey of situation awareness requirements in air-to-air combat fighters. *International Journal of Aviation Psychology* 3 (2): 157–168.

Endsley, M.R. (1995). Toward a theory of situation awareness in dynamic systems. *Human Factors* 37: 32–64.

Endsley, M.R. and Garland, D.J. (2000). *Situation Awareness Analysis and Measurement*. Mahwah, NJ: Lawrence Erlbaum Associates.

Endsley, M.R. and Kiris, E.O. (1995). The out-of-the-loop performance problem and level of control in automation. *Human Factors* 37: 381–394.

Endsley, M.R., Selcon, S.J., Hardiman, T.D., and Croft, D.G. (1998). A comparative analysis of SAGAT and SART for evaluations of situation awareness. *Proceedings of the human factors and ergonomics society annual meeting* (October), Los Angeles, CA. Sage Publishing, 42(1): 82–86.

Hosey, G. and Melfi, V. (2015). Are we ignoring neutral and negative human–animal relationships in zoos? *Zoo Biology* 34 (1): 1–8.

Kahneman, D. (2011). *Thinking, Fast and Slow*. Macmillan. ISBN: 978-1-4299-6935-2.

Occupational Safety and Health Administration Act (1970). s 5 (b). https://www.osha.gov/pls/oshaweb/owadisp.show_document?p_table=OSHACT&p_id=3359.

Ward, S.J. and Melfi, V. (2013). The implications of husbandry training on zoo animal response rates. *Applied Animal Behaviour Science* 147: 179–185.

Box C1

Training Animals in a Group Setting

Kirstin Anderson-Hansen

There are few institutions that are fortunate enough to have one trainer per animal and therefore in most facilities, it is necessary for the trainer to train multiple animals at once. This is both challenging and difficult, both for the trainer and for the animals.

One of the primary reasons for training our animals is to increase the animal's well-being by creating situations for the animal to be mentally and physically challenged. This can be compromised if the guidelines set by the facility are inconsistent with how to train animals in a group setting. When working one-on-one with an animal, the animal has the trainer's full attention but this all changes when there are multiple animals and only one trainer. For these animals, the concept of sharing resources, i.e. the trainer's attention and reinforcement, is not something they are naturally willing to wait patiently for whilst another conspecific is being fed especially with social animals living in a hierarchy. However, it is possible to train them to allow other individuals to receive a reward whilst they patiently wait, through establishing consistent guidelines.

To make sure we do not comprise the individual animal's welfare whilst working multiple animals together, certain considerations need to be made:

1) Is it possible to separate, either spatially or physically?

To be able to separate spatially more than one person is required, where one person has one animal whilst the other person has the other animal or animals in a group training situation (see Figure C1.1a). To train the animals to comfortably separate, the animals should be trained near each other at the beginning then slowly increase the distance between the trainers with small approximations. Since this is usually a counter conditioning situation, the approximations should always be small enough to not cause the animals to be nervous and tense and the sessions should be kept short and positive. This same method can be used to physically separate animals (see Figure C1.1b).

2) Is the species a social or solitary species? When working with a social species, for example primates, separations can cause more harm than good by creating a stressful situation for the entire group. Therefore the animals may be more comfortable being worked as a group. Whilst other species that don't necessarily live in groups, for example tigers, may be more comfortable being separated and worked individually (see Figure C1.2). Thus it is always important to know the species social behaviour and adapt your training to meet the needs of the species you are working with.

Zoo Animal Learning and Training, First Edition. Edited by Vicky A. Melfi, Nicole R. Dorey, and Samantha J. Ward.
© 2020 John Wiley & Sons Ltd. Published 2020 by John Wiley & Sons Ltd.

(a)

(b)

Figure C1.1 It is possible to train multiple animals at the same time by either spatially separating them (a) or physically separating them (b); both of these regimes can be seen here in grey seals at the University of Southern Denmark, Kerteminde, Denmark. *Source:* Kirstin Anderson-Hansen.

3) Is there a special social situation? Is there a hierarchy in the group?

Since we always want to set the animals up for success, it can be an advantage to respect the hierarchy in the group when training them together. This can be done by making sure certain individuals are spatially or physically separated during the training sessions. One technique that trainers use is stationing (training animals in different places within the enclosure using different individual stimuli (e.g. different shaped targets) to identify the different stations. For exam-

ple, if you have a dominate male then you might want to station him separately from the rest of the pack. However, if you have a more complicated hierarchy. For example, a group of grey seals with a dominant male, a dominant female, and three subordinate females at different rankings, then we would typically station the dominant male on the far right side of the group next to the dominant female, and the most subordinate female would be stationed at the far left side with the two other females in the middle (see Figure C1.3).

Figure C1.2 Whether the multiple animals you're trying to train are social or not, might impact whether you want to train all the animals in a group or individually. *Source:* Odense Zoo.

By placing the animals far apart at the start and reinforcing them *consistently* for patiently and calmly waiting at their station, they learn that it is not necessary to compete with the others. With time, the distance between the animals can be decreased, making it possible to reduce the number of trainers necessary to work with the entire group (see Figure C1.4a and b).

Reinforcement

Animals will learn through a strong positive reinforcement history to wait patiently and accept another animal to receive reinforcement. Therefore, at the start of the training programme, the animals are just reinforced for stationing patiently. Once this is properly conditioned, the trainer can begin to slowly work individual behaviours for a short period. However, when a trainer is working with one individual, they should never forget that the others are also working – waiting patiently.

For example, during a group training session, if the trainer asks Animal #1 to perform a behaviour (i.e. mouth open) and Animal #2 is asked/expected to wait patiently, the animal waiting patiently is having to do the more difficult behaviour – waiting whilst the

Figure C1.3 Considering the social hierarchy in a group is essential when training animals together; pictured here the grey seals at Hel Marine Station, Poland, stationed in hierarchical order. *Source:* Kirstin Anderson-Hansen.

(a)

(b)

Figure C1.4 The distance between animals which are being trained together can impact the training session; here a group of grey seals are stationed and trained with a larger space between animals (a), but animals can also be reinforced for being stationed and trained when in close proximity to one another (b). *Source:* Kirstin Anderson-Hansen.

other animal gets attention from the trainer. This calm and patient behaviour needs to be reinforced and maintained.

A common basic behaviour that most animals are initially trained to do is a 'hold' behaviour. By asking the other animals to do a 'hold', for example, the animals that are expected to wait are given something to do. This will allow the trainer to work with one individual for a short period and

reduce the risk of the other individuals becoming frustrated.

Anthropomorphism

Anthropomorphism can contribute to poor group training practices. For example, animals could get frustrated or even aggressive if a trainer make assumptions of what other animal(s) are thinking. Ultimately the behaviour could break down and the animal could stop paying attention or refused to participate in future training sessions. A better solution would be to reward each animal for either waiting patiently or for performing a behaviour. With this method, the animals all receive attention and reinforcement and it is likely to increase their calm and patient behaviour whilst stationing in the locations the trainer would like them to be in.

Options

When training animals in a group setting, there are several possibilities for how this can be accomplished (Ramirez 1999):

1) *Location*: each animal has a location (station) that is specific for them. This is where they need to be to receive reinforcement (Figure C1.5a).
   Advantage: the animal learns quickly where it needs to be for the start of the session.
   Disadvantage: can be difficult to move the animals to different locations if necessary

2) *Position*: each animal is placed (stationed) in a specific order in relation to each other (Figure C1.5b).
   Advantage: the animal learns quickly where it needs to be for the start of the session.
   Disadvantage: can be difficult to move the animals to different locations if necessary,

(a) (b) (c) (d)

Figure C1.5 Group training sessions can be facilitated by: reinforcing animals only in a specific location (station) (a); placing animals in a position for training (b); animals are trained using their individual 'tag' (c); and training starts wherever the animals happen to be, but during the session the trainer moves them into positions which better serve the training session (d). *Source:* (a) red panda, Odense Zoo; (b) harbour porpoise, Kirstin Anderson-Hansen; (c) dwarf alligators, Randers Regnskov; (d) bottlenose dolphin, Kolmården Zoo.

and changes in the hierarchy can cause problems with stationing.

3) *Name targets:* each animal is assigned a 'name tag', which works as a target and can guide them where they need to be throughout the training session (Figure C1.5c).

Advantage: the animal learns quickly where it needs to be for the start of the session, gives the animal a task to do (targeting on its name tag) whilst waiting, and allows for lots of flexibility during the sessions. For example, if animals need to be moved around or shifted.

4) *Animal Shuffle:* the animals start where they choose, and the trainer then moves or sends them to where the trainer wants them (Figure C1.5d).

Disadvantage: can require a lot of training time to first get them where they need to be.

Reference

Ramirez, K. (1999). *Animal Training: Successful Animal Management through Positive Reinforcement.* USA: Shedd Aquarium Society.

Box C2

This Generation's Challenge

Gary Priest

The nineteenth Century witnessed the dawn of the industrial revolution and within the span of 100 years, efficient machines were developed that harvested the earth's abundant resources at increasingly unsustainable rates. During the twentieth century, two World Wars and the growing demands of expanding human populations saw the curtain close on the earth's truly wild places. By the twenty-first century, virtually all the earth's animals, to some degree, were managed by humans.

In a perfect world, every human would have the opportunity to appreciate the diversity of life that has existed on earth, cherish it and protect it, as our common and irreplaceable biological inheritance. But, as our population approaches eight billion people, we must recognise the situation for what it is and not what we wish it would be. As a result, the mission of zoos and aquariums has never been more pressing than it is right now. Today, in view of the challenges faced by the earth's animals, zoos and aquariums have two primary roles; 1. increase human's collective appreciation of animals and our need to protect and conserve them; 2. apply every conceivable energy and technology to maintain sustainable captive populations and work collaboratively with in situ conservation projects.

The World Zoo and Aquarium Association reports that annually over 700 million people visit zoos and aquariums (Gusset and Dick 2011). As I see it, we have two strategic priorities. First, we must change hearts and minds and sound the alarm. Without popular support, our second strategic priority, to protect and conserve endangered species, will become moot. We simply cannot allow the failure of either priority. Without wide popular support, our second mission to protect and conserve endangered species will become a moot point. We simply cannot allow the failure of either mission.

Fortunately, something *is* being done and it is amazing. Concerned young people all over the world are responding to the alarm and are dedicating their lives to reversing the course we have been on.

In the past 25 years, captive management programmes (Species Survival Plans, SSPs; European Endangered Species Programmes, EEPs, and a host of comparable programmes around the world) have been developed and adopted by zoos and aquariums globally (Che-Castaldo et al. 2018). The goal for each plan is to preserve the maximum genetic diversity of the founding population and maintain a genetically diverse and demographically robust population. Plans for over 600 species exist with more being added each year. This work involves geneticists, population biologists, reproductive physiologists, zoologists, and thousands of dedicated animal care workers.

We know that animal behaviour is plastic and is modified by experience (Wong and Ulrika 2015). It is this same adaptive mechanism that allows animals to exploit changes in their environment. Using well established positive reinforcement techniques, a huge

Zoo Animal Learning and Training, First Edition. Edited by Vicky A. Melfi, Nicole R. Dorey, and Samantha J. Ward.
© 2020 John Wiley & Sons Ltd. Published 2020 by John Wiley & Sons Ltd.

(a) (b)

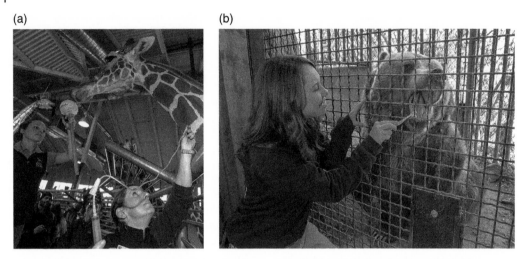

Figure C2.1 One of the many benefits of integrating training into zoo animal management is that ability to maintain and support healthcare; as evidenced here by the ability to take blood without the need for anaesthetics (a) and brushing a bear's teeth (b). *Source:* Steve Martin.

variety of animals in zoos and aquariums are now being conditioned to facilitate husbandry aimed to ensure their care and long-term survival. During the past two decades, bypassing the risks of anaesthesia, scores of endangered species from giant pandas (*Ailuropoda melanoleuca)* to orangutans (*Pongo sp.),* to tigers (*Panthera tigris*) have been trained to voluntarily accept ultrasonography, radiography, the giving of blood and tissue samples, as well as a wide variety of other routine health care procedures (e.g. see Figure C2.1a and b). From the radiographs and biological samples, geneticists and reproductive physiologists determine genetic diversity and better understand the reproductive biology of the species. The technology is also applied to minimise the effect of stress that may be associated

with our captive management of animals, such as computers, televisions, and radios (Clay et al. 2011). Biologically, the reduction of stress is a critically important component in reproduction. In two short decades, the advance of the application of conditioning through positive reinforcement has become a zoological institutions best practice and its application is rapidly spreading across all taxonomic boundaries.

The challenge faced by this generation regarding the preservation of the earth's animals and their habitats is unique in all of human history. The time is now. There will not be a second chance and the consequences of failing to do all we are able, using every tool at our disposal, is inconceivable. No better, more ethical, or responsible option is available.

References

Che-Castaldo, J.P., Grow, S.A., and Faust, L.J. (2018). Evaluating the contribution of North American zoos and aquariums to endangered species recovery. *Scientific Reports* 8 (9789): 1–9.

Clay, A.W., Perdue, B.M., Gaalema, D.E. et al. (2011). The use of technology to enhance zoological parks. *Zoo Biology* 30 (5): 487–497.

Gusset, M. and Dick, G. (2011). The global reach of zoos and aquariums in visitor numbers and conservation expenditures. *Zoo Biology* 30: 566–569. https://doi.org/10.1002/zoo.20369.

Wong, B.B.M. and Ulrika, C. (2015). Behavioral responses to changing environments. *Behavioral Ecology* 26 (3): 665–673. https://doi.org/10.1093/beheco/aru183.

Glossary

antecedent Environmental events that occur before a behaviour.

antipredator behaviour Mechanisms developed through evolution that assist prey organisms in their constant struggle against predators.

aversive conditioning Any stimulus, event, or condition whose termination immediately following a response, increases the frequency of that response.

backward conditioning Respondent conditioning in which the conditioned stimulus follows rather than precedes the unconditioned stimulus.

classical conditioning The modification of respondent behaviour by stimulus–stimulus contingencies, also referred to as Pavlovian or respondent conditioning.

cognitive map A mental representation of one's physical environment.

concept learning A learning task in which a learner is trained to classify objects by being shown a set of example objects along with their class labels.

conditioned place avoidance A form of Pavlovian conditioning used to measure the motivational effects of objects or experiences.

conditioned stimuli A stimulus which evokes a response or alters some other condition of behaviour only because of a history in which it has been paired with a stimulus (often unconditioned) having the same effect.

conditioned taste aversion Refers to when the subject associates the taste of a certain food with sickness.

continuous reinforcement When every emitted target behaviour is followed by a reinforcer.

cultural transmission The way a group of people or animals within a society or culture tend to learn and pass on information.

desensitisation Any form of counterconditioning that reduces an inappropriate negative response to an event.

differential reinforcement of incompatible behaviour (DRI) Reinforcement is provided for one behaviour that is incompatible with another behaviour.

discriminative stimulus The events that precede operants and set the occasion for behaviour.

establishing operations An environmental event, operation, or stimulus condition that affects an organism by momentarily altering (a) the reinforcing effectiveness of other events and (b) the frequency of occurrence of that part of the organism's repertoire relevant to those events as consequences.

event marker A signal used to mark desired behaviour at the instant it occurs.

extinction burst A rapid burst of responses that occur when extinction is first implemented.

Zoo Animal Learning and Training, First Edition. Edited by Vicky A. Melfi, Nicole R. Dorey, and Samantha J. Ward.
© 2020 John Wiley & Sons Ltd. Published 2020 by John Wiley & Sons Ltd.

extinction A procedure in which the reinforcement of a previously reinforced behaviour is discontinued.

fixed interval A schedule of intermittent reinforcement in which the first response occurring after a fixed interval of time is reinforced.

fixed ratio A schedule of reinforcement in which a fixed number of performances (counted from the preceding reinforcement) are required for reinforcement.

forward conditioning A type of motivating operation in which the pairing of two stimuli such that the conditioned stimulus is presented before the unconditioned stimulus.

free-contact The direct handling of an animal when the keeper and the animal share the same unrestricted space.

habituation A form of learning in which an organism decreases or ceases its responses to a stimulus after repeated or prolonged presentations.

imitation An advanced behaviour whereby an individual observes and replicates another's behaviour.

imprinting Any kind of phase-sensitive learning that is rapid and apparently independent of the consequences of behaviour.

intermittent reinforcement Reinforcement that does not follow every response.

intervention training The addition or change of several independent variables at the same time to achieve a desired result, without testing the effect of each variable individually.

local enhancement The attention of an individual is drawn to a specific location or situation.

magnitude of reinforcement Refers to the quantity, intensity, or duration of the reinforcer provided for responding.

mal-imprinted Suffering from a defect in the behavioural process of imprinting.

match-to-sample A procedure in which the choice of a stimulus that matches a sample stimulus is followed by a reinforcer.

motivating operations Environmental variables that alter the effectiveness of some stimulus, object, or event as a reinforcer; and alter the current frequency of all behaviour that has been reinforced by that stimulus, object, or event.

negative punishment The removal of a stimulus decreases the target behaviour.

negative reinforcement When a response results in the removal of an event, and the response rate increases.

observational learning Learning that occurs through observing the behaviour of others. It is a form of social learning which takes various forms, based on various processes.

operant conditioning Arranging the reinforcement of a response possessing specified properties, or, more specifically, arranging that a given reinforcer follow the emission of a given response.

positive punishment A stimulus is added, but the rate of the behaviour decreases over time.

positive reinforcement A stimulus following a behaviour is added and increases the likelihood of that behaviour.

Premack principle A principle that states that contingent access to high-frequency behaviours ('preferred' activities) serves as a reinforcer for the performance of low-frequency behaviours.

primary reinforcement Reinforcers that are not dependent on their association with other reinforcers.

protected-contact The direct handling of an animal when the keeper and the animal do not share the same unrestricted space.

proximate cause A cause that is close in time or sequence to the thing it is causing.

rate of reinforcement Reinforcers per unit time.

ratio strain A reduction in the rate of a target behaviour and an increase in emotional behaviour resulting from an increase in the ratio of behaviour to reinforcement.

reinforcement Any event which maintains or increases the probability of the response it follows.

reinforcers A consequential stimulus occurring contingent on a behaviour that increases or maintains the strength (rate, duration, and so on) of the behaviour.

schedules of reinforcement A rule governing the delivery of reinforcers.

secondary reinforcers Reinforcers that are dependent on their association with other reinforcers.

semantic ability Language skills refer to an understanding and appropriate use of meaning in single words, phrases, sentences, etc.

sensitive periods A time or stage in a person's development when they are more responsive to certain stimuli and quicker to learn particular skills.

shaping The reinforcement of successive approximations of a target behaviour.

stationing Refers to teaching an animal to go to, and stay on, a home base of some sort.

stereotypies Rhythmic, repetitive, fixed, predictable, purposeful, but purposeless movements.

stimuli A physical object or event that has an effect on the behaviour of an individual.

stimulus control The control that stimuli in our environment acquire over the behaviour we emit in their presence.

syntactical ability A metalinguistic skill, which concerns the ability to consider the structure rather than the meaning of a sentence.

theory of mind The ability to attribute mental states – beliefs, intents, desires, emotions, knowledge, etc. – to oneself, and to others, and to understand that others have beliefs, desires, intentions, and perspectives that are different from one's own.

ultimate cause The cause you can trace back to the beginning: the very first event in a chain of causes.

unconditional stimuli A stimulus, the capacity of which to elicit a response does not depend upon its having been paired with another stimulus possessing this capacity.

variable interval A schedule in which a person is reinforced for the first response after a varying period of time has passed since the previous reinforcement.

variable ratio A schedule of reinforcement in which one response is reinforced after a variable amount of time has passed.

working memory The part of short-term memory that is concerned with immediate conscious perceptual and linguistic processing.

Index

Note: Glossary items are in bold.

a

abilities *see* skills and abilities
accommodation (eye)   70
Acinonyx jubatus see cheetah
acoustic function *see* hearing
active training   119–141,
    158–160
  birds   234
Adelaide Zoo   175
adult life   87–94
Africa, presentations and
    performances   253
ageing, cognitive function and
    44, 111
  see also lifetime
Ailuropoda melanoleuca
    (giant panda)   334
alarm calls   15, 24, 36, 91, 92,
    124, 298, 299, 302
Alzheimer's disease   44, 45,
    46, 78
ambiguity   313
antecedents   12, 131,
    140, **335**
anthropocentrism   4, 217
anthropogenic changes,
    learning in response
    to   27
anthropomorphism   4, 217,
    330–331
  group training and
    330–331
  presentations and
    performances   255

anticipatory behaviour
  food   107, 256
  positive   283
antipredator behaviour *see*
    predator
Antwerp Zoo   175
Aonyx cinereus (Asian small-
    clawed otters)   88, 162
apes, great
  event markers and   127
  hand-rearing   85
  life experiences   219
  weaning   85
  see also bonobo;
    chimpanzees; gorillas
Aphelocoma californica)
    (Western scrub
    jays)   236
Apis mellifera (honeybee)
    16, 90
aquariums   239–242
  legislation   319, 320, 322
  reptiles   221
aquatics *see* marine (aquatic)
    species
Ara spp. (macaws)   122, 130,
    257
Arctic fox (*Vulpes lagopus*)   40
artificial behaviours *see*
    unnatural behaviours
Asia
  legislation   320–321
  presentations and
    performances   253

Asian small-clawed otters
    88, 162
assessment (husbandry
    programme)
    144–149, 157
Assiniboine Zoo   125
associations, learning   235,
    276–277
Athene noctua (little
    owl)   299
Atlanta Zoo   175
attitudes
  personal (zoo
    professional)   315
  visitors   263
Auckland Zoo   223, 224
audition *see* hearing
Australasia, presentations
    and performances
    253
Australia, legislation   319
aversive stimuli and
    conditioning
    16, 54, 57–58, 59, 62,
    63, 292, **335**
  predator recognition
    297
  taste   92, 292–293, **335**
awareness
  elephant self-awareness
    192
  of mental state of another
    animal (theory of
    mind)   236, **337**

Zoo Animal Learning and Training, First Edition. Edited by Vicky A. Melfi, Nicole R. Dorey, and Samantha J. Ward.
© 2020 John Wiley & Sons Ltd. Published 2020 by John Wiley & Sons Ltd.